Advance praise for *The Resilient Farm and Homestead*

"In *The Resilient Farm and Homestead*, Ben Falk gives us a delightful and inspiring description of his years developing a 10-acre permaculture farm in the Green Mountains of Vermont. Readers from regions outside New England, however, should not assume that Falk's practical, hard-won knowledge will not apply to them. His discussions invariably transcend the specific applications revealing principles that should be useful to homesteaders everywhere."

—**LARRY KORN**,
editor of *The One-Straw Revolution*
and *Sowing Seeds in the Desert*
by Masanobu Fukuoka

"Ben Falk calls his book about reviving a worn-out hill farm in Vermont an example of resilience and regeneration; I call it pure natural magic. Grow rice in New England? Yes. Heat water to 155°F on cold winter days at a rate of a gallon a minute by piping it through a compost pile? Yes. How about dinner tonight of your own rack of lamb garnished with homegrown mushrooms? Yes. Your choice of scores of different vegetables and fruits even in winter? Yes. Plus, your own dairy products from your own sheep. All the while, the soil producing this magic, on a site once thought little more than a wasteland, grows yearly more fertile and secure from natural calamity."

—**GENE LOGSDON**,
author of *A Sanctuary of Trees*
and *Small-Scale Grain Raising*

"*The Resilient Farm and Homestead* is a terrific book. Simultaneously inspiring and practical, Ben Falk takes you from the why to the how. . . . a journey where you will create a present and future filled with optimism and joy.

—**SHANNON HAYES**,
author of *Long Way on a Little*
and *Radical Homemakers*

"Imagine. Honoring biodiversity in a place we each commit to for the long haul is what it takes to address a rapidly changing climate. Problem solved! Plant trees, let greenness thrive, learn the ways of fungi, be joyful. Ben Falk provides the encouragement and critical know-how to create your own food-producing sanctuary in *The Resilient Farm and Homestead*. The time is now to engage in healing the land and secure an ongoing future for generations to come."

—**MICHAEL PHILLIPS**,
author of *The Holistic Orchard*

"Ben Falk extends the conversation about resilience to deep resilience—resilience from the level of personal attitudes and skills to the design and creation of the maximally resilient homestead. *The Resilient Farm and Homestead* weaves together permaculture theory as modified by actual practice on a 10-acre Vermont farm with a thorough preparedness guide for times of climate change and greater uncertainties of all kinds and sizes. The book is greatly enhanced by numerous glorious photos of permaculture plantings as hedgerows, rice paddies, people swimming in swale-enclosed ponds, fruit and vegetable harvesting, and foraging sheep, chickens, and ducks. I particularly appreciate that Falk tells us what didn't work as well as what did. This book will be essential reading for the serious prepper as well as for everyone interested in creating a more resilient lifestyle or landscape."

—**CAROL DEPPE**,
author of *The Resilient Gardener*

"With *The Resilient Farm and Homestead*, Ben Falk has definitely planted the seeds of a positive, abundant legacy. This book outlines the process of designing one's homestead with not just the future in mind, but the imminently practical NOW! This one is going on my shelf next to Helen and Scott Nearing."

—**MARK SHEPARD**,
author of *Restoration Agriculture*

The Resilient Farm and Homestead

The Resilient Farm and Homestead

An Innovative Permaculture and Whole Systems Design Approach

BEN FALK

Illustrations by Cornelius Murphy

CHELSEA GREEN PUBLISHING

WHITE RIVER JUNCTION, VERMONT

Project Manager: Patricia Stone
Editor: Makenna Goodman
Copy Editor: Eileen M. Clawson
Proofreader: Helen Walden
Indexer: Linda Hallinger
Designer: Melissa Jacobson

Printed in the United States of America.
First printing May, 2013.
10 9 8 7 6 5 4 3 2   13 14 15 16 17

Our Commitment to Green Publishing
Chelsea Green sees publishing as a tool for cultural change and ecological stewardship. We strive to align our book manufacturing practices with our editorial mission and to reduce the impact of our business enterprise in the environment. We print our books and catalogs on chlorine-free recycled paper, using vegetable-based inks whenever possible. This book may cost slightly more because it was printed on paper that contains recycled fiber, and we hope you'll agree that it's worth it. Chelsea Green is a member of the Green Press Initiative (www.greenpressinitiative.org), a nonprofit coalition of publishers, manufacturers, and authors working to protect the world's endangered forests and conserve natural resources. *The Resilient Farm and Homestead* was printed on FSC®-certified paper supplied by QuadGraphics that contains at least 10% postconsumer recycled fiber.

Library of Congress Cataloging-in-Publication Data
Falk, Ben, 1977–
    The resilient farm and homestead : an innovative permaculture and whole systems design approach / Ben Falk.
        p. cm.
    Includes bibliographical references and index.
    ISBN 978-1-60358-444-9 (pbk.) — ISBN 978-1-60358-445-6 (ebook)
1. Permaculture. 2. Permaculture—Vermont. 3. Agricultural ecology. 4. Agricultural ecology—Vermont. 5. Restoration ecology.
6. Restoration ecology—Vermont. I. Title.

    S494.5.P47F35 2013
    631.5′8—dc23
                    2013004638

Chelsea Green Publishing
85 North Main Street, Suite 120
White River Junction, VT 05001
(802) 295-6300
www.chelseagreen.com

For my parents, Marcia and Stephen Falk, with gratitude.
And for the generations after me, I hope this helps.

Contents

Chapter One

Creating a Positive Legacy while Adapting to Rapid Change

It is not the strongest animal that survives, nor the fastest, but the one most adaptable to change.
—LEON C. MEGGINSON,
paraphrasing Charles Darwin

Regeneration involves seeing things as they could be, while resiliency requires dealing with things as they are. This book shares principles, strategies, and components being tested at the Whole Systems Research Farm (WSRF) and homestead that are helping to transform a beat-up old Vermont hill farm into a highly biodiverse and productive human-supporting ecosystem. These systems, we have learned, must be simultaneously regenerative and resilient, for without regeneration health and production are limited. This landscape, like much of the world, is a damaged place, and without enhancing the health of soil and water (and the human body-mind as a result), one cannot increase productivity in durable ways. Without increasing productivity of the land in a durable manner, one's resiliency is not bolstered. Yet without also focusing on specific near-term needs such as having a plentiful fuelwood supply, backup lighting, fuel or tools, and basic skills, one's ability to do regenerative work is limited—the "long" work being too easily interrupted by bumps in the short term.

Along with my family, friends, and colleagues, our goal at the homestead is to implement and maintain biological and built systems that yield intergenerational value. The work undertaken at the WSRF references the most abundant, durable, and longest-term human settlement strategies developed across the globe and is intended to leave a valuable legacy for multiple generations into the future. In this way our work is intensely optimistic—we are planning for a more viable and thriving future in this place ten years down the road, an even more abundant one in a hundred years, and, ideally, an Edenic garden lasting centuries beyond that. If I hadn't experienced directly the possibility of this, I would think of this as unreasonable. But I have seen with my own eyes that human hands in partnership with seeds and fungi, animals and rainfall can, over relatively short time periods, transform sickly land into thriving living communities. Human work, it is safe to say, can *speed* the healing of more-than-human systems. As is said in the permaculture community, "We are nature working."*

Our goal as participants in the land must be to do better than "less harm." Why focus on doing less bad when we can actually improve, actually regenerate?

---

* This quote is most often attributed to Penny Livingston Stark, from whom I believe it was originally sourced.

The highest possibilities of human presence are staggering—I would consider the idea of a Garden of Eden a fantasy if I weren't confronted with evidence to the contrary in my daily life. The Whole Systems Research Farm shows clearly how infertile land that once supported only fern, moss, blackberries, and white pine can turn into a lush multilayered landscape of grasses, flowering herbs, fruits, nuts, berries, mushrooms, and livestock supporting one another, wildlife, and people. The work here in the past ten years also shows how seasonally inundated, compacted, oxygen-deprived soils and abused, abandoned, eroded land that's been clear-cut multiple times, bulldozed, stumped, and brush-hogged mercilessly can be transitioned into a place of deepening soils, of balanced moisture, of increased wildlife, of food growing in every corner, and of the unique beauty that emerges out of the synergy between land and people.

It's hard work at times, indeed, but the process is an enjoyable, vitalizing one, and the results are staggering and humbling, and they have come to define "gratefulness" for me. Perhaps as importantly, I have in the past few years begun to see my own vitality (both mental and physical) be enhanced as this landscape enlivens. So since the resurgence of health, both of Earth and of ourselves, is so plainly possible, it becomes the primary design imperative. The Garden of Eden is both a practical goal and one truly worthy of our efforts.

Unfortunately, however, the early part of the twenty-first century on Planet Earth happens to be the

Early autumn sun breaks across the integrated landscape outside the WSRF studio. Shelter, fuel, water, food, and medicine permeate a warm microclimate built from the soil up. Photograph courtesy of Whole Systems Design, LLC

antithesis of humanity's cultivating the Garden of Eden. The world today seems dominated not by a process of cultivating health and value, not by a process of supporting living systems, but by the erosion of life systems—by the mining of value, the ransacking of future possibilities for overuse and abuse in the present moment. Indeed, industrialized societies the world over seem more bent on war and consumption than on leaving any type of living legacy for their children. Sadly, the idea of posterity seems almost absent in much of the industrial world today. Simply considering the next generation, let alone the next handful of generations, seems remote from the predominant paradigm among people today.

So how do those of us working toward regeneration and the highest possibilities for human and land health deal with this striking paradox—that while we may plant chestnuts and build stone barns for the next seven generations, we are a tiny margin of the world's people? It's as if a group walks through a desert planting trees, while another moves through behind them chopping them down. This is not a workable situation. It is not viable to be working for one future while disregarding completely those around us—numbering in the millions—working (consciously or not) toward the opposite future. This is a bizarre paradox of this age, and it's where resiliency and adaptation come in.

Regeneration: Enhancing Life Systems

Regenerative actions harness the unique force omnipresent on Planet Earth—the power of living systems. These are the miraculous organizing forces that transform bedrock, sunshine, and rain into lichen, lichen into soil, soil into plants, and plants into animals. And with each cycle in the growth-decay process a deeper layer of nutrients and organic matter accumulates on the earth's surface and in her waters, ever-greater levels of biodiversity emerge, and the vitality of the waters and organisms alike increases.

A force is regenerative when it accelerates the process of transforming mineralogical matter into complex living organisms. The basic design tools to do this are simple: bedrock, water, sunshine, atmosphere,

> **A force is regenerative when it accelerates the process of transforming mineralogical matter into complex living organisms.**

and humans. The resilient homesteader is less a creator than a facilitator, ensuring the presence of the components (water, plants, animals, wood, people, etc.) and relationships (spatially and temporally) between them to maximize the rate at which dead matter becomes life and the rate at which bedrock and subsoil become living soil, living soil becomes plant cells, plant cells become animal tissue, animals become soil again.

My good friend and fellow designer-maker Chris Shanks uses the term "accelerate succession" to indicate this imperative and process. I like to add "*steer* and accelerate succession," to show the need not for simply any succession of life unfolding but for those scenarios that most rapidly increase biomass and biodiversity, two key metrics of regeneration. Humanity now needs to work land in such a way as to undo the damage we've already done, then extend this recovery to a state of health further than it has ever been. If we "steer" Earth toward adaptive regeneration, we'll see true resilience much sooner than by just letting it develop "naturally." It is our responsibility, in other words, to make up time for a long history of damages. An increasing number of people are answering this challenge to Earth's ecosystems by cultivating systems that produce as much food, energy, materials, medicine, wildlife habitat, water purification, carbon sequestration, pollination, and other ecosystem services as possible in the smallest amount of space possible for the longest amount of time possible.

Two inextricable ideals—regeneration and resiliency

Resiliency: Becoming an Adaptive Animal

Insanity: doing the same thing over and over again and expecting different results.

—Variously attributed

Before addressing the actual solutions that form the focus of this book, it is helpful to understand the patterns of mind that inform such solutions, the mental framework from which we make effective—or poor—decisions. The unsustainability of modern agriculture and society seems to stem from a poor understanding of cause and effect. Humanity's various failings on both the individual and collective level can be traced back to this basic phenomenon: misunderstanding how one's actions affect an outcome, or oftentimes, not recognizing that one's actions have *any* effect on an outcome whatsoever. In evolutionary biology this is called a *maladaptive response*. Acting in this way gets you booted out of the great wheel of life pretty quickly. The resilient homesteader is interested in the opposite: how do we fit in, respond to, adjust and adapt to constantly changing conditions? How do we do so with grace and joy? Indeed, should adaptation not be uniquely mastered by an animal as conscious as *Homo sapiens*?

Understanding the mechanics of maladaptation is important to clearly navigate the terrain of adaptation. So what creates a disconnect between our understanding of cause and effect? This forms the basis of many lengthy philosophical discussions, but for the purposes of this book, I will keep it very brief. How is one able to do *x* and ignore the fact that *y* results (e.g., defecate uphill of your water source, then get sick). Two reasons seem clear enough: The first is not recognizing that an action performed could actually have an effect on

What was once a steep slope has been transformed into terraced gardens of vegetables, perennial flowers, and fruit trees.

HOW THE WILDERNESS CONTEXT INFORMS A RESILIENT LIFESTYLE

When I trace my inclination toward preparedness, I always find a link to my time in backcountry travel and living experiences—living life on expeditions—from a summer on skis crossing an ice field in Alaska to weeks climbing mountains or paddling on lakes. All of these experiences have a commonality: During journeys in the backcountry, one is in constant anticipation of conditions that can change *at any point in time*. Such change is simply a given—whether it's a storm that will test the soundness of your tent's rigging or your retreat plan off the mountain you attempted to climb. In such situations you learn basic and crucial skills in observation and awareness, judgment and decision making, decisive action and determination.

Although true "wilderness" experiences are less a part of my life today, their role as a central theme in my own development has been crucial. Anyone, especially those in college and younger, stands to gain immeasurably by wilderness experiences—and any other expeditionary-like journeys in which humans push their comfort zone, rely upon themselves, test themselves, and ultimately learn what they are made of. Through encounters with elements in the more-than-human world, we can see the reality of life on this planet, which is often hidden from us but remains a basic fact of existence: The world around has always changed, sometimes rapidly, and it will do so again. Life in a tent, out of a backpack, and in the open exposes us to such change—brings it to the fore. When living outdoors we have few buffers separating us from these adverse influences.

Our daily life in the front country, back in "civilization," is filled with conveniences that, while enjoyable, distance us from the changing nature of the world around us. This affords us a certain degree of comfort, time savings, and other advantages but comes at a severe cost, one of which is losing a certain degree of personal self-reliance and the skills needed to be highly self-reliant. These include acute environmental awareness and response. In civilization much of those and other skills simply aren't needed. You don't need to watch the weather because the weather is totally irrelevant to the question of where you're going to sleep for the night or how you're going to prepare dinner. Expand that irrelevance to many spheres of your life and you can see how rapidly and profusely life in the comforts of "civilization" dumbs down our capacities simply by removing the need for many of them. What you practice you become good at; what you don't you gradually lose competency in.

This is not to say that some lifestyles in the front country don't demand certain and highly developed skills and awareness—some do. But for the most part our homes, cars, computers, and other comforts retard the development of fundamental awareness capacities that are necessary to think about, plan for, and act on a future that is guaranteed to be saturated with change. The take-home message here is that we become proficient at what we practice, and we tend to practice most intensely based upon needs. We can intentionally place ourselves into environments where meeting basic and constantly shifting and challenging basic needs is required of us. This is, in part, what a "wilderness" (read unmediated by human systems) experience offers all of us. The learning and growth opportunities lying within them are deeply practical and endlessly rewarding.

something seemingly unrelated (going to the bathroom and staying healthy). Notice the example here is not touch a hot coal and get burned, because humans seem good enough at understanding very *simple and immediate* cause and effect, but we are particularly poor at grasping this process when there is a time delay or when the system contains complexity, such as when numerous agents are acting upon it. The second possibility is a lack of understanding how action in one sphere can affect a seemingly unrelated area—pooping on the ground and drinking water from a well.

How do we overcome these two tragic misunderstandings? The first seems easy enough: recognize that all actions incur a result, whether we understand the result or not. The world is connected—a web—and we cannot act in any way that does not affect this

> Should adaptation not be uniquely mastered by an animal as conscious as *Homo sapiens*?

web. Basic stuff. The second is more challenging and involves a degree of observation, intuition, analysis, and critical thinking. It's at the core of what it means to be a conscious being, to be human. It can be called critical thinking: reasoning a problem through in its entirety, design thinking, problem solving, or systems thinking. Not that this is a simple undertaking, especially in today's world of increasingly dulled, technofied minds. Cultivating systems-thinking humans, however, is not the focus of the book but I would encourage the reader to seek the many great resources existing on the subject including works by Fritjof Capra and Peter Senge among many others.

A self-reliant nation is built upon a citizenry living in resource-producing and relatively self-reliant communities. Self-reliant, tenable communities are composed of self-reliant households. And relatively self-reliant households are the basic building block of any culture that is viable over the long term without requiring war (stealing of resources) to sustain itself. No democratic civilization can last long if it is built upon a citizenry that consume more than they produce; that's debt and debt is inherently unsustainable and ultimately undemocratic. If our goal is a peaceful, just society, self-reliance at the home and community levels must be a central focus of our lives.

Using This Book

The Resilient Farm and Homestead is written to share approaches that have been rewarding in my own life and have resulted in the sometimes stunning revitalization of the beat-up old hillside I call home. It is written to share the regeneration and resiliency strategies I have been employing, what's worked and what hasn't. In certain areas of the book, I share general information about a topic that may go beyond what I have worked with at the Whole Systems Research Farm. However, in general this is a personal and direct account of my experience of working with a specific piece of land. This book is not a rehashing of information found elsewhere but only of direct experience. I have found that much of what I've read in the literature around homesteading, permaculture, and ecological

restoration often conflicts with the reality of these disciplines when practiced on the ground.

This should not be surprising, I suppose, as too often what makes it into books is theory, not practice translated into words. I aim to achieve the opposite. In this way it's a story, and a collection of experiences that readers can adapt to their own lives. It is written with the hope that people the world over will find value in it as they take back control over some measure of their own lives, empowering themselves and their families in the pursuit of resilience and regeneration and revel in the health, freedom, and fulfillment that is a natural outgrowth of such a life.

But this book is written not only for today but for a time when we look back at today. It is written for a time when we realize how fragile the means of our existence had become—a time when we might have the opportunity to choose what to rebuild and how. Most importantly, this book is a personal story of aiming to thrive in changing times necessarily, involving two inextricable approaches—regeneration and resiliency. Regenerative strategies enhance the health of the systems we live and work within, while resiliency strategies enable the system—and ourselves as members of that system—to adapt to constantly shifting conditions. It is these imperatives we must constantly work from and toward.

This book offers various angles on the challenges posed by regenerative and resiliency imperatives. It is in part:

▶ **A case study in homesteading and farming**—what's worked and what's been most challenging, especially useful for those homesteading and farming and wishing to do so successfully in a changing future

▶ **A resource for the professional or the student of ecological design**—articulating principles, concepts, strategies, and language in performing whole systems design, with particular use for architects, landscape designers, planners, engineers, builders, and community organizers

▶ **An ecological restorationist's clue book** of ideas—especially useful in the evolving field of restoration agriculture, of speeding the transformation of degraded landscapes into productive, fertile, and biodiverse systems that humanity and the earth increasingly need urgently

▶ **A preparedness manual**—especially helpful for parents who want to invest in an adaptable future for themselves and their family

The lessons learned here are applicable to those interested in enhancing the vigor of their own homestead and farm, person, family, and community. It is not a prescription for what to do in all other locations and at all scales, as all projects contain a multitude of unique and differing characteristics, from climate to landscape size to social factors. However, many of the strategies we've employed are transferable to cold-temperate climates of the world in particular. Chapters on water and soil management, tool selection and use, and the social considerations of using land well are applicable to all areas of the world and projects of all scales. Chapters focused on the challenges we face at the dawn of the twenty-first century—including toxic contamination, economic insolvency, soil loss, human health challenges, and the like—are also relevant, sadly, to all peoples in all parts of the globe. Readers in cool or cold-temperate regions of the world can safely assume that most of the approaches shown in this book are of high relevance to their own lives, whereas readers in other climates should pick and choose among these techniques and experiment to see what is most applicable to their own landscapes. This book should be used like a cookbook: not as a prescription but as a resource of ideas to get you thinking and acting. In the same way that no great chef ever confined herself to a recipe, no great land tender should ever confine herself to another's ideas. Each of us working with land must ultimately listen to the clues continually emerging from our own direct interaction with the land under our feet, if we are to find the ways that work best.

The lessons learned here are applicable to those interested in enhancing the vigor of their own homestead and farm, person, family, and community.

Permaculture

Throughout this book I refer to locations on a site as being in a specific numeric zone from 00 to 5, with 00 representing the human body/mind/self and 5 being unmanaged "wild" areas of a landscape. These zones have emerged from a discipline called "permaculture," a design approach and framework for problem solving developed most notably by Bill Mollison and David Holmgren. Their work emerged in Australasia and combined some of the smartest pattern-based and ecological design approaches thus far developed around the world. In the past forty years Mollison and Holmgren, along with other pioneers such as Geoff Lawton, John Todd,[*] Penny Livingston Stark, Sepp Holzer, the Bullock Brothers, Mark Shepard, Dave Jacke, and many others have developed and articulated these pattern understandings into a set of principles and design approaches that are manifest as a set of land use (and other) strategies.[†] A permaculture can accurately be said to be any system in which "the whole function of each part is fully realized."[‡] This often takes the form of a landscape in which fertile topsoil is being produced, water quality enhanced, and wildlife promoted while humans also garden valuable yields of food, fuel, fiber, or medicine from the system.

In essence, permaculture as a land use approach (the aspect of permaculture I use and refer to in this book) is a system of ecological regeneration in which the production of products for human livelihoods is also a key component—a marriage of ecological restoration and gardening, if you will. Permaculture land systems always promote biodiversity, not simply biomass (output), and are heavily focused on increasing ecosystem health and reducing mechanical energy

---

[*] Though more associated with general ecological design by many, Dr. John Todd so deeply advanced the field of ecological design (in which permaculture is rooted) to such a significant extent that the language and ideas he has developed are inextricably part of permaculture today. A similar statement can be made about the work of Allan Savory.

[†] The original groundbreaking permaculture work most referenced is *Permaculture: A Designer's Manual* by Bill Mollison, published in 1988 by Tagari Publications.

[‡] Tanya Srolovitz, student in a permaculture design course the author organized in 2001 at the Island School in Eleuthera, Bahamas

Elderberry harvesting in the currant microswales during our permaculture design course. This is zone 1 in midsummer abundance. Note the five-year-old black locust in the left foreground, which is already fence-post size and has broken its Tubex tree tube. The free-ranging ducks are searching out slugs amidst the mulch.

expenditures and off-site inputs into the system at every point possible. Permaculture seeks always to allow a land system to perform its own functions when possible instead of the human management performing them, even when these are slower or lower yielding in the short term. These approaches are extensively explored in the large permaculture literature available.

Permaculture design has more clearly articulated the zonation concept of laying out a landscape according to frequencies and ease of use than any other modern design discipline. This zone approach is essential to the proper layout of components of a site, and I refer to these zones throughout this book. Since permaculturists vary in what they mean with each number (although zone 5 always means unmanaged land), I want to define how the specific zones are used in this book.

Zones and Their Definitions

Zone definitions vary significantly with overall occupants per acre of the site. For example, a family on a quarter-acre lot will often visit zone 3 multiple times per day, whereas zone 3 on a rural homestead of ten acres with a handful of people, like the WSRF, often only sees a visitor a few times per week or less. Variation with the season is also enormous, and the definitions above are for growing-season periods: I visit zone 3 in the winter less than once a week and sometimes only once a month. Of course, the given zone of an area is also in constant flux, with zone 3 becoming zone 2 over time and vice versa as a site is developed and maintained. For use in this book I consider zones as follows:

oo The human being, physical, mental, and spiritual: body, mind, self. This is the space you occupy every moment of every day.

o The home shelter: house and kitchen area especially. These are the spaces you occupy many times each day. In a well-designed homestead, this includes outdoor living space near the dwelling.

1 The most often used landscape areas: vegetable garden, barn, parking areas.

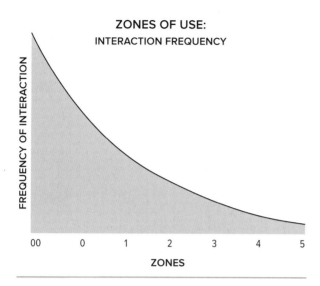

Interaction drops rapidly as we move from zone 1.

2 Lesser-used areas that are frequented once to twice a day or so: typically, some staple annual crops such as potatoes; perennials, such as fruit trees and berries; grazing areas; mushroom yards.

3 Areas frequented a few times a week: typically, grazing paddocks, most orchards, mushroom yards, and nutteries.

4 The least frequented areas under management: woodlots, some grazing areas, some forest garden zones.

5 Unmanaged land: "wild." Land that is left for observation alone—one visits but does not cut, harvest, or forage in any significant way. Not all or even most sites may have a true zone 5.

Who Are "We?"

At points in the book, you will see the term "we," referring to users of this site. While the opinion shared in this book is solely that of the author, the work performed on-site has included far more than my labor and input alone. At the time of this writing, the Whole Systems Research Farm has included interns for seven years of its existence, group participation in projects ranging from permaculture design courses to college and local groups of five to twenty people in size, hired help for 15 to 25 percent of this period during the growing season, and input from work-trading volunteers

sporadically over the last nine years. The site's general development, while always under the direction of Ben Falk, has benefited from the input, and especially the labor, of dozens of people over the years.

WSRF Site Specifics

The Whole Systems Research Farm comprises ten acres of hillside land perched above the Mad River in Central Vermont. The site has been home to at least half a dozen families since Europeans first settled this part of the world about three hundred years ago. So, too, has this site been host to various land uses, from logging to maple sugaring, the pasturing of beef and likely dairy cows, to the more recent perennial cropping and grazing systems we have employed. The following specifics should be referenced when comparing the lessons learned in this book with your own site.

▶ Scale: ten acres, perimeter approximately 3,600′, elevation change ~150′.

▶ Average slope: 10 to 15 percent, many areas of 15 to 20 percent, almost no truly level areas except created terraces

▶ Global climate classification: cold-temperate

▶ Frost-free days per year: average (mid-May to mid-/late September). This varies by one to five weeks per year, however.

▶ Coudiness: Very cloudy most of the year with generally sunny summers for a total of 49% annual average sunshine of total possible, making this area one of the cloudiest in the lower 48 United States comparable with much of the Great Lakes and Pacific Northwest.*

▶ Aspect: primarily westerly with some southwest and northwest.

▶ USDA hardiness zone: 4 to 5

▶ Coldest temperature recorded in past nine years: −26°F (−32°C)

▶ Warmest temperature recorded in past nine years: 92°F (33°C)

* NOAA climate data on percentage of sunshine out of possible: http://www.ncdc.noaa.gov/oa/climate/online/ccd/pctposrank.txt

THE WHOLE SYSTEMS RESEARCH FARM—
PRIMARY FEATURES

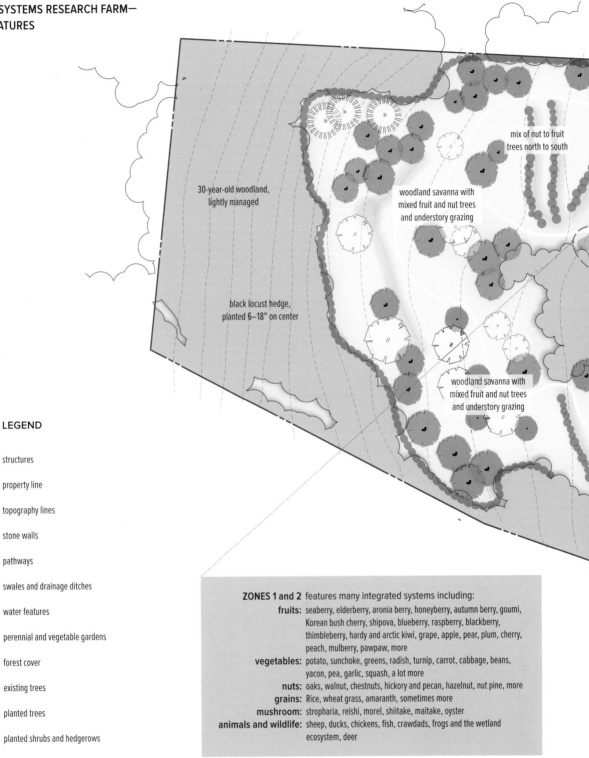

mix of nut to fruit
trees north to south

30-year-old woodland,
lightly managed

woodland savanna with
mixed fruit and nut trees
and understory grazing

black locust hedge,
planted 6–18" on center

woodland savanna with
mixed fruit and nut trees
and understory grazing

LEGEND

structures

property line

topography lines

stone walls

pathways

swales and drainage ditches

water features

perennial and vegetable gardens

forest cover

existing trees

planted trees

planted shrubs and hedgerows

ZONES 1 and 2 features many integrated systems including:

fruits: seaberry, elderberry, aronia berry, honeyberry, autumn berry, goumi, Korean bush cherry, shipova, blueberry, raspberry, blackberry, thimbleberry, hardy and arctic kiwi, grape, apple, pear, plum, cherry, peach, mulberry, pawpaw, more

vegetables: potato, sunchoke, greens, radish, turnip, carrot, cabbage, beans, yacon, pea, garlic, squash, a lot more

nuts: oaks, walnut, chestnuts, hickory and pecan, hazelnut, nut pine, more

grains: Rice, wheat grass, amaranth, sometimes more

mushroom: stropharia, reishi, morel, shiitake, maitake, oyster

animals and wildlife: sheep, ducks, chickens, fish, crawdads, frogs and the wetland ecosystem, deer

Plan view of the Whole Systems Research Farm site in Central Vermont

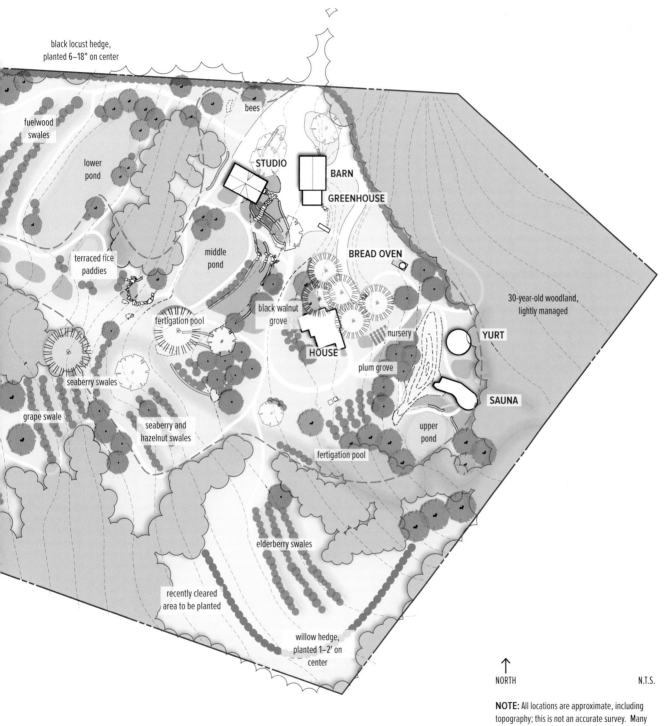

black locust hedge,
planted 6–18" on center

fuelwood
swales

lower
pond

terraced rice
paddies

middle
pond

STUDIO

bees

BARN

GREENHOUSE

BREAD OVEN

30-year-old woodland,
lightly managed

black walnut
grove

fertigation pool

nursery

YURT

seaberry swales

HOUSE

plum grove

grape swale

SAUNA

seaberry and
hazelnut swales

fertigation pool

upper
pond

elderberry swales

recently cleared
area to be planted

willow hedge,
planted 1–2' on
center

↑
NORTH

N.T.S.

NOTE: All locations are approximate, including
topography; this is not an accurate survey. Many
features, such as plantings, minor earthworks, and
other infrastructure are not depicted.

- Latitude: 44°N, 73°W
- Elevation: Approximately 850′ to 1000′ above sea level
- Wind exposure: low, maximum winds on-site approximately 65 miles per hour, typically very low with maximum wind stress occurring in thunderstorms two to four times per year, with winds of 30 to 45 mph. Site is prone to cooling down-valley (katabatic) breezes nightly from upslope.
- Soils: silty and gravelly clays, 0″ to 4″ of topsoil present on-site upon my arrival atop a subsoil that ranged from 10+ feet to less than a foot deep. Roughly a third of the land we work intensively is less than two feet to bedrock and some cropped areas have 1″ to 10″ of subsoil atop bedrock.
- Groundwater: depth to water table is within 2″ of the surface for most of the year over at least ¾ of the site. The water table falls 3′ to 5′ during most summers.
- Existing vegetation cover at time of my arrival: 50/50 forested/abandoned "old field" with saplings of birch, poplar, and *rubus* species becoming predominant.
- Occupants: 1 to 6 over the ten years I've been on-site.
- Machinery used in development, at various times: 21 horsepower tractor backhoe (original swales), full size tracked excavators (primary ponds), compact excavator of about 8,000 to 9,000 lbs (most newer swales, pools, terraces, and paddies, and a 56 HP tractor on rare occassions.

Site History

- Geological/glaciological
  - Located in the Green Mountains physiographic region, composed of metamorphic bedrock, schist. Soils tend to become acidic over time as a result but are relatively well mineralized.
  - Was buried in ice thousands of feet thick during Holocene glaciation ten to twelve thousand years before present (YBP), then a lake for some hundreds of years.
  - Northern hardwood/boreal forest mix began on-site nine to ten thousand YBP.
- Pre-European settlement
  - Little seems to be known about this area of Vermont, but it was likely frequented by bands of gatherer-hunters that may have either altered the ecosystem significantly or left little mark on it. While it's impossible to know for sure, the research I have done for the past fifteen years on Vermont land uses indicates that this area of the state was used lightly by first peoples—confining their activities to the milder and more productive areas around Lake Champlain. That's where I would have liked to be as well if I were living here hundreds and thousands of years ago! However, since we cannot know the presence and impact of past human settlement on this site and this impact is usually more than we know and realize in the present day, it may be safe to assume that human impact on this site is greater than I realize.
- Post-European settlement
  - 1800–1900: Former dairy cow pasture; likely cleared completely about two hundred years ago, initially. Likely grazed for fifty to seventy-five years before abandonment.
  - 1900–1950: Partial abandonment; half the site grows into northern hardwood forest composed of maple, beech, ash, hop hornbeam, yellow birch.
  - 1940s/1950s: Complete abandonment; the remaining field areas of the site succeed into white pine (*Pinus strobus*); red maple (*Acer rubrum*); aspen (*Populus tremuloides*); paper birch (*Betula paperifera*); grey birch (*Betula populifolia*); brambles; fern, especially sensitive, bracken, and ostrich; and goldenrod.
  - 1970s: Second-home development of the site for skiing-related recreation. The site was cleared with heavy machinery for a view. This represents the second, third, or fourth major assault on the health of the site's ecosystem, with likely no seed spread after machinery was used to clear-cut the forest.

Site Future: What's Possible?

Given an understanding of the history of this (and your own) location, what visions can reasonably be drawn about the future of the site? Although the past is often

Some of the sheep herd taking in the view west above the rice paddies, pond, and lower swales. Photograph courtesy of Whole Systems Design, LLC

prelude to the future, today we stand in stark realization that we cannot afford to continue repeating the mistakes of the past—most notably the loss of topsoil and water from the hillsides, the dependence on distant resources to meet our basic needs, and the lack of human vitality that past systems have too often yielded. It's clear that a newer, much more effective presence in our places is now necessary. To envision such a future, it's helpful to look back at the long history of human land use, what's worked and what has not*, and to also reference the more recent trends of human settlement. To highlight potential answers to these questions, I offer the following brief exploration of what it means to flee, dwell, and invest in a place to be successful over the long haul.

---

* The most useful book I have found for offering answers to this primary question has been *Farmers of Forty Centuries*, by F. H. King.

Fleeing

The average American moves 11.7 times in a lifetime.

—United States Census Bureau,
Geographic Mobility Report 2006

The best time to plant a tree is twenty years ago. The second best time is now.

— Japanese Proverb

It's not surprising that we North Americans still call this continent the "New World." Relative to the first peoples in America, who have lived here for between four thousand and fifteen thousand years, we just got off the boat. It's new to us, and so far we don't seem intent on staying. I was taught in school that the American frontier closed in the nineteenth century, yet the same boom-bust cycle has continued into the twenty-first, shifting from the Appalachians to the Prairie to the West to the Rust Belt to Silicon Valley and the Sun Belt. Now—*finally*—we're almost out of both places to live and places from which to extract our living. Our distant sources of labor, food, energy, water, and rare-earth elements are running dry. Africa won't feed China for very long, nor can Canada and the Amazon feed and fuel the United States for more than a handful of decades—the land base simply is not big enough or productive enough by any measure to feed the surging populations. Though we fled from distant lands to America, we continue to live much like refugees, constantly moving from one place to another, never staying long enough to cultivate the richest values possible in a specific place. In doing so we've traded uniqueness for the generic, culture for commerce. Even those of us who can afford to usually don't stick around long enough to harvest the fruits of our labor—nomads not seeking safety but "success."

We need the opposite kind of culture, a people that mean to stay. Strangely, running out of places to go and resources to plunder may be what we need most to convince us that a regenerative presence is called for. It's easy to wreck a place when you know you can move on to the next; without another place to go, might we finally be forced to open our eyes to what's at hand? To gaze not at a distant horizon but at the ground beneath our feet? Then might we ask, "What can I do here? What can I make of this place?" This transformation is inevitable and will happen whether we engage it or not; the earth is finite, and we're spectacularly overshooting our resource base. This shift will not be just personal but cultural.

"Staying" seems to be one of the key ingredients to a resilient and adaptive culture and to any civilization that can last beyond a few centuries, especially in the modern age. Rootlessness is simply not a viable operating system in a high-tech (high-footprint) world with billions of humans, and it begets a mind-set of conquest, a broken chain of cause and effect, not of accountability. Indeed, the concept of "life, liberty, and the pursuit of happiness" seems hinged upon close feedback loops between action and consequence. But true "staying" can only happen in a settled society, in cultures where "home" and community are central, where the individual is embedded in a long chain of generations, inheriting from those before, leaving for those who will come after.

Fortunately, this pattern is hardly new. The instances in which human groups have sustained themselves in specific places for millennia occur where cultural and economic (resource) systems were organized not to maximize wealth for the individual but to grow and transfer value *across* human generations. *Not* moving to the next place has been the only way we've built wealth enduringly. This kind of value takes decades and centuries to develop: barns spilling over with the autumn harvest, apples stacked high to last through a winter, disease-resistant crops from hedgerow to hedgerow, towering groves of nut trees, abundant herds of game, lush pasture and sturdy animals, vigorous people mastering their work, and vibrant cultural memory. Human culture can create all of these conditions—even thriving ecosystems. But it takes generations of people skillfully committed to each other, and to a place, to do so.

Our task, then, at the dawn of the third millennium, is to transition from a society based on mining the most value as quickly as possible to a long-haul culture living not on the principal but on the interest. So how do we develop perpetual, interest-bearing systems from which we can live? We can start by looking at those places where human inhabitation has lasted millennia—and at those who dwelled and did not despoil their homes.

Dwelling

In difficult dry regions of the Iberian Peninsula, a complex agroforestry system based heavily on the interactions between an oak-and-chestnut overstory and a grazed understory (using pigs and small cows especially), called the dehesa system, was devised, likely in the first millennium AD. Grazing animals were rotated through the woodlands, with animals thriving primarily on the produce of the trees. The nuts offered a wellspring of fat and protein from year to year, with no pruning, no fertilizing (other than animal rotations), little disease pressure, no irrigation, no bare soil, no erosion, and complete groundwater recharging/moisture retention. This kind of land use is the opposite of desertification. The productivity of the dehesa system has been found to be higher per unit area than any version of modern agriculture in Spain, when accounting for all inputs and outputs. At the same time the quality of the systems' outputs is superior to those of modern agriculture: Chestnut-fed swine has long been regarded as one of the finest meats in the world, as flavorful as it is dense in nutrients, beyond comparison with grain-fed meats. The savanna-mimicking dehesa silvo-pastoral systems were so widespread, evolved, and practiced for so many centuries that until the twentieth century many ecologists did not recognize the anthropogenic origins of these ecosystems. As the agroforestry practices of planting, cutting, pruning, and grazing waned in the modern era, so, too, has the diversity of "wild" life in these woodlands. While springs dried up, soil building slowed, and the region has become more arid and brittle and less productive. It is probable that, as in many abandoned or untended places, diversity dropped when beneficial human management was removed: Lack of good grazing removed fertility from the cycle—while poor grazing eroded land, and trees stopped being replanted.

In what is now California, the Sierra Miwok, Yokuts, Chumash, and at least a dozen other first peoples developed perennial, fire-managed ecosystems that grew a stunning abundance of game along with medicinal plants in the understory of black oak–dominated woodlands. Peoples in California also developed systems based around sugar pine, hazelnut, and other masting and often exceptionally long-lived plants, using fire, transplanting, and selective cutting rather than grazing (having none of the domestic-able animals that were available in Eurasia). In the Sierra Nevada Mountains individual sugar pine groves were often tended to by single clans, climbed and harvested from for a dozen or more human generations (sugar pines can yield rich pine nuts for three hundred to five hundred years).

Imagine harvesting food from a tree that your great-grandfather planted, that your grandfather then climbed to harvest nuts from, that your father climbed and rested beneath, whose seeds your mother made a flour from to nourish you, that your son will feed your grandchildren from, that your grandchildren, when the tree dies, will use the wood from for shelter, the inner bark for medicines, the resin for fire starter, the needles as incense in a ceremony for the tree and for the lives that the tree made possible. Such is the life of a people who live close to trees, intentional in their legacy.

Over a period of at least a hundred human generations, those dwelling in eastern North America guided the development of vast food forests. The Wabanaki, Algonquian, and Mahican peoples, the Abenaki, Huron, Iroquois, Manhattan, Massachuset, Narragansett, Penobscot, Seneca, Shinnecock, and others promoted an intergenerational food, fuel, fiber, and medicine ecosystem whose foundation was the mast-bearing tree: oak, walnut, hickory, chestnut, butternut, pine, beech, hazelnut (they did not yet have the apple from Asia).

The earliest European accounts of this land describe it as having an open understory and being full of oak and walnut trees. These visitors thought they had encountered an unusually beautiful wilderness. But as has become clear, this was no wilderness but a continental-scale forest garden whose crops were trees, the game they sustained, and harvestable understory plants. As in other regions of the world where cultures figured out how to dwell for thousands of years in a single place, the tools and techniques of choice were fire; hunting; selective cutting; promoting the largest, most useful seeds; and dispersing them (think Johnny Butternut); and a deep awareness of seasonal cycles to properly time these activities.

Energy Cycling

Why are trees—especially nut trees—at the basis of these regenerative land-use systems and highly adapted human cultures? In the simplest terms it has to do with inputs and outputs. A nut tree is simply more effective *and* efficient at converting sunlight and precipitation into value, over the long term, than any other technology humans have yet designed. This becomes clear when comparing biological systems in general with nonliving technologies. Consider a photovoltaic panel or wind turbine, for example. Each requires large and

Lupine—a superplant in that it both puts nitrogen into the soil and is a nutrient hyper-accumulator—building soil quickly.

damaging inputs to generate single outputs. What are the inputs for a photovoltaic panel? For one thing, Bauxite from which to smelt the aluminum frame, as well as silicon and numerous other minerals (many only found in a dwindling number of difficult-to-access places on the planet). These all must be mined, transported, refined, transported again, then fabricated, then shipped again. All for one output: electricity.

What are the inputs required for a nut tree? At most an exchange between breeder and planter, transporting of the seed or seedling, some wood chip mulch, rain, and sunshine. And *time*. What are its yields? Oxygen, soil, wildlife habitat, moisture retention, carbon sequestration, air and water enhancement, human food, stock feed, building materials, shade, windbreak, and beauty, to name a few. The former resource path of the photovoltaic panel—the abiotic—provides us with a practical service at great cost. The latter, biological (or "soft") path creates an enduring and generative legacy of positive value. And whereas a solar panel, wind turbine, or green building offers diminishing yields over time, a nut tree's output actually increases, for at least the first century or two of its lifetime.

Such is the power—and imperative—of biological systems: They are the only means we have of sidestepping entropy, at least for significant periods of time, on this planet. That's what tips the balance; it all comes down to capture, storage, and transfer. The most functional human-land arrangement is the one that can harvest the most sunlight, moisture, atmospheric fertility, and biological energies, then accrue that value for the longest period of time while converting some of it into products and services that other living things, such as humans, can feed on. Biological systems do this very well, while nonliving mechanical systems cannot.

In the modern era enough research has been done to quantify the advantage of cropping with trees over annual crops. Accepted yields for chestnut, for example, are eight hundred to fifteen hundred pounds per acre. That rivals modern corn production on deep-soil land. However, corn only produces such a crop with constant labor and fertility inputs each year, while reducing the land's capacity to produce because of its erosive forces on the soil. A chestnut orchard, on the other hand, actually improves the land's (and climate's) capacity from year to year *while* it yields; it requires no bare soil or off-site fertility inputs, and it produces hundreds of crops from each plant on marginal, shallow-soiled land (far more of the earth's cover type than deep-soiled land), while taking up less space than corn. And you can crop the same area with other species simultaneously; for example, a chestnut orchard is *also* a pasture, *also* a game preserve/farm, *also* a place for understory berries and medicinal crops.

All in all, you can grow three to eight times the product value (protein, fat, carbohydrate, Btus and other nutrients/values) via a tree crop system than with an annual, input-dependent crop such as corn, and you can do so while improving the land from decade to decade.* Indeed, tree cropping and ecological restoration can be performed simultaneously. Annual cropping the same land, year after year, however, usually leads to a ruined soil and culture, even on flat lands (and always on steep lands unless it's rice). Mesopotamia, much of Greece, and many other empires were once forested; now, they are deserts.

Despite abundant human cleverness, we haven't invented a better way to store energy than a stack of firewood. We haven't yet devised a more effective

---

* See J. Russel Smith's classic *Tree Crops: A Permanent Agriculture* for more on this whole aspect.

means of capturing solar energy than by putting up a cow and hay in a barn through the winter. Biological energy harvesting and storage is what has allowed us to survive to this point, and our experiments of replacing biological systems with mechanical and chemical systems have at best been delayed catastrophes. We must rely on some nonbiological aspects (the barn in the previous example), but wherever we do we compromise the system and our own returns in the long term. The minute a barn is built, it begins to decay. The famous comparison of a tractor with a draft horse highlights the entropy principle at work here: A tractor and horse are comparable in the amount of work they can achieve on a small piece of land, yet after a time the tractor dies and the horse makes another horse. Only life processes are regenerative. Hence, our prospects for thriving on this planet depend on our ability to partner with life forces.

Reinvestment

Life, however, can be slow. Who can wait decades for a return on investment?

Actually, most of us do already: pensions, Social Security, mortgages. But a nut tree beats an IRA, hands down, on a strictly monetary basis alone (not counting all the side yields). Indeed, one could consider such an investment a "collective retirement account," maturing in ten to thirty years and yielding ever-increasing returns for its first one hundred to two hundred years at least. Stone (nut) pines (which cover huge swaths of the Siberian taiga and are adapted to the world's cold-temperate climates) often bear for four to five hundred years. Try gambling on Lehman Brothers for just a one-hundred-year return. Your apple tree, however, can easily do that. Planted for $100 and tended to at a cost of $50 per year (in your time), the tree will yield roughly fifty thousand pounds ($150,000 worth at $3 a pound) of fruit in its first century—a total return on investment (ROI) of 2,841 percent and an annualized rate of return of 7.1 percent (almost exactly the same as a 50/50 bond/stock portfolio over the *last* hundred years). That's *not* counting any wood/timber value from the tree upon its harvest, which can be enormous in the case of a nut tree such as black walnut, oak, and

chestnut. If you didn't count your time pruning and harvesting, and chalked that up to family fun, your overall ROI would be 150,000 percent in a hundred years. Over fifty years your APR would be 15.8 percent—*not* slow money.

Trees are one of the only financial instruments we can rationally depend on for long-term returns on investment. Perhaps this is why humans have invested in trees for millennia and in banks for a mere moment in time. Unlike an IRA or Social Security, barring a lightning strike, your family's nut tree carries a guarantee that the US Treasury simply can't make (even if it weren't bankrupt); it simply hasn't been around long enough. One can find mature nut trees today that started yielding before the United States existed, and one can plant a tree today that will likely be bearing after this nation's lifespan is over. On the thousands of pounds of value falling from your tree year after year, you will pay not one cent of tax. The value is all for you—and for the squirrel, the owl, the soil, the groundwater, the climate, and your children. Imagine inheriting a food forest. Imagine creating one. Planting season begins when the ground thaws and ends at leaf out. Your intergenerational legacy can begin today.

The Green Distraction and the Political Black Hole

We are now a few decades into the Green Dream. Sometime in the latter half of the twentieth century, upwardly mobile, socially conscious, academically educated professionals—those who could afford to—began to drive the commercialization of products and services that were healthier, less cruel, and more conserving of natural and cultural resources. The intent behind this movement was, and is, well meaning. It grew out of an increased awareness of the destruction wrought by global consumerism and has sought to change that; in the words of the movement itself, to "make the world a better place through conscious consumption." People set out to reverse the course of destruction wrought by consumerism, through a different type of consumerism.

Decades before the Green Movement emerged, a similar political movement was embraced by even larger segments of the population. Progressive politicians and activists worked through the political process, legislating for increased social justice, revamping laws to clean up waterways, and regulating the processes of modern industry to better protect biodiversity and do less damage to Planet Earth in the creation of "products." Indeed, much of modern "progressive" politics can be seen as an attempt to minimize the damage wrought by the increasingly destructive ways citizens of this nation make their living. And with each decade Americans have moved further away from domestic production toward an ever more globalized, colonial resource relationship, all the while exponentially increasing the take-make-waste capacity of each citizen.

No doubt this movement toward no-VOC paint, ecotourism, green building, compact fluorescent lightbulbs, organic foods, fair-trade goods, low-flow fixtures, hybrid vehicles, and more stringent regulations slowed the rate of cultural- and natural-resource obliteration, but it has not reversed the trend. These progressive consumer and political movements of the late twentieth century failed to change the underlying structure that gave rise to massive human-ecological unsustainability in the first place. Radical consumerism and its transference of value from two-thirds of the world's humanity to the richest third continued unabated, further bankrupting Earth principal (biodiversity, soil, fresh-water and clean-air reserves), mining human capital (physical, intellectual, and emotional health of individuals and societies), and looting value from distant places and from future time periods. Thus, despite these movements the scope of human destruction continues to expand rapidly into the twenty-first century with

▶ Greenhouse emissions of nations that ratified the Kyoto Protocol still on the rapid increase
▶ Tropical deforestation accelerating
▶ Nuclear-waste production increasing
▶ Species extinction accelerating
▶ Resource-related warfare on the rise, with concomitant waste in money, energy, and lives
▶ Overall biospheric toxicity increasing faster than at any other time in the past 400,000 years, at least

Confronting the fact that the social justice and green movements (let's call them "surface movements") have not succeeded in changing the human trajectory away from perennial emergency toward a positively evolving, healthy, peaceful world forces us to recognize the structural forces that are at work. We start to see how surface movements have served largely to distract us ("Let them have green products" instead of "cake"). The most meaningful forces determining the resource relationships between humanity and Planet Earth operate largely beyond the influence of these movements. So how do we effect meaningful change, recognizing that our choice of dish detergent or fair-trade goods is not going to change the underlying drift toward deepening catastrophe?

Exodus from Consumer Society

Sometime in the early twenty-first century, the systems that had concentrated wealth in the hands of the few—the same systems that had become the most dominant social-organizing systems on the planet—began to slowly become unhinged. A few generations of accumulating instability from the system's sheer scale and

Table 1.1: Typical Consumer-Based Actions and Solutions That Address Problems

Issue	The "Less Bad" Consumer	The Producer
Food	Buys organic groceries	Grows a vegetable garden, maintains food trees and berries, raises animals
Waste/ Energy	Recycles	Buys less and produces, processes, and stores more
Social Justice	Donates to a national charity	Organizes neighbors to alleviate a local problem
Energy	Conserves electricity	Produces electricity with solar, wind, water, or wood, or doesn't need it
Water	Buys a water-conserving appliance	Harvests rainwater and greywater and cycles it on her land
Policy	Votes once a year	Organizes with neighbors, meets with elected officials, holds town office

depth of injustice will at some point, as it always has historically, overwhelm the system's capacity to contain its own fallout. What if the same cultural process that stimulated the social-justice and green causes coalesced into a massive force and began to replace consumer society itself with a society of producers based in decentralized, egalitarian, human-scaled, smaller units of organization? What if crises stimulated this process out of necessity? This shift is beginning to happen, especially at the home scale.

Ask yourself what actions you can take to harness this transition away from a consumer society that belittles your own humanity to an organizing force that fosters individual empowerment—a liberating and enlightening cultural revival that replaces consumers with producers, hyperdependency with self-reliance. Table 1.1 highlights the relationship between typical consumer-based actions and solutions that address

problems (classified as "Issues") at a deeper, more systemic level. The categories are *not* mutually exclusive: Actions defined as "Less Bad" often support the regenerative "Producer" action but by themselves usually will not result in meaningful, long-term change at the societal level or empowerment at the individual level. This is just the tip of the iceberg. Starting down this road opens the door to scores of other possibilities. The lifestyle of the producer can actually be far more stimulating, complex, and interesting than a consumption-oriented way of living.

Becoming Useful in the Transition

I am often asked at the end of presentations a question that goes something like this: "I have limited time, money, and skill. What would be the most important things to focus on in becoming more adaptive to the

HOMESTEAD ENERGY FLOW

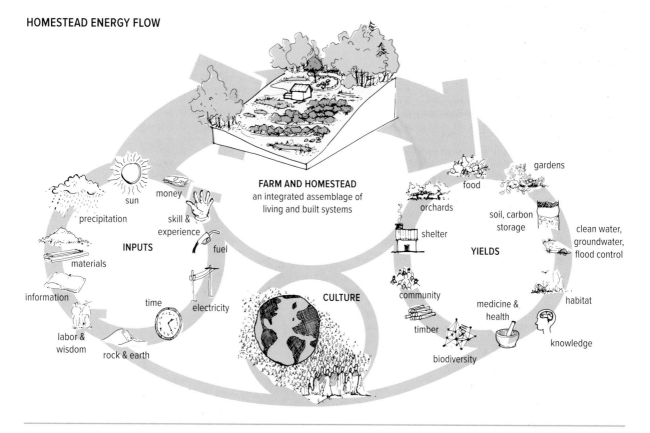

The overall flow of energy and materials through a regenerative and resilient homestead and farm is complex but designed and managed to continually generate value on- and off-site, while requiring fewer off-site inputs over time. A functional society of free people is necessarily built upon a foundation of such productive systems.

changes now underway in the world?" This is a difficult question, and the answers necessarily vary by region, one's existing skill set, the area of interest, and the physical and social context of the transitionee. However, it seems useful to at least offer an attempted distillation of answers to this question. This list, of course, is not exhaustive. It should be useful in helping us ask the right questions and begin or continue to think more clearly about our resources and what remains to learn, acquire, and develop as we attempt to become more helpful and resilient members of a rapidly changing world. The steps toward a more resilient lifestyle can be thought of in the following order and categories: Empower/Mind Shift; Root/Community; Harvest and Cycle; Shelter; Feed and Vitalize.

1. Empower Yourself: Reskill and Reattitude

Master something people *need*, not simply what they want: food, clothing, shelter, information, tools, wellness. Remember, skills are one of the only things no one can ever take from you. Skills also tend to accrue, and many of them only accumulate over life, rather than wither. We may at times be rich or poor, socially connected or relatively alone, but for the most part our skills are ours to keep. Our attitude manifests everything else. If you don't believe you can do something, you certainly won't. Those who learn fast, adapt to challenges in their lives, and are generally "successful" in any way, shape, or form in this world tend to have some traits in common with each other: They believe in themselves and act from a place of confidence, not fear; they tend to treat change as an opportunity; they do not bemoan challenges or loss; they look ahead; they look practically at the future—sober in what challenges it may bring and hopeful in what opportunities may emerge. All the other steps described below to becoming useful and adaptable in the transition can be achieved with relative ease if you empower yourself in skill and in attitude.

2. Establish a Land Base and a Community: Put Down Roots

After a solid skill set and adaptable attitude, land and community are the two most fundamental tools for leading a resilient life. How will you connect with a piece of land in a long-term way? What legal or other agreements will allow you to stay on that piece of land and work toward its empowering you? Who are your people? How will you find them or attract them? Do you need to change location, career, outlook, or even paradigm to find them? Do you need to lead for others to follow? These questions must be addressed and pondered, if not necessarily answered, before you can proceed to the next step.

For wanderers who think one can be highly resilient in a nomadic way, I do not mean to discount that path. It seems plausible that a group of people—not an individual—could indeed develop some level of resiliency within a nomadic lifestyle. The challenge, however, becomes one of starting from scratch every time: Unless you're a scavenger or a raider and live on the produce of others, you need access to land to be a producer yourself (barring the truly nomadic tribe lifestyle of ages past that hunted and shepherded). So at best, the nomadic path requires one to constantly be investing in establishing relationships with new places, new people, and new land. That approach hits the reset button constantly. While the nomad is all the time reinvesting, the person who chose to set down roots is growing shoots, putting up branches toward the sky, connecting with those who have rooted around them. This does not mean that one should not be ready and willing to pick up and move if necessary, but that certainly would be a major blow to the lifestyle of one pursuing resiliency and regeneration.

3. Harvest and Cycle Energy, Water, Nutrients

The productivity of the place you dwell is next in the order of establishing a viable lifestyle for the long haul. The land's ability to produce is dependent upon its ability to capture sunlight, rain, snow, wind, atmosphere, and other forces and transform those forces into food, medicine, fuel, and other yields you need in a particular location. That transformation depends on sunlight's being processed through functional water, soil, plant, fungal, and animal systems. It takes time to establish these fertility systems. Years. Best to get started now. These systems include compost, humanure, greywater, and nutrients salvaged locally. Because most other systems depend on the productivity of a site, making your place as fertile as possible should begin to happen as early as possible when you establish a home in a new place. It is this step and the following two that are the focus of much of this book.

4. Develop Passive Shelter

Also at the beginning of establishing a resilient lifestyle is the need for shelter. Ensuring that this shelter goes up quickly and is as nonreliant on off-site resources as possible is a baseline part of the whole system. Keeping things simple here pays off in spades. You can always add comfort—read complexity and often expense—later. A highly functional nondependent shelter is most easily achieved if it is relatively small and very well insulated; has wood heat, gravity-fed water and a hand-pump backup; and has a cool/cold location built into it for food storage. Shelter fit for a dynamic future is fixable and adjustable by its user over the long haul.

5. Learn to Cultivate and Wild-Harvest Food, Medicine, and Fuel

Food production and foraging is fundamental; so too, for an increasing number of people is the production and processing of medicine and fuel. The more self-reliant and skillful one can be with all the fundamentals of life, the more one can thrive through the period of change the world is entering. To survive and thrive in this changing future, one must be fluent with most or all of the basic production systems, including raising vegetables, foraging and hunting, animal husbandry, and growing perennial crops, such as fruits and nuts.

When Systems Fail: Emergencies and Resiliency

In my work the lines between planning a landscape and planning a lifestyle are becoming increasingly blurred. From increasing climate shifts to global economic insolvency, and the various instabilities these set in motion, sound planning for both land and lifestyle looks forward and aims to respond ahead of the curve—where response is most strategic (before the flood, before the well dries up, before the dollar tanks, before a gallon of gas is six bucks). It's becoming increasingly clear that both our habitats (land and infrastructure) and our lifestyles (and the community they are connected to) must adapt to increasingly rapid economic, social, and political shifts that ultimately determine what ways of living will be more or less viable.

And viability or—more accurately—*resiliency* is what we're after. That means responding ahead of the actual event, swinging the bat before the ball whizzes by. Responding deftly to changing conditions is at the core of successful adaptive responses in all creatures great and small on this particular planet. This holds true both for individuals and for species. To be adaptive we must extrapolate current conditions into specific future conditions that guide our planning. Since we can't accurately predict the exact future, we must entertain diverse scenarios and plan around a selection of them—including those that present acute challenges—emergencies.

Any rapid changes in living conditions, whether from a job loss, an injury, or a global catastrophe, can be adapted to far more easily when a group of people can rely upon one another.

Elderberries producing heavily in their fourth year. Preparing is always best done well ahead of time since systems take time to establish—food systems often require years to yield, no matter how much money or effort is thrown at them.

SCENARIO PLANNING AND ACTING

Planning for the future is greatly aided by organizing around specific events and their particular consequences—aiming to avoid the "wish I'd done that" pitfall common to generalized planning. While we cannot go through all scenarios that are worthwhile planning for, we can "mock up" some aspects of almost all of them. For instance, you can't simulate a nuclear power plant spewing radiation into your neighborhood, but you can run a fire drill in which you do what you'd need to in such a situation: communicating with family and friends, sealing up the home and garden, potentially evacuating to a predetermined location, and so on. Only through activating these scenarios can

you learn the particulars—often crucial details—that will empower a more successful response when the event is not a drill. Running event scenarios, however, takes time, and not everyone has the time or resources available to run such drills. Therefore, a combination of drills (event acting) and event planning is practical for most people to carry out as one decides what to prepare for and how.

Since we do not know which events and which particulars will come to pass, we should think through many of them; scenario planning with diverse possibilities is most helpful. Once scenarios are laid out before us, we can decide which of those are most important to actually practice around. It is most effective to start scenario planning and acting with foundational questions specific to each of our living situations. These questions are identical from location to location, but the answers vary (sometimes) with location, people involved, and other context aspects.

SCENARIO MAPPING: PLANNING QUESTIONS

What events can happen globally, regionally, and locally that will impact me in significant ways? What would it mean, and what will the results be in each situation? What is currently in place in my life that can improve my ability to deal with each of these events? What aspects of my life currently present the biggest challenges to dealing with these events? Who would I turn to for help, and who would need my help in each event? What tools, equipment, skills, and other resources would be utilized in each event? And finally, what resources would most likely be lacking?

As you go through this mapping process, be sure to note, especially, all strong points and missing links or weaknesses. Write them down, and make an action plan for addressing each of these. It's also very easy to know something is a weakness but forget to do anything about it. For instance, I always knew in the back of my mind that my lighting situation for dealing with power outages was not ideal, involving only multiple headlamps, batteries, and candles. When I did lose power a few years ago, it became apparent that the light from my Coleman Lantern was so much more helpful than candles or headlamps when a group of us gathered together to cook and eat a meal. Although I had the lantern, I did not consider it part of my lighting plan and only had a few ounces of old white gas for it. Now, I know its value and have put up a couple of sealed jugs of white gas, which, unlike gasoline, lasts a very long time, and have procured another lantern, since two is one, one is none.* And one sure is useful.

Working through each of these questions is no small task and will probably take multiple years to do thoroughly. It is most helpful to record all of these on paper as you go and, ideally, go through them with at least one other person. It is easy to miss key aspects, and individuals also vary in their needs and desires, which affect the answers to the above questions.

There are many other important emergency preparation approaches we utilize on the home and farm and the details of them could (and likely will) fill another book. However, due to space limitations this topic is out of the purview of this book, but is explored in great depth during our workshops on the farm.

---

* I first heard this phrase from Jack Spirko who runs the vast and valuable resource of the Survival Podcast and its forums.

Chapter Two
The Design Process and Site Establishment

The design process is the starting point—if there ever is a discrete point of initiation—for making a regenerative and resilient place a reality. It's probably more accurate to say that the design process is the period when we spend more of our time focused on what we should do (goals), what's possible (analysis), and how to do it (the plans) than at any other part of the thinking-making continuum. The boundaries of the design process include any observations, drawing, photographing, conversations, and hard plans that one engages in during the development of a place.

It is essential to remember that this process does not stop once the shovel hits the ground. Designing is a constant state of being, and when engaged in the world as a problem solver, you never turn off the tendency to notice a suboptimal situation and think systematically about how to improve it. Design processes can take many forms, and no one approach can be prescribed as the best for all people in all places at all scales. However, it can be said that any effective design process is rooted in intense engagement with the problem at hand and the world in which that problem resides. This is often where "modern" design fails; while the problem may be engaged, the world in which the problem lives is often ignored. This, of course, is a logical consequence of the modern "designer" being someone that works inside a building called a studio. The designer, if she is worth the tracing paper she uses up, is someone digging in the dirt, talking with the neighbors, sitting quietly waiting for the sun to rise at daybreak, and embedding herself in the subject at hand (which is always a *place*) in countless ways. This section offers a brief account of the approaches and frame of reference I have employed to think clearly about and activate functional human habitats.

Planning and Design: Observation before Action

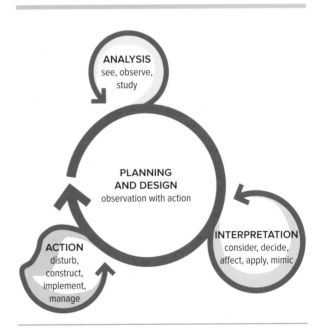

Site planning should be continuously fed by a never-ending process of analyzing, interpreting, and acting. Illustration courtesy of Whole Systems Design, LLC

When beginning to plan a project and the establishment of systems, the primary task is imagination. We must envision what the site could be like—and should be like—in one . . . five . . . twenty . . . a hundred years hence. Taking the time and mental space needed to do this both in the beginning and repeatedly as the site unfolds is crucial; remember, site analysis is never complete because the site is always becoming something new with each passing week, month, year. So when I say "should" be, I mean an intention that is open to new information coming in as the site unfolds. We must be able to both intend a site with will power and listen to the site as it evolves—both, not one or the other alone.

The Designer's Set and Setting

Successful problem solving is based to a large extent on the designer's learning environment. As those seeking to be adaptive and successful in the face of a world in constant flux, we are all designers by definition. We are all problem solvers. As problem solvers we must be several things, including:

> **In the realm of living systems no design is ever complete.**

▶ **An observer:** We must notice what is happening in the world around us, both immediately at hand and globally. This awareness must observe patterns in both physical conditions (space) and events (time). And the extent to which we notice and "see" is highly dependent on what we know and have language developed for. Seeing clearly is also dependent upon acceptance and a constantly repeating initial nonjudgment of what is happening around you. Clear analysis depends on seeing things as they are—jumping to thoughts about why something is present or why a process is happening the way it is clouds a clear vision of what it is.

▶ **An interpreter:** As we notice patterns we must be able to translate that raw data into intelligent hypotheses and strategies for action. The ability to perform this "if-then" part of the design process is enhanced especially by associative reasoning skills

Students in the Whole Systems Applied Permaculture Design Course working on their schematic designs.

and direct familiarity with ecosystems and how they function across time. One can think of this part of the process as *ecoliteracy** *in action*. It's also important to understand leverage points and how to get the most result for the minimum effort—ecosystem management tai chi or utilization of Buckminster Fuller's "trim-tab" principle.† Since there is always far more adjustment to the system than we have time or energy resources to accomplish, the designer's task occurs in a constant state of prioritization—of tweaking the trim tabs to adjust the movement of the entire ship using as much leverage as possible. If, for a moment, you think the system is complete, done, perfect, you haven't looked hard enough—there is always optimization that can be facilitated.

► **An activator:** Following identification of a need or challenge and the interpretation of how it might be met comes the action part of the process—the digging, planting, hammering-nails phase. This requires mental and physical skills, patience, good judgment, pacing, the ability to work smart and apply our physical force via our bodies or other tools (including machinery) to the project. As we engage in this work, it is important to do so with great awareness; don't fall into the easy trap of lowering the head and forgetting to look up from time to time to see how the work unfolds.

Actually, it is only during this activation phase that the true accuracy of the identification and interpretation phases begin to become known. As we hammer nails and raise the walls of a house, we start to see if the design was a good one. As we look back at lines of plants in the ground, we begin to know more completely if they are truly in the right locations. So begins the cycle of think-design-do-reflect-redesign-do. It's

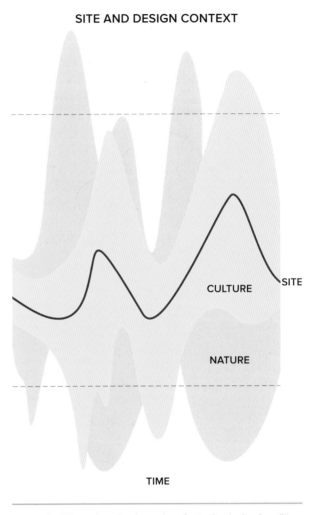

SITE AND DESIGN CONTEXT

CULTURE — SITE

NATURE

TIME

The site should be evaluated as the product of natural and cultural conditions across time. Graph courtesy of Whole Systems Design, LLC

a never-ending cycle—in the realm of living systems, no design is ever complete. Plants grow, soils build and change, animal habitats change, and people shift in interest and available time and capacity. The entire system is always in a state of flux, with all parts moving relative to both the larger world and to each other—like one giant mobile rotating upon itself in the breeze. To function intelligently and with some level of mastery, we must always be willing to look up, stop, take a breath, and reassess.

Indeed, the clearest measure of the value of our approach in regenerative systems development only fully emerges over the course of years, as tree crops

---

* The concept of ecoliteracy can most directly be traced first to Fritjof Capra's work of the same title but is now a widely used term.

† "When I thought about steering the course of the 'Spaceship Earth' and all of humanity, I saw most people trying to turn the boat by pushing the bow around. I saw that by being all the way at the tail of the ship, by just kicking my foot to one side or the other, I could create the 'low pressure' which would turn the whole ship. If ever someone wanted to write my epitaph, I would want it to say 'Call me Trimtab.'"

mature, as soils become what they can be, as populations of insects, fungi, and animals interact with the system. In year one we know little; by year five we know something about the system, but not until year ten or twenty can we start to draw conclusions about much of how the system functions. And a few decades in, if we are ready to be surprised, to be wrong, we will continue to learn. Eventually, if you are actualizing yourself in your place, you become the world's foremost expert on how to help make that place work—on what it was, is, and can be—and how to get there.

The ideal design for a site is never static and is always evolving according to such constantly changing variables as

- ▶ Human resources available
- ▶ Climate
- ▶ Soil conditions
- ▶ Succession stage of ecosystem on site, including maturity of plants and animals
- ▶ Outside resources available off-site: money, materials, energy
- ▶ Off-site social conditions: stability of society, needs and desires of neighbors, people in the region and beyond—movements in "the market" at large

Site Establishment Leverage Points

Since time, money, labor, and other forms of energy are always limited in some capacity, we must always be seeking the most powerful ways of effecting regenerative change in a landscape. The regenerative land worker's routine must always be guided by finding those methods that offer the most influence for the least input, since it is these inputs that are limited. Compiled below is a list of the top five leverage points I have found in cold-climate regenerative agroecosystems establishment and maintenance. These are the strategies that offer the most potent positive impact relative to the smallest energy and time input (most of these usually happen most optimally in concert with one another). These actions should be seen as the land worker's primary tools for change making:

- ▶ Swale construction and fertigation via swales
- ▶ Grazing animals, especially large ruminants and chickens
- ▶ Broadcasting seed and planting live plants
- ▶ Cutting and clearing trees
- ▶ Scraping and tillage: mechanical bare-soil creation by hand or via bulldozer, excavator

Recall these primary approaches as you attack any land-based problem on your site, and remember, too, that they are only those I have found most essential. There are other important succession-altering tools I have found extremely useful, such as fire, but that are not essential enough to fit within the first edition of this work. Stay tuned to the next edition for elaboration on these techniques and the likely inclusion of others that are emerging on this farm.

Ecosystem Management: Steering Succession

The area of the world I live in wants to be forest. Fueling a merino wool gold rush, a timber rush, and the charcoal (metalworking) industry, 80 to 90 percent of my home state of Vermont was cleared of its forest only a century and a half ago. Yet without replanting it is nearly entirely reforested today. We are fortunate to live in a place and time where the right combination of moisture consistency across the year and temperature patterns promote the establishment and growth of trees on all surfaces of the landscape except on the steepest cliffs and open water.

From the homesteading and resiliency perspective, however, today's forest cover is as optimal as its composition and health are natural; that is, simply because the forest is here does not mean it is inherently healthy or optimized in terms of species composition, stocking density, species health, or soil and water health. Today's New England forest is simply the result of the succession taking place in the wake of the nineteenth-century abuse and abandonment. Species such as chestnut were replaced with white pine in this landscape change. Precolonial forests also contained a much higher concentration of black walnut and other mast-yielding trees, in all likelihood because of the native human

OIL TO SOIL—USE IT OR LOSE IT:
LEVERAGING THE CHEAP-OIL WINDOW FOR MAXIMUM EFFECT

When I started developing the WSRF about ten years ago, I did so with the idea in mind—an academic and ideological one—that it would be best to do the work necessary without using fossil fuel in the process. However, in establishing the Whole Systems Research Farm, we have found several actions that apply liquid oil energy more directly than others for the maximum result. These actions accelerate the rate of site development for the better and take maximum advantage of oil energy. Could these actions be accomplished by hand? Of course (and they once were), but the site's rate of development would lag behind the current trajectory by many years.

Given that cheap oil will obviously continue to be used and abused by various groups—most notably the US military and citizens for driving and space heating—we have made the conscious decision to take advantage of the small window of time still remaining with which to develop intergenerational land and infrastructure systems, which greatly enables long-term production of the site without any oil input for hundreds if not thousands of years.

Indeed, the earthworks built literally overnight by using cheap oil energy will last until the next glaciation, which we hope will be many millennia from now. I would not want to explain to my children or grandchildren when pressed why with all of the cheap energy and machines

we didn't choose to optimize the shape of their farm, construct infrastructure, and do the heavy work that now they must do with a strong back and years of heavy toil.

► **Cutting trees and processing firewood with a chain saw:** No other use of liquid oil compares in bang for the buck with crosscutting wood with a chain saw.
► **Earthmoving** and **digging** via excavator and bulldozer, for making swales, paddies, terraces, foundations, roads or paths, and so on.
► **Stone wall construction** with boulders, not small stones, with excavators for retaining walls and terraces.
► **Making electricity via generator for power tool use**; most notably, for ripping boards on a table saw. Ripping boards (cutting lengthwise) by hand is enormously labor intensive and slow. Note: Most conversion into electricity using liquid energy is *not* a sensible conversion.

The least productive, highest-entropy ways to convert liquid fuel into work includes the following:

► Light mobility: especially, moving people
► Space heating: that's simply insane
► Making electricity

presence on the land, during which specific species were promoted—those that fed people (directly) and wildlife (indirectly) the most were selected for. Think of Johnny Appleseed's work being applied across a range of productive species by a million human beings over at least a few thousand years. In this perspective it's not hard to see how the forest was so vastly different from what it is today.

> Nut trees produce more fat and protein per acre than other trees and therefore can grow more wildlife per acre and distill sunlight into a more potent value.

For example, the Northern Forest cover type today contains a lone nut-bearing species, the American beech (*Fagus grandifolia*). Nut trees produce more fat and protein per acre than other trees and therefore can grow more wildlife per acre and distill sunlight into a more potent value. For those interested in producing the most resources (outputs) with the fewest inputs possible, the imperative here is clear: We need a greater percentage of the forest to be composed of nut-bearing trees. We also need to develop those land-use systems that may in fact not be forest, if and when they convert more sunshine and rainfall into potent values than forested land does.

It turns out that intensive rotation grazing has a uniquely high ability to convert subsoil to topsoil and, in the process, sequesters more carbon from the atmosphere into the geosphere (as organic matter) than any other cropping system. We are beginning to realize synergies between such grazing strategies and permanent tree cover and recognizing the wisdom inherent in millennia-old food systems such as the Spanish dehesa system, in which animals are grazed in and around an open forest (woodland), with yields of timber, nuts, soil, meat, and even milk all overlaid on the same piece of land. We now call this approach silvopasture to indicate the simultaneous growing of trees and animals. It is becoming increasingly clear that much of the cold-temperate climate landscape should be in some form of permanent silvopasture. (See chapter four on grazing systems for a more in-depth look at silvopasture strategies we are employing in particular.)

Simply because a piece of land is currently forested does not mean that such a forest is optimal for the health of the ecosystem or most productive for the human members of that ecosystem. This idea is strong

Permaculture can be thought of as applied disturbance ecology.

and pervasive—and one that I had for many years when I spent more time in the academic study of land than in daily interaction and participation with it.

We must confront the myth of the "natural" and the tendency to call something natural simply because it seems to exist free of human management. A regenerative agent in the system must take his vision beyond this snapshot and romantic image and see the current landscape as simply one ecological succession unfolding, one scenario among many possible, which happens to be the result of past disturbances and the particular events occurring immediately after these events. Permaculture, in this view, can be thought of as applied disturbance ecology. What disturbance forces have been at work

Every site is a response to past disturbance forces. Our task is to make conscious those disturbances and steer the system's response to them in such a way that the outcome is as biologically complex and productive as possible.

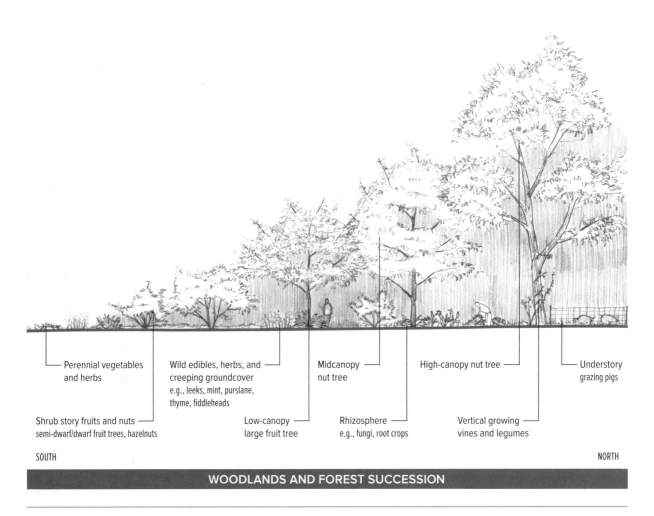

Perennial vegetables and herbs

Shrub story fruits and nuts
semi-dwarf/dwarf fruit trees, hazelnuts

Wild edibles, herbs, and creeping groundcover
e.g., leeks, mint, purslane, thyme, fiddleheads

Low-canopy
large fruit tree

Midcanopy
nut tree

Rhizosphere
e.g., fungi, root crops

High-canopy nut tree

Vertical growing
vines and legumes

Understory
grazing pigs

SOUTH

NORTH

WOODLANDS AND FOREST SUCCESSION

Foundational to the design process is an understanding of the inherent potential of the existing ecosystem and what the site ultimately "wants" to be. Illustration is not to scale. Illustration courtesy of Whole Systems Design, LLC

on this landscape in the past? What species have been filling niches as they open and why? What would the most optimized landscape here look like in terms of ecological and human health—what is configuration of water flow, species composition and their arrangement? What changes (disturbances) are needed to evolve the current system to the more optimized one?

Disturbance forces include:

► Cutting vegetation (low or high on the trunk) or uprooting vegetation
► Grazing, human foot traffic
► Tillage/excavation/removal/scraping (by machine or animal)
► Fire
► Seeding
► Flood
► Drought
► Heat and cold
► Wind
► Pollution, chemical
► Pollution, radioactive
► Pollution, genetic (disease, GMOs, mutational)

These are most of the tools at our disposal with which we can modify the direction of succession in a place—not that we want to use the last few, mind you. Those of us in a forested region of the world must start site development with design questions that include these: What aspects of the existing forest cover are functioning toward resilient, regenerative, and productive ends? What species, what characteristics of the water flows, human interaction with the system are most limiting to the promotion of a more diverse and biomass-producing condition?

Resiliency and Regeneration Principles

While giving a tour of the research farm recently, I stopped by the rice paddies, as I do on most walks of the farm. A student in the group asked questions I hear often: "How often do you water the paddies, how deep should the water be, and how do you know when to add water?" I began to think of the answers, and within seconds was inundated with various potential answers. I almost began to provide what have been the varied answers to these questions, including the state of the rice; the temperature of the day; the past days' weather and the forecasted weather ahead; the time of the season; the condition of the rice as evidenced by color, size, and overall vigor; how busy I am; and many other factors. Then I realized that all I could say was, "It depends."

What to do in a given situation when working with land always depends on the conditions one is facing, and the conditions are always myriad. Those conditions are also in a state of dynamic flux. This is acutely challenging for most people who are products of the industrial schooling—often an unlearning—system, which trains people to follow discrete sets of instructions when addressing a problem. Life, people, and the relationships between them, however, are far too complex, dynamic, and nuanced for rote instructions to be effective most of the time. Habit is not sufficient to solve problems, though it can be useful; it must be coupled with awareness and novel responsiveness to novel conditions.

The land system is not a machine—it doesn't function in merely mechanical ways, though it is in part mechanical. This is probably why people are easily confused and end up habitually managing land as they would a machine. The rub is, however, that it

> **Habit is not sufficient to solve problems, though it can be useful; it must be coupled with awareness and novel responsiveness to novel conditions.**

also functions in far more complex ways beyond the patterns of a machine, or nonliving system. The land system is *alive*; thus, in a constant state of flux, evolving, responding, adapting, adjusting. It is never the same thing from one month to the next, one day to the next. Thinking it is the same thing leads us to conclusions that are at best ineffective, at worst dangerous. Relating in a way that truly appreciates and accounts for the complexity of the living land system is not mysterious or difficult—it is no different from relating to another human being. Healthy people recognize the complexity and changing nature of other humans: We wouldn't say the exact same thing to one another every morning over breakfast, act in the same manner to one another each day, year to year. Of course not, as people's needs, desires, and overall contexts change. Healthy interaction is *responsive*—always based on the conditions of the moment and on past patterns and future goals.

Healthy interaction is dynamic, elegant, soft, improvisational, but not robotic. Most of us know this on a human-to-human level. Yet when it comes to land interactions, we tend to think repetitive actions are appropriate, as if the land system is the same from day to day, year to year. Truly, at its root, the idea that we can figure out some aspect of the land system and think that management needs should stay the same from year to year is insane. No complex system works that way—and such an approach is completely blind to the in-flux nature of reality as a whole. Modern industrial schooling and the unlearning process it tends to facilitate are highly effective at patterning people to act in this way, however.

The following principles and strategies represent some of the guiding directives that I have identified in the work I have been practicing. The list is not complete—therefore, please do not be limited by them. There are dozens more that apply less often in my practice or that I have left out for brevity. There will likely be many directives that need to be identified in your own endeavor that do not appear here—especially if your climate, scale of work, focus, and other contextual aspects vary significantly from my own. The process of discovering these directives is a rewarding one, and I encourage you to continue that journey in your own life; it's a personal one, and the only way to amass plenty of clues in this process is by getting your hands dirty. Enjoy the process!

RESILIENCY AND REGENERATION DESIGN

1. Maximum Outputs for Minimum Inputs
The optimal system yields the most value—in quality and quantity—with the fewest inputs. Our task always involves maximizing the ways to grow the most value while reducing dependence on off-site resources to do so.

2. Transform Dead Matter into Living
Regeneration relies upon the upcycling of matter, ultimately based upon a foundation of bedrock, atmosphere, water, and the elements. Our task is to facilitate the conversion of rock to soil, soil to plant, plant to animal, animal to soil. And the cycle continues, each time accumulating a net gain in value—with more organic matter existing than the round before. In this way we

can think of the entire regenerative land-use practice as an attempt to transform inert material into as much living, breathing, organic material as possible—the earth itself as one huge compost pile upon which we grow.

3. System Establishment versus System Maintenance
The needs of a site during the early months and years of establishment vary greatly compared to the needs of the system over decades and centuries. It is often sensible to use tools, materials, people, and other resources in the present moment for system establishment that may not be available or desirable in the future for system maintenance; for example, an excavator to make rice ponds or paddies.

4. Biological Complexity, Technological Simplicity
Resilience is greatest when living aspects of a system are complex, diverse, and connected, while the nonliving aspects of the system are simple. This is rooted in the fact that technical systems are constantly prone to entropy and are always moving toward failure, whereas living systems actually tend to build higher levels of order over time. Living systems amass sophistication, durability, and productivity. As resiliency seekers we aim for a system in which ever less time and energy are spent on infrastructure maintenance so that time can be dedicated to cultivating and optimizing the living systems on-site.

5. Resilience = Diversity × Redundancy × Connectivity × Manageability
The ability of a system to recover from disturbances is highest (1) when the system is composed of a high diversity of elements; (2) where there are backup elements to all crucial components of the systems; (3) where the connections between each component form a web with as many connections and modularity as possible, but (4) where the system is simple enough to be legible, manageable, and accessible for human participation—where the system's needs for optimization do not overwhelm the capacity of the human occupants to help meet those needs.

6. Regeneration Metric = Biomass and Biodiversity
Though impossible to simplify into any single formula, the most concise way I have been able to define

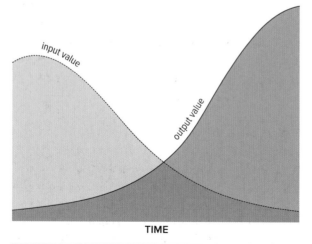

REGENERATION AND RESILIENCY DESIGN GOAL

input value

output value

TIME

Maximizing outputs while minimizing inputs across time

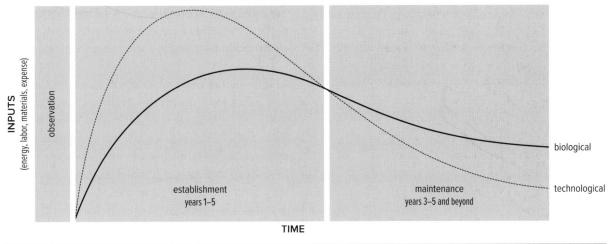

A high level of inputs is needed during the establishment phase to create a sharp rise in the biological activity and infrastructure that will last and add value over time.

whether an action is regenerative or not is to evaluate the answer to the following question: "Is the action increasing biodiversity *and* biomass?" Conventional farming, of course, is focused solely on biomass production, while conservation biology and ecological restoration is focused on biodiversity preservation and increases. Permaculture aims to increase the interdependency upon each other.

7. Facilitate the Vital Force

No hard line exists between living and dead matter. There is an animating presence in all living beings that seems impossible to isolate. This animating force makes regeneration possible, and our work is always to encourage its expression and to align ourselves with it, to receive its gifts—for the force is abundant and what it provides are most accurately described as gifts.

8. Human Management = Primary Limiting Factor

I have found that even on just ten acres, space, soil, water, infrastructure, and skills do not constrain the optimization of the site more than the capacity of the human inhabitants of the site to provide time, labor, and awareness services in the development and maintenance of the site. This principle is closely connected to Bill Mollison's "yield is theoretically unlimited" statement. Indeed, the yield of a system, since it is

the synergistic product of air, water, soil, and many other components, is not limited by any one, and only one component leverages them all together—human management.

9. Stress as Stimulus

We are after the rhythm between rest and stress that promotes the most biodiversity and biomass.

10. Responsiveness, Not Habit

As mentioned in other areas of the book, the most effective actions, though sometimes stemming from positive habits, are not limited by habit but are informed by habit. Such habits promote awareness. This awareness allows us to see what novel ways of responding to conditions are required. Responding to new and emergent conditions (always the case) demands that we can act in ways we have never acted previously.

11. Human Resource × Site Characteristics = Ideal Site Design

An "ideal" site design, though hypothetical, is a useful goal. This selection of elements and their positioning in the landscape is never the same from site to site because they must be wholly responsive to conditions (of both humans and nonhuman components) that always vary from location to location. Therefore the "ideal" design

SITE AND CLIENT

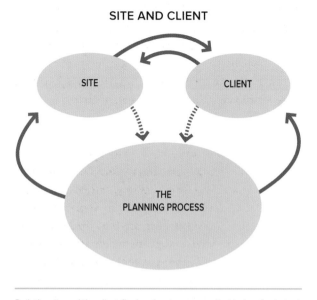

Both the site and the client feed a planning process that in turn feeds back into the site and the client through implementation.

must be completely referenced from the site's characteristics and the human resources in place, which include number of people, interests, skills, ages, and other factors. Give five groups of people the exact same site, and their five designs should not be the same—no one design is correct in such a situation—and the best one is so because it is suited in part to the humans inhabiting the place, not just the other site conditions. In this way there is no "correct" design for any one site. What's true for you is going to be different from what it is for others. Plan on it. So don't simply copy another's approach thinking it will give you the same results, as, likely, it won't.

12. All Design Should Be Modular

Since the future of both land and person(s) is unpredictable and guaranteed to change, good design and developments must be able to be added to, subtracted from, moved around, and adjusted constantly over time.

13. Structural Diversity Begets Biological Diversity

Biological diversity is often most limited by the physical three-dimensional structure of a space. This is a commonly cited principle in marine ecology that is highly visible in the example of a sunken ship landing on a bare

sandy sea floor. Life flocks to such a situation where little existed before. So, too, is this phenomenon in effect in terrestrial systems seen especially at the edge between field and forest, in swale-mound systems versus flat fields, and other situations where three dimensions of complexity exist rather than simply two. Promoting this positive effect encourages us to develop structure, whether it is buildings, swales and other slope elements, trees within fields, or other such spatial changes. Organisms exploit edges and structure constantly—when you add structure you see the results quickly.

14. Habits of Mind

The outcome of any action is highly determined by the mental frame of reference used by the actor. One has the power to shift this without dependence on outside events, people, money, or other resources, so one's own attitude management is highly empowering.

15. Spread Pulses

The most resilient systems spread intensities across time for maximum value absorption into the system. Examples of this include slowing, spreading, and sinking water via swales, terraces, ponds, and paddies; thermal absorption in high-mass materials such as stone and water for release when ambient temperatures are lower; fertility spreading via keyline ditches from concentrated areas such as barnyards to low-fertility areas; and delayed processing in the harvest season when crops can be put aside for processing because time is more plentiful.

16. Disperse and Extend Fertility

On all sites there are zones of nutrient concentration and/or high productivity. It is the job of the regenerative designer-maker to spread such fertility from areas of concentration to areas lacking fertility. This strategy includes both dispersing fertility across space and extending fertility across time; for example, space: biomass harvested from a productive pasture moved to feed animals while they are in less fertile paddock; and time: using humanure/urine generated in the winter to water and feed plants in the growing season. This principle ties in closely with the principle of spreading pulses.

17. Land as Value Distillation Tool

The land system as a whole and all elements in it (including people) is useful to view as a tool for concentrating the most beneficial yields, including medicine, food, fiber, materials, and fuel. A landscape should be thought of as a net that is constantly being cast, through which the gardener–farmer–solar energy angler reaps the most positive interactions among sunshine, soil, rain, wind, plant, animal, and fungal activity. The regenerative and resilient designer-maker is a facilitator in this interplay between forces, the overyield of which can be stored, shared, and accrued as fertility to be cycled (banked) back into the system to continually bolster the principal over time. With the principal constantly being increased, the interest (yield annually) offered by the system can also continually increase over time.

18. Multiple Functions from Single Expenditures (Always Do or Get Two or More Results)

A primary permaculture principle—all elements and actions/processes ideally always yield more than one desired result: A duck fertilizes, reduces pests, and makes eggs and meat; urinating outside allows you to see the status of a plant, health of an animal, or a pest eating your basil, while offering fertility to the site. If anything you are doing seems to yield a single result, closely evaluate if there is not a better way.

19. Moving Things Is Entropy

It's easy to get multiple results from some single actions—think of planting a tree: You're fertilizing the soil, weeding, inoculating, being healthy, and putting a new plant in the ground all at once. Moving something from one place to another, however, tends to yield less value than it costs in time, energy, or materials. Hauling a bucket of water from one spot to the next or driving a cord of wood from one area to another offers little benefit except the result of a material in a new location. Simultaneously, it carries a cost of energy, time, and usually money being spent while compacting soil where one walks, potentially hurting one's back, killing something in the path or road, and so on. The most optimized sites

Every element of a system should serve multiple functions; a duck fertilizes, reduces pests, and makes eggs and meat.

> **The most optimized sites reduce wherever possible the need to move materials from one place to another.**

reduce wherever possible the need to move materials from one place to another. Where moving needs do exist, they are done as passively as possible. Granted, we all need to move things actively: I move a lot of firewood, but every time I move firewood, I realize that the same effort I put into such a task could be applied to tree planting, soil inoculating, plant or animal tending, sowing, or innumerable other regenerative actions. Given this reality, it is important to continuously evaluate what you're moving and why, and how to reduce moving needs so that energy spent there can be applied to more regenerative actions.

20. Value across Time
The most potent values in a system are yielded across the greatest length of time; for example, nut pines such as *Pinus koraiensis* and *Pinus cembra* take twenty or more years to begin bearing but yield for four hundred or more years. Plums and peaches bear within five years and yield for thirty to forty years. Nut pines are made up of mostly rich fat and protein; plums and peaches are tasty but offer mostly only sugar and basic vitamins. The best soils in the world weren't built overnight but over thousands of years. A chestnut can outyield a cow in terms of nutrients without needing any food to be provided for it, but it takes a decade or so to begin bearing, whereas a cow starts bearing quickly.

Examples of durable abundance and vigor in human cultures are always most manifest in examples of people living in close contact with one another and their physical places for many generations. The best things usually require a wait. Working on a longer time horizon than is typical in the early twenty-first century on Planet Earth is crucial to developing individual, community, and land health.

21. Essential Functions Provided by Multiple Elements
If it's essential, ensure that you have multiple ways of provisioning that need; for example, I can get water from the well via pump in normal conditions; via the well via pump via generator when the grid is down and the generator or pump is broken; via a spring and tubing when the grid is down or via rooftop catchment into barrels; and via a 50-gallon storage tank in an attic or ponds and buckets if all else fails. Barring something too extreme to plan for, I am going to have water.

22. Simplest Solution Is the Best Solution
There are numerous ways to solve almost any challenge, but the simplest approach involving the fewest steps and least energy, materials, and time is always the most effective, long-term, viable solution.

23. Efficiency Does Not Equal Resiliency
Simply because a system transfers energy or materials quickly or with little waste does not mean that such a transfer is durable in the face of shifting conditions. For example, watering multiple five-hundred-foot rows of tomatoes with manufactured fertilizer injected into water and distributed via drip lines is highly efficient at what it does *on the farm*. Take any one of the inputs needed to make this system work out of the equation, however, from shipment of fertilizer to pumping of water and you'll see how brittle the system is. Highly efficient systems often actually come to us courtesy of compromised resiliency. Often, we must make a choice between durable, adaptive but somewhat inefficient systems (in the short term) and systems that offer extreme efficiencies in the short run but at the cost of brittleness in the long run. Resiliency necessarily carries with it an extended time horizon.

> **Resiliency necessarily carries with it an extended time horizon.**

24. Increase Diversity, Don't Reduce it
The task of the resilient homesteader is nearly always one of promoting diversity. This becomes particularly challenging when pest issues arise. The conventional response is to remove a biological element when a

pest problem occurs—in permaculture we generally try to figure out what to add instead. Asking the question "What eats this?" is often one of the most useful approaches to such challenges.

25. Quality-Quantity Relationship

In general the smallest production system can produce the highest quality yield, while quality is usually reduced as production scale increases.

26. Scale and Proportions Are the Most Difficult

In the design of any space, remember that it is most often the overall size of a space and its proportions that are chosen badly, not the quantity, type, and position of components in the design. It is often very difficult to get the proportions of a space optimal, and a desirable result is a dynamic product because no two spaces are the same: You can't just copy the proportions from a space you know and have the system work out in exactly the same way in a different location. This is where the "art" of design often comes in. It pays off in spades to get experience into the equation at these junctures in the process.

27. Oil Intervention

As discussed earlier, rather than not using currently available and inexpensive fossil fuel, one can use the existing flow of such a resource to establish systems that do not require fossil energies to operate perpetually. Think of a swale as an example: It requires forty hours of labor to dig, say, two hundred feet of swale or thirty minutes of excavator use and eight ounces of diesel fuel. Not using this eight ounces of fuel (or call it a hundred ounces of fuel with a pro rata of embodied energy in the machine) won't undo the fact that such fuel will be burned up in other avenues such as by the US military or your neighbor in his SUV making an unnecessary trip to go shopping. The point is simply this: Use it or lose it. Intervene in the oil flow, and apply the potent energy to establish long-term systems. Choosing not to simply means that you'll be digging for many years and the systems' overall development will take many times longer than would otherwise be the case. Imagine our children digging paddies and ponds and swales and wondering why we didn't establish these systems while the digging was so easy.

28. Probability × Impact = Risk

The likelihood of an event's happening times the severity of that event if it occurs defines risk; for example, economic recession versus a comet striking Planet Earth or getting the flu versus contracting the Ebola virus. Sound planning is risk based.

29. Niches in Time

Good planning and action always make use of an opportunity in time—the "moment." There is always an optimal time to perform any action—never is it as good to perform an action "whenever." Often, an action is *only* appropriate in very small windows of time.

30. Zone 1 Site Mimic

Given that awareness, time, and labor are the limiting factors to maximum realization of abundance and health on most sites, and that many sites are simply too large or complex to tend with complete diligence, mimicking the entire site as much as possible in zone 1 enables the human inhabitants to survey the entire site by inspecting zone 1; for example, planting a few of every species in zone 1 lets you know what is fruiting in more distant zones of the landscape that are easily missed. I often miss the first honeyberries ripening because I have no honeyberry in my zone 1—the only plants are in a zone 3-ish area that I do not frequent. A few "barometer" honeyberries such as I have with seaberry, currant, and mulberry tell me when I need to walk a few hundred yards into zones 2-3-4 with a bucket to harvest.

31. Past Is Precedent

Exceptions to the general pattern of history are rare. When planning actions the regenerative and resilient designer-maker must aim for the highest and most beautiful possible outcomes while simultaneously being aware of the most likely future scenarios. This resilient designer-maker uses history, rather than pie-in-the-sky fanciful visions as a guide in this work.

RESILIENCY AND REGENERATION HABITS OF MIND

32. Good Design Always Empowers

Any system that promotes regeneration, resiliency, and adaptability empowers the human beings, plants, animals, and other forces acting in the system. All good design facilitates the free, conscious, and subconscious actions of members within the system and encourages the *manifestation of instinct* by all members of the system. Good design allows rather than restricts, encourages rather than suppresses. Good design is suspicious of "rules" and sees regulation as an indicator of an area in need of attention and improvement. This does not always mean that an optimal system is completely nonhierarchical, though in general it is relatively nonhierarchical. Good design always facilitates the manifestation of all the genius latent in each member of the system, and usually, rigid hierarchies do hinder that.

33. Passive versus Active Observation

Permaculturists are fond of the dictum, "Observe, then do." While passive "observation before action" is necessary, so, too, is observation through action. Much of what is necessary to learn in human habitat development and management can only be learned through experimentation with various approaches over time. Many aspects cannot be learned through passive observation alone, and attempting to figure out a challenge without doing is often impossible. We must begin planting, building, *acting* to learn about the system more thoroughly. I learned far more about the soil and its variations on this site in two weeks of planting trees than in three years of observation through walking and looking. In retrospect, this should not be surprising—getting one's hands on the material at hand (soil, plants, water, and so on) is the most direct route to actual seeing. This should not discourage deep detective inquiry through reading the landscape's plants and other ecological indicators, but should remind us of the limits of indirect contact.

34. Observation Action Chronology

In reference to the previous principle, it is important to note that the most irreversible actions should be conducted following the most passive observation while the most changeable actions should be conducted earlier in the development time line.

35. Two Is One, One Is None

Elements fail; crucial elements must always be backed up. Things fall apart. Rust never sleeps. Entropy is. If you need it, back it up. Ensure that it is redundant and, ideally, alive; for instance, the most durable food storage is chickens in the yard, cows in the barn, vegetables growing in the garden, kimchi in pots underground.

36. Character of Work over Time of Work

When deciding how to allocate time, it is easy to forget that the type of work involved in a task is often more important than the time needs of the task; for example, digging holes for a couple of hours is less desirable work for most than splitting wood for four hours; pruning for a day is more practical a task than hauling slash for just four hours. In general, heavy, dirty, or toxic jobs should be planned out of the equation as much as possible in place of jobs that may require more time but less brute strength. Personally, I'd rather mow with a scythe for half a day than run a weed whacker for two hours, for instance, all things being equal.

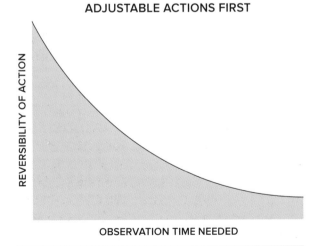

ADJUSTABLE ACTIONS FIRST

REVERSIBILITY OF ACTION

OBSERVATION TIME NEEDED

The most reversible or changeable actions should be carried out earliest in site establishment, while the most irreversible should be carried out after more passive observation.

37. Immerse in Abundance

The vital force tends to produce abundance. Since our task is to facilitate this abundance, we must immerse ourselves in it to know it, to encourage it. Immersion in abundance also serves to expand our perspective and allows us to work from a frame of "enough," not of scarcity. When we work from a reference point of scarcity, there is never enough to go around. Without harnessing abundance as our guide it is unlikely that we will cultivate truly regenerative and resilient systems.

38. Maximize Site Awareness

Take advantage of all opportunities to increase sight-line distances and clarity of that distance. The area that you can hear on a site and the acuity with which you can hear activity in this area is of crucial importance to managing a site well. And all ways of sensing what is occurring on the site at a given moment are helpful. See expanded explanation of this principle in the grazing section of chapter four.

39. Embedding Skills and Practice in Daily Routine

Life gets too full and time too limited for all the important skills of a land-based lifestyle to be practiced enough if they require many hours of practice in addition to daily needs and chores. Therefore, integrating the skills we'd like to develop into the regular daily rhythm is crucial to actually practicing them. This occurred to me clearly one day as I found myself in the middle of a few-week period of making my morning coffee on a mini wood cookstove that I use for camping and emergencies. I made the fire with a fire steel using no match or lighter. This took maybe a minute longer than it would have with a lighter but allowed me to practice an important fire-starting skill while accomplishing a normal daily task. Such rituals can also be highly satisfying.

I have found similar opportunities to embed skill development in my daily routine and always seem to find that they often take a little longer than the easiest, least skill-intensive method but far less time than dedicating specific nonroutine time to them; for example, taking an hour to work out when I can haul firewood by hand instead of using my truck or jogging

to the mailbox instead of driving. This principle is also connected to the concept of designing in challenges to one's daily life, such that vigor and skills are maintained during the day: I use ladders instead of proper staircases to go between floors in my studio-shop. Sure, stairs would be easier, but the steep ladders keep me more nimble, healthier, and probably happier, too. Ease should not always be the goal, and it is often actually counter to maintaining the most vigorous, aware, and satisfied existence. This is not to say that having easier backup ways of getting around in the case of injury or friends and family to help would not be a good idea.

40. Skills = Most Durable Resource

Beyond land, tools, money, even friends and family, your own skills—including those soft and hard—from growing a potato to making a friend, are your most dependable asset. Your land may be taken from you or your job downsized. Some of your friends and family will certainly leave this life. Your tools will rust or break and can be stolen or lost. But your own aptitudes are there for you to rely on no matter the condition of the world in which you find yourself. And with the right skills you can make every other resource from the world around you. Skills are the foundation on which the rest of your life value is cultivated.

41. Awareness Limits Action

Action is only as productive as one's awareness allows. You can't do more than you can see, hear, feel, know. Therefore, sensing as much as possible is key.

42. Environment Limits and Manifests Action

In the same way that ecological succession is informed by the seedbank available, disturbance forces acting upon the site, and other factors, so, too, does personal mental evolution depend on factors in the environment. We need to stack those factors such that our design aptitudes are enhanced over time. In this way we need to be our own continual health-care (mental included) practitioners. Our surroundings limit or empower our minds. The resilient and regenerative homesteader therefore must enhance her own surroundings for her to actually be able to carry out the work of imagining and implementing positive solutions. A poor environment retards this ability.

43. Solutions = Alignment

Solutions tend to emerge from alignment with, not opposition to, forces—not from resistance but from transformation. This is how water works.

44. Figure It Out: Try Stuff

Many of the most needed solutions and approaches have not been figured out, and if they have been, they have been in different places, periods, and groups of people. We each need custom solutions specific to the uniqueness of our lives and places. These always vary. Because they vary there is no instruction book we can follow to gain all the necessary solutions. Trying a wide variety of approaches is crucial to finding the best solutions specific to your unique situation. Fear and lack of confidence retards this. Be confident. Try stuff. Those who have figured out important approaches most often happen upon them by simply trying a variety of tactics.

45. Miracles Everywhere

Stepping back for a moment amid our daily routine, we can sometimes see that each flower, each animal, raindrop, and breeze, is itself actually a miracle in disguise. I say "in disguise" because our minds tend to quickly come up with answers as to why and how something exists—whether it's the moon, a rainstorm, or a snowflake. Yet when we probe and trace the lineage of anything as far as we can, we indeed find that each thing is linked to every other thing and a source for each thing cannot be determined. Everything arises, emerges. At its core this seems clearly unexplainable, both in scientific and other terms. The practical implication here is clear: the moment we believe that we have the full explanation of why something happens is the moment we begin limiting our vision as to the phenomena at hand. Remembering that we are only seeing part of the vast and mysterious processes underway in each phenomena around us actually allows us to see, learn, and do more as designers. The concept of miracles, therefore, is a highly practical one.

FOOD AND FERTILITY

46. Constant Organic Matter Accumulation

Aiming to build soil fertility and nutrients, suppress weeds, decrease drought vulnerability, and produce a constant stream of garden soil, the homesteader and farmer should be in a constant state of collecting organic matter. We harvest from neighbors' driveway edges when the leaves build up in the fall, from under pine stands for the blueberries, from local arborists when they've got a truck loaded with chips. Any and all sources of organic matter are good as long as you ensure a low level of toxicity in the material.* *You can never have too much compost.*

47. Paths as Biomass Producers

At more than very small scales in cold-temperate climates, paths will eventually be grass and other herbaceous plants—even if they start as gravel, woodchips, or even pavement. Such pathways should serve as biomass and soil production for other areas by composting and mulching with harvested plant matter and then composted or used as mulch in specific areas around perennial plants.

48. Seed Often and Lightly

Since successful germination of seed in land renovation requires consistent moisture (not a deluge or a drought), the most successful strategy we've used is to seed lightly but very often in land renovation/enhancement work. This means we seed starting in late March—frost seeding of clovers—even while snow melts back. We continue to seed areas that are in poor condition (abused and abandoned) from three to ten times per season, aiming to seed especially before July, when seed can best establish and moisture is most reliably available.

49. Passive Forage-ability

The entire landscape should be managed as an intensive foraging zone—this means "stocking" the site with the most multifunctional beneficial plants (and fungi,

---

* Avoiding leaves from oft-used roads is key to reducing collection of benzene and other toxins. When collecting bagged leaves in a local town, we avoid yards with Norway maples or black walnuts in them—those trees produce leaves with toxic compounds We also avoid manure from farms whose practices we don't know and trust to preclude pharmaceuticals in the material.

sometimes even animals) (a) for which the habitat is made, structurally and biologically; (2) that are managed to balance as an overall system and maximize biodiversity and biomass; and (3) that are harvested passively while one moves through the landscape doing active or recreational activities. The most productive sites relative to the amount of inputs needed into the system offer yields to the forager within the landscape; one does not need to find food only in a garden bed or planted perennial zone. As an edible ecosystem matures, human food self-seeds and begins to colonize the site even in areas where it was never planted or sown. In this way there is human-ecosystem coevolution occurring. This relationship should be fostered wherever possible.

50. Plant as Densely as You Can Afford To

Because plants sometimes die and you can't go back in time, in addition to the fact that trees are much faster to cut down than to grow (you can get intermediate yields from dense systems before thinning), plant as densely as possible, with thinning happening later on. This is no different from seeding two to three or more times the vegetable seed you need in a bed and thinning the extras. Seed is cheap and you can save your own.

Plants can be as well. You cannot go back in time and put a thirty-foot walnut in where one died, in between two walnuts twelve years after planting when they were spaced forty feet on center to begin with. We try to plant at two to three times the "horizon"/final desired spacing with the intent to thin if no one dies in the meantime. We'll also see yields as the trees reach maturity but before they get too crowded. For example, we plant a bur oak silvopasture system that is ideal with spacings of thirty to fifty feet on center when fully grown (years fifty to three hundred or so) at ten- to twenty-foot spacings. We are already getting nut yields in years five to six and will continue to get some yields, most likely, for another ten to twenty or more years before thinning is needed. And then we'll have nice oak posts or other building materials.

51. Animals above Plants

Because plants are limited by nutrients—especially nitrogen—provided in large part by animals and their by-products, and because gravity never ceases, placing animals above the elevation of crop plants in a landscape is a primary approach to maximizing productivity. The best sites have an access road high in the landscape where a zone 0/1 exists, including home and barn. These nutrient sources are then easily fed via gravity and water (fertigation) downhill to plant-based cropping systems. At the WSRF we have positioned our rice paddies, many veggie gardens, and nut tree systems downslope from residences and the barn. Ponds, too, if you are growing fish, represent a nutrient source that can fertigate plants below. Animal by-products should flow into plant systems so that these by-products (N, P, K, and many other nutrients) fertilize plants. The direction of material flow should never be animal to animal, as nutrient overconcentration occurs in this configuration. Plant-into-animal-system flows are to be avoided because fertility is exhausted quickly in this scenario.

52. Pee on Plants (or Next to Plants)

Closing the fertility cycle between humans and the systems that feed us is fundamental, and there's no other opportunity to reciprocate the giving nature of plants so readily as urinating. Doing so is simply returning what was given in the first place, cycling, giving back, being in reciprocal mutualistic relationship.

53. Swales Everywhere

Swales are fundamental to a landscape that aims to reduce the constant effects of erosion and entropy. Swales stop the flow of water downhill, force the water to be infiltrated, make more soil-air interface (on the mound), and make more land, literally.

ECOLOGY AND MANAGEMENT

54. Disturbance Stimulates Yield

Resilience and regeneration tend to be highest when the evolution of the system is stimulated by disturbance (stress) events combined at the right interval with rest events to build biological vigor. An example of this is clearly illustrated by the way intensive rotational grazing works: Plants grow tall and deep, grazing occurs rapidly, plenty of rest is applied to allow full plant recovery. The biological deepening of the system is maximized by a correct tempo of rest-stress–rest-stress, similar to the way fitness and muscle building occurs in the human body, not from all exercise all the time, not from all sitting on the couch all time.

55. Succession Determined by Disturbance and Its Aftermath

Because disturbances open niches for life to inhabit, ecosystems shift quickly immediately after disturbance. Disturbance creates opportunity and challenge: opportunity if the opening in the ecosystem is seized, challenges if the ecosystem is left to succeed "randomly."

56. Fill Open Niches Immediately

Whenever disturbance is applied, be ready to fill open niches created; for example, plant, graze, seed. Biology must follow technology (where technology is the excavator, fire, or the chain saw, and biology is plants, animals, seeds).

57. Systems Establishment Overshooting Management Capacity

There is a strong tendency for humans to develop systems that are too large or complex for successful management over time because of the future availability of time, labor, finances, intelligence, skills, or energy. In other words, the system-establishment phase tends to bite off more than the system-operation phase can chew.

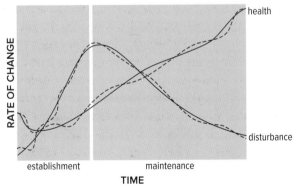

DISTURBANCE MANAGEMENT

RATE OF CHANGE

TIME

health

disturbance

establishment

maintenance

An increase in disturbance during the establishment phase often helps system health increase rapidly.

58. Biology in Place of Technology

Wherever possible replace mechanical elements with biotic, living components—they last longer and yield side benefits such as soil, meat, milk, and fiber. They also reproduce and make more of themselves, whereas nonliving elements in a system are always decaying and need replacement.

59. Annual-Perennial Balance in System

In general, the more growing space available, the more brittle the climate, the steeper the slope, and the less fertile the soil, the more crucial the role of perennial food plants; whereas, the less space available, the less brittle the climate, the lower the slope, and the more fertile the soil, the greater the role annuals should play in an agroecosystem, if they play a role at all.

60. Modularity and Agility

Remember that planning and planners are inherently imperfect. We can only imagine and solve for so many possibilities. Lives always unfold in unpredictable ways: Conditions change (climate, economy, society, family), goals change, our health changes. The future is always being made in ways that vary from our vision of it. Therefore, wherever possible develop systems to be adaptable over time: Insulate the house with cellulose instead of foam so you can remove a wall and add on; stub out plumbing in that wall just in case; expose the wiring and plumbing when you can (you *will* renovate, nearly all of us do eventually if we stay in a home long enough); photograph and map utilities in the ground or in the walls; and so on.

People tend to hold tight to a specific vision, especially when real effort has been made to plan a project carefully. But the same planning that can empower can and does often blind. I witness this in the professional world of planning all the time. It's especially easy for planners and designers to cling to a specific vision, for they've thought it through. But we'll always only have a partial picture. Therefore, respect the changing nature of the future. Hedge your actions, and keep systems as adaptable as possible. Decide and make only as much as you need to. Defer decisions when you can reasonably. When wrestling with a difficult design decision,

Designing and constructing elements in a human habitat is the easy part; integrating them to optimize the function of the whole system and doing so in a manageable, not overwhelming manner is the real challenge. Photograph courtesy of Whole Systems Design, LLC

"We don't have to decide that now" is one of the wisest things people can say.

61. Ecosystem Partnering, Not Stewardship

Stewardship implies dominion, whereas partnership implies coevolution; mutual respect; whole-archy, not hierarchy. A partner is sometimes a guide, always a facilitator, always a coworker.

62. Partnering with Vigor

One of the regenerative designer-doer's primary tasks is to facilitate vigor and vitality in the ecosystem she is partnering with, identify that vigor, bolster vitality in areas where vigor is low.

63. Sculptable Landscape

As an edible, multifunctional ecosystem matures into a multilayered annual, perennial, and grazed system, the need to prune back plants becomes significant, both to continue allowing sunlight penetration and for optimal soil building. We need to plant systems very densely to do the rapid soil/water/site enhancement necessary and to promote maximum yields. Root dieback events caused by pruning/coppicing/grazing are crucial in this regard. Grazing can provide this service in the understory, but to cut back plants too significant to graze, we need to prune/pollard/coppice. Developing a landscape as a three-dimensional sculpture of sorts becomes a clear need as the system moves into "the pruning phase." We primarily prune black locust, alder, willow, and seaberry in this capacity. In the tropics there are dozens of species—mostly N-fixers—used in this way.

64. Native to When

When using the term "native," what year do we use to determine whether a plant is "from here" or "alien"? If we choose European contact, we ignore a multiple thousand-year history of anthropogenic plant dispersal that was highly active before Europeans began to settle the "New" (actually, very old) World. Using the term "native" without indicating a date of arrival is to ignore the vast majority of human-plant history—the multi-thousand year legacy of dispersal and coevolution.

65. Cheap Tools Are Too Costly

Only high-quality, well-made tools and materials are worthwhile unless the goal is short-term, poorly done work. It's hard to achieve a higher quality work result than the tools you use; for example, you can't cut to $1/32''$ if your marking tool is only accurate to $1/16''$ marks. The precision of the tool begets the precision of the result. Buying the highest quality tools you can afford is the only affordable option in the long run, and using them is far more enjoyable anyway. The only tools or design decisions I've ever regretted buying or making were those I was trying to save money or time on. Over the long haul—and long haul is the point here—only high-quality tools and materials function well enough and for long enough to be both effective and economical. It's also important to understand how quickly the quality of tools and materials has deteriorated over the years. In just the past ten years, the quality of basic hardware

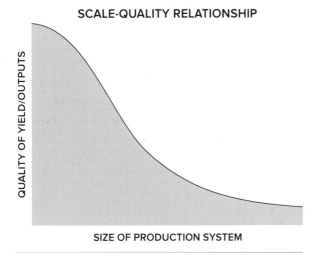

SCALE-QUALITY RELATIONSHIP

(y-axis: QUALITY OF YIELD/OUTPUTS*)*

(x-axis: SIZE OF PRODUCTION SYSTEM*)*

In general the quality of yield is reduced in proportion to the scale of the system in which it is produced.

store items has gone markedly down, let alone the "old steel" put into our parents' tools, which now can only be found through very high-quality makers that must be sought out. In a day and age when junk with planned obsolescence is the norm, one has to go out of the way (often times *far* out of the way) to find something actually worth buying.

66. Quality of Work Affects Labor and Management Capacity

The type of work, not simply the amount of it, influences to a large extent the capacity of human beings in the system.

67. Apply Present Resources Now

A great example of this strategy is buying and spreading organic seed now, such as alfalfa, which is currently under GMO research and likely dissemination (if it's not already in the food supply). This means that future alfalfa carries with it a high likelihood of contamination, even if it's certified organic. A similar example can be found in many tools—whether an axe, a flashlight, a firearm, or an excavator—they are extraordinarily cheap (relative to their true, past, and likely future costs) today and won't be forever. In fact, many will likely become relatively unavailable down the road. Fossil fuel is a prime example.

68. Storage Always Runs Out

Stockpiling of energy or materials, while often a valuable strategy, must always be tempered with the need to renew such stores. All physical resources eventually perish, and therefore, renewal is crucial. The balance between storage and renewal is in constant flux and should always be recognized and adjusted for.

69. House as Water Tower

The need to store water high in a site such that it can be gravity fed to points of use is fundamental. In a cold climate we must bury this storage below frost, ensure high flow volumes, or store it in a warm space. On most sites without access to a significant spring high on a slope above zone 1 and other areas of the site, using the home itself as a water tower is the most practical solution. This can easily be achieved with a small- to medium-size cistern of plastic or steel positioned in the top floor of a home. (See chapter three for more on this approach.)

70. House as Dehydrator

In cold, humid climates the major harvest period often coincides with cool, wet weather, making drying difficult. Given that dehydrating is a fundamental food storage strategy, we must have multiple methods of drying foods when possible. Using the home itself for such a task is often the most practical approach, since it must have a heat source anyway and should be well ventilated. It is easy to make the home serve dehydration needs with a woodstove, adequate stack effect for ventilation,* and space for hanging crops to dry.

71. Clarity Points and Leverage Points in Time

Throughout the year there are periods punctuated by intense windows in time when one can see how the systems around them function. These are often during the swing seasons and in times of contrast, such as the first rainstorm after a dry stretch, the first warm day in spring, or the first frost. These periods can come unexpectedly and offer crucial insight opportunities for learning about our places and how to fit within them synergistically. Spontaneous action to able to access the value of these periods is crucial. You must see them coming or quickly note their arrival and be able to drop everything at a moment's notice to take advantage of these opportunities.

Such instances sometimes entail a long walk at night, getting dirty, lying on the ground in the mud to see something occur, listening carefully in the night and getting out of bed to see the event, and so on. Such events will come, and you need to be ready to listen, see, feel, maybe even taste. Windows of alignment with a place through ever greater realizations about the place are there to utilize if you are open to them and are willing to be uncomfortable or joyfully spontaneous when necessary.

Squash harvest at the WSRF—one of our primary storage crops

72. Principles Are Only Useful if Actually Followed!

With principles and strategies in mind (I hope you've used the above directives to remind yourself of those you have noticed already), we are now ready to move into the next phase of thinking about the place we will be cocreating: a systematic process for decision-making—the pencil-meets-paper step in the design process.

Understanding Your Site and Finding the Synergies

At the outset of the journey down the resiliency and regeneration path, one is often overwhelmed with questions—the unknown simply outweighs the known

---

* Stack effect is the passive flow of air or water (or any other fluid) from warm to cold and low to high in a space; also called a convection loop.

at this point. A structure for sifting through the seemingly endless variables is needed. Enter a *process*. Its beauty is its ability to narrow down options; its danger is in missing solutions that may be important. It is important to begin with two foundational elements: (1) you and (2) your place (or intended place). The rest of the design process can flow effectively from these two starting points but only if it is informed by the *existing conditions* of you and your place. At this point in the design process, you may be looking for a piece of land on which to develop a home or farm, or you may already reside in that place. Either way, the process can be used to bring to the surface your own goals (and those of others living or working on the site) and the characteristics of the land on which you are considering residing on or already do. The design process is successful when it finds synergy between these two variables at work.

"You" means your goals, your skills, your vision, your past, and your intended future. It also means your family, your pets, and your stuff. "Your place" means the physical site—your homestead—on which you live or intend to live. The goal in the beginning of the design process is to find as synergistic a fit between you and your place as possible. Ideally, a place matches your goals before you have arrived at the place, but it doesn't always happen so neatly. Much of the time people already have a place and must work to understand it,

learn what it can do well, and decide how to adapt their own goals to meet their place in synergy.

To successfully develop a durable homestead, farm, or any other land-based endeavor, we must partner with processes that aid, not discourage, us. That means we must be willing to put our own original goals aside and allow our minds to be changed. It is important to detach from expectation; we're always in a state of adapting ourselves to our places–learning what, truly, our place can manifest most fully, what it can and truly "wants" to do. For example, even if it has always been your dream, you don't want to build a massive fish-growing infrastructure if you have little water, or plant a large elderberry farm if you're on droughty soil, or anything else that is misaligned with the place you find yourself. Creating a successful farm or highly productive homestead is hard enough—even if you're *not* fighting the natural tendencies of the site.

In our design practice we tend to work with professionals that are relatively successful in their field—they are good at what they do and are used to achieving positive results for their efforts. They buy a piece of land, usually coming to the process with several clear goals of what they'd like to accomplish. They tend to have the goals in hand but the inherent characteristics of the site out of mind. This often leads to problems. It's not rare that we've seen a client attempt projects

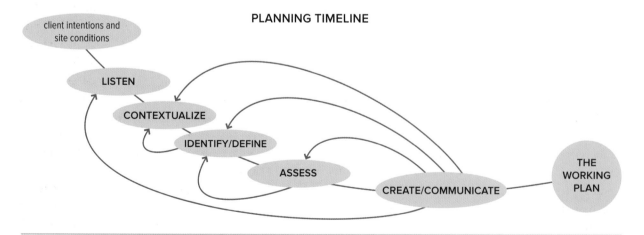

Our planning process transforms the client's intentions and site conditions into an optimal plan. Without a process rooted in analysis and feedback, quality spaces are not achieved. Illustration courtesy of Whole Systems Design, LLC

that are unsuited to the site he happens to live on or has recently acquired. This seems to happen most commonly when people have experienced success at something and attempt to extend that success into a realm in which they have less experience or in which the context (and rules of the game!) have changed.

This reminds me of a quote from Jamshid Gharajedaghi that a good friend often uses to describe such situations—these difficulties being common in the world around us and sometimes quite tragic:

> The most stubborn habits which resist change with the greatest tenacity are those which worked well for a space of time and led to the practitioner being rewarded for those behaviors. If you suddenly tell such persons that their recipe for success is no longer viable, their personal experience belies your diagnosis. The road to convincing them is hard. It is the stuff of classic tragedy.

Land plays by its own set of rules, and it can be surprisingly unyielding. Failure on land is much more common than success. Indeed, the history of humans and land is most often a legacy of spectacular failure. Our job as designers of land and human habitat systems must therefore be as ecologists first and foremost—to understand how the land system is and wants to function, before inserting our own human goals and program into the system. We need the land's tendencies on our side, which is best accomplished by adjusting our goals, rather than the characteristics of the landscape (though that can be done to significant extents as well). That means we must sometimes be willing to put our own original goals aside and allow our minds to be changed in the process of building a land relationship. We must be open to possibilities other than those we originally envisioned, for the genius of a particular place to manifest. Our task is find the best insertion points toward meeting our goals. Ideally, we flex those goals constantly to best meet the site's abilities, and if we're fortunate we do not choose a site until we've thoroughly assessed the

> **It is often those with the strongest convictions about their own goals and destiny that are often bound for the biggest struggles and disappointments in land and in life.**

characteristics of various landscapes in which our project could occur.

I came to my landscape thinking I wanted to plant an orchard, grow some vegetables, and perhaps raise some fish. Fortunately, I had little else specifically and clearly in my mind in terms of intentions. In retrospect this was a blessing; I have since come to learn that it is often those with the strongest convictions about their own goals and destiny that are often bound for the biggest struggles and disappointments in land and in life. I am not suggesting that having clear goals is a bad thing. Clarity of goals is key, and the willpower to make the goals a reality is crucial. But unchanging, stubborn effort toward such goals is dangerous when the universe (or a piece of land) is conspiring to help you do something very different.

For example, my first project on the site was to plant a small orchard. In the spring after I moved in, I had identified a very sunny area that seemed good for a grouping of apples and pears, so I walked down one evening and started digging holes. The next morning I went down to begin planting the few trees I had managed to buy thus far for the project. All the holes were filled with water, but curiously, it hadn't rained. I knew enough at the time, barely, to realize that without major modifications this was not a good site for an orchard. The following year I had saved up enough money to hire an excavator to build a pond—figuring that such a pond would help dry out the field below by catching the water. I was still fixated on *that* orchard in *that* location.

By the next year a pond was built and full, but the field below was still just as wet. Only sometime that summer while swimming in the pond did it fully sink in that this area of the site just *wanted* to be wet, and there was little if anything I could do to change that.

It clicked at that moment that I needed to embrace this fact. Small holes filled with water weren't very useful, but big holes were—I began building more ponds and eventually rice paddies; farming these wetland conditions—not fighting the land's tendency but working with it.

Embracing what the land we are working with can truly do best is a large part of learning to inhabit a place well—the site assessment and analysis process is aimed at uncovering the basic nature of these land tendencies. The site design process then helps us decide what elements harness the landscape's tendencies and how to scale, arrange, and manage them. While all landscapes tend toward certain functions more than others, these tendencies are not necessarily innate and can be altered, even to a large extent. I am not suggesting that we do not alter these tendencies significantly in our site development work (in the site establishment phase). However, success in any project is most economical and fastest to achieve if we tap into the existing tendencies in a site rather than change them radically. That said, however, urban areas, especially, are areas in which a radical change in site *conditions* is often called for.

Goals Identification and Requirements of the Design

To aid in goals articulation, it is important to come up with a list of questions—and it shouldn't be brief! After all, there's a lot to hash out, and chances are, your goals are not simple. In my work with Whole Systems Design, we have developed a comprehensive goals-articulation sheet. It is ideal for those at the outset or continuation phase of a project to make such a sheet specific to their own lives and projects.

Once the goals articulation process is in full swing (it's not something that is ever "completed" by the way) land can begin to be identified for the project using the site assessment tools described on pages 49 and 50. This is also the time when a *program* can be developed. My work as a resilient designer is not program intensive as is common in most architectural approaches; it is to aim for a more organic process where the program—exact needs of the development in numbers—emerges over a period of time. In most architectural firm approaches program development is primary—the first meetings are all about it. From this program everything else follows suit. In my perspective that leads to dangerous design approaches which we see manifest in the world today—environments that are not appropriate to or taking into account of their contexts, the living places in which they exist. Such a program-first approach puts the human goals at the center of the process not placed as one link in the web that is the place in which the human project will be happening. It is inherently un-ecological. It is for these reasons that the program should usually emerge over the analysis phase, not be set out initially as the set-in-stone goal for everything else to bend around. Often, there are specific program criteria that must be met, however, like a school needing to expand to house x number of new students or a farm needing to graze x number of new acres. These program specifics should always be inspected and evaluated for value, however, in light of the place for the project—often they are inappropriate and countless developments occur each year in which goals are outlined that are completely incompatible with the site chosen.

For those already on land they'd like to stay on, I recommend that a detailed site assessment be done initially—before any goals articulation. This "what is" assessment explores the features and processes of the land to identify what it has a high capacity to produce and what it will trend away from, ecologically. Goals and assumptions should be completely on hold in this phase—as we seek only to ask questions, not draw conclusions. This is an incredibly important aspect in any design process—that the assessment phase be unpolluted with judgments and unclouded by an agenda, bias, or any kind of desired outcome. Let the land speak for itself—just listen. For if your own agenda talks more loudly than the characteristics of the land, you will be in for a lifetime of struggle and frustration. Ultimately, the farms and homesteads that truly succeed over the long haul are those on which the land's innate capacities are being tapped and harnessed, never resisted or forced into something they are not.

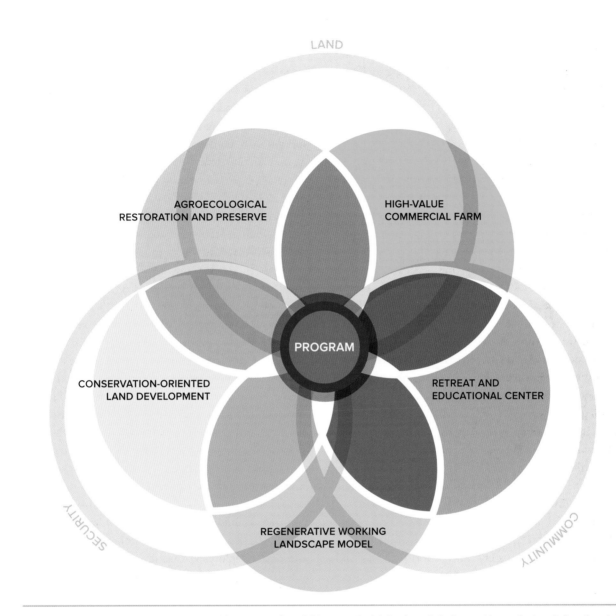

A graphic depiction of a program. The program should emerge out of possibilities recognized during the analysis phase, not imposed onto a site from the outset.
Illustration courtesy of Whole Systems Design, LLC

Assessing the Site

Site assessment is performed both when one is looking for land and home and when one already resides in a place. This part of the process involves overlaying criteria on prospective properties—or the one you live on—to find good matches with the identified goals. The goals-articulation work performed above is used to generate a list of criteria that is applied to various landscapes of interest for those seeking land. A gradient of criteria is helpful in this process, starting from primary criteria, which are "nonnegotiable," to secondary, tertiary, and so on, ordered by importance. For example, we often work with people whose primary criteria include all-day sunshine (solar access), two-wheel-drive vehicle access, abundant water, and decent soil; with secondary criteria being walkable to a school and views; with tertiary criteria such as great neighbors and an existing barn. The following criteria are commonly identified in our process with clients as we help guide them toward sites that are most suited to helping them reach their goals.

- ▶ Location within area of interest: county, town, neighborhood
- ▶ Site access: Length and grade of driveway? Condition of town road?
- ▶ Solar access: How many hours of sun?
- ▶ Aspect: N-S-E-W facing?
- ▶ Size (acres) and shape of parcel
- ▶ Elevation change (feet)
- ▶ Views: positive and negative, night and day
- ▶ Water: Amount of rainfall, streams, seeps, springs, wells, ponds
- ▶ Soundscape: road noise, especially
- ▶ Soils: Any prime or quality agricultural soils? Inundated soils, restrictive layers, or shallow bedrock?
- ▶ Slope: Steep areas?
- ▶ Vegetation: Dominant cover type—field, hardwoods, softwoods, maturity, species, values, unusual species?
- ▶ Microclimate (within region): Cold pocket? Snow belt? Warm? Dry?
- ▶ Privacy
- ▶ Schools
- ▶ Taxes
- ▶ Regulations: setbacks, zoning, and so on
- ▶ Mineral resources: quarry, sand, clay
- ▶ Overall beauty

Land Analysis

What will nature allow us to do here, what will nature help us to do?

—WENDELL BERRY

Site—land and infrastructure—analysis is performed when land is obtained, or before goals articulation when land is already available for the project. Site analysis involves directed observation and on-paper map drawing of the features and processes of the site and those that influence the site from elsewhere. There's an important distinction here: *Features* are what people tend to think of identifying—rock outcrops, soil, water, buildings, trees. But *process*—phenomena that occur across time—are just as important: wind, sun/shadow, animal movement, people movement, noise, water tables rising and falling, erosion, and so on. Site analysis involves direct observation of as many of these conditions as possible within a reasonable time frame and uses indirect tools such as mapmaking to aid in the process of discovery.

If you are fortunate enough to have a piece of land already with which to utilize your goals, you should be, in part, directed by the site—its characteristics, what it can accommodate well, and what it is not well suited for.

The site analysis process includes assessment of the existing conditions of the features and processes on-site that are of most influence on the design and arranged according to scale. The analyses are not complete until the implications of each assessment are understood—ideally, through *implication statements* made on each analysis drawing. An example of an implication statement would be, "Because the site is exposed to cold northerly winds, sensitive plantings and heated buildings should be protected from the north." And "Because the home on-site is listed on the National Historic Register, effort should be made to preserve exceptional aspects of the home should any renovation occur," and "Because of a seasonally high water table over the northern half of the property, buildings, plantings, and other developments that require well-drained soil conditions should be avoided there unless the condition is modified."

The phrase "of influence" at the top of the previous paragraph is crucial because you can't analyze every existing condition. But selecting what not to assess is always a judgment call and is where experience comes into the process—it's not easy to decide what is of influence, for example, if you've only done site analysis a handful of times before, so those not experienced in doing site analysis on multiple projects/landscapes should err on the side of including more features/

processes than fewer. *Existing conditions* here, is the operative term—as *future conditions* are the design (intending/proposing) phase—analysis is the "what is" phase. Arranging the analysis across scale is important to get a sense of how various factors influence the site from large scale context-level conditions such as bedrock, soils, winds, and even regulations, cultural or legal aspects, and mid-scale site characteristics such as existing vegetation to microsite scale conditions such as gallons of water available per minute in a spring. Such existing conditions include but are not limited to the following.

CONTEXT
▶ Geology and soils
▶ Climate
▶ Ecology, forest cover, wildlife
▶ Legal/social/cultural/economic

SITE SCALE
▶ Slope and topography
▶ Aspect
▶ Microclimate
▶ Soils
▶ Vegetation
▶ Wildlife
▶ Views
▶ Water/hydrology
▶ Access and circulation
▶ Infrastructure
▶ Soundscape
▶ Historical and exceptional features

GLOBAL ANALOGUES

similar latitudes

cold-temperate climates
approximate—as defined by Koppen climate map

N.T.S.

Climate analogous regions to the WSRF site. These areas contain the most useful design strategies that we reference in our work. Illustration courtesy of Whole Systems Design, LLC

GLOBAL AND REGIONAL CONTEXT

THE LARGEST GROUP OF ISLANDS IN THE ATLANTIC OCEAN, the Bahamian archipelago is composed of more than nine hundred islands and cays laying in the easterly trade winds. Home to the world's deepest limestone deposits and just three hundred thousand inhabitants spread over 11,400 square kilometers, the Bahamian islands have accumulated at the edge of deep abyssal plains creating a seemingly endless land-sea edge environment.

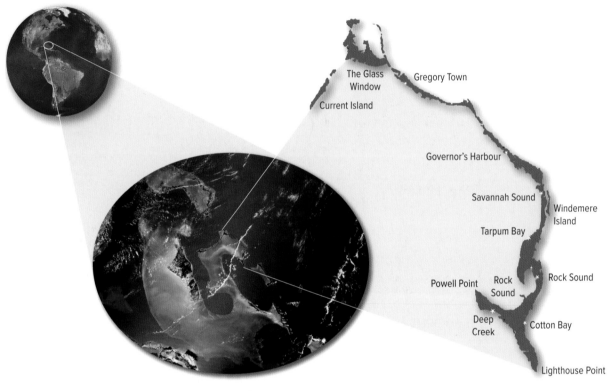

TEMPERATURE VARIATIONS IN THE BAHAMAS are minimal because of the buffering effect of the surrounding sea with average daytime air temperatures ranging from 80 to 90°F. The subtropical dry climate is dominated by easterly trade winds for a majority of the year with a monsoonal weather pattern bringing fifteen to thirty-five inches of rain to the islands, mostly in the summer and autumn months. Accumulating from millions of years of sedimentary deposition, the islands' highest point lays less than two hundred feet above sea level and the nation contains no surface fresh water.

ELEUTHERA, GREEK FOR "FREEDOM," lies sixty miles east of Nassau and stretches more than a hundred miles north to south. The original population of Taino people rapidly declined after European landfall in the fifteenth century. Since then, the island has had a history of farming, primarily pineapple, dairy, beef, and vegetables. The construction of large sawmills in the nineteenth century allowed for the almost complete deforestation of the island's ninety-foot mahogany and teak hardwood canopy, leaving the majority of the island covered with the early successional forest typical throughout the island today. Eleuthera's marine ecology, especially its coral reefs, are also experiencing rapid decline due to both local and global influences.

In the twentieth century Eleuthera became home to major international resorts, many of which were closed in the 1990s. However, today many new resort developments are planned for the island. Major settlements include Governor's Harbor, Rock Sound, Harbor Island, Gregory Town, and Tarpum Bay.

↑
NORTH　　　　N.T.S.

Our process starts by taking a step back, looking at what systems are in play on larger scales than those within the boundaries of our site. At this scale we examine social, cultural, and environmental history, as well as broad climate conditions and land types. Illustration courtesy of Whole Systems Design, LLC

LOCAL CONTEXT

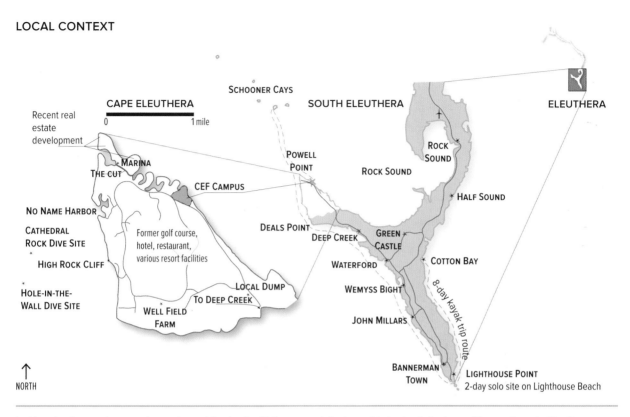

Looking at a closer scale, we again examine social and cultural influences and dig deeper into land and climate conditions, such as plant hardiness zones, natural communities, and historic land-use patterns. Illustration courtesy of Whole Systems Design, LLC

BIOPHYSICAL REGIONS OF VERMONT

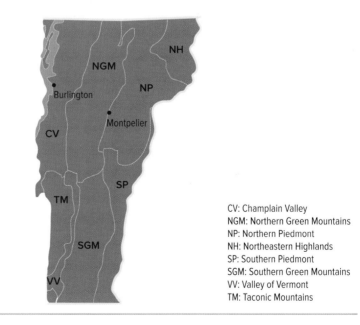

CV: Champlain Valley
NGM: Northern Green Mountains
NP: Northern Piedmont
NH: Northeastern Highlands
SP: Southern Piedmont
SGM: Southern Green Mountains
VV: Valley of Vermont
TM: Taconic Mountains

An example of regional design context: Biophysical regions help frame the site's place within the larger landscape by defining broad regional characteristics, such as plant communities, geologic and physiographic history, animal communities, and climatic attributes. Illustration courtesy of Whole Systems Design, LLC; Source data from Thompson and Sorenson, 2000

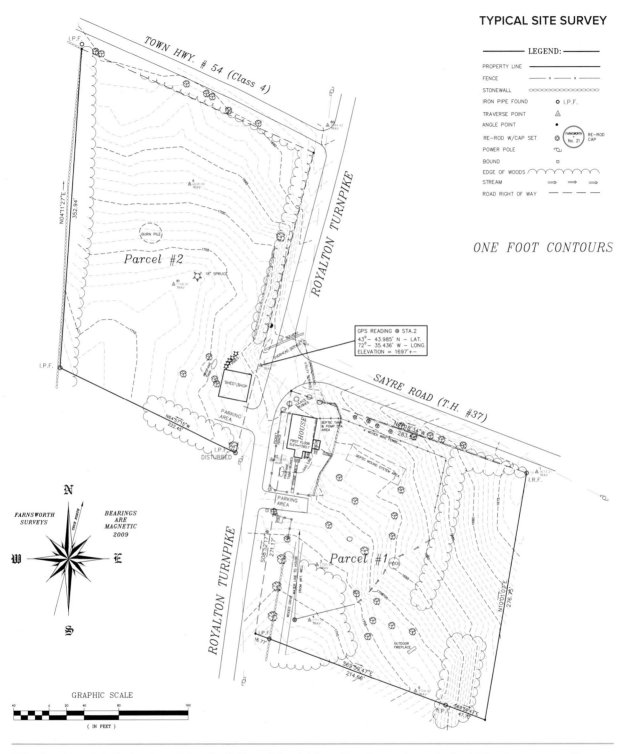

At the site scale we start with a base layer of information that is usually heavily informed by a site survey by a professional civil engineer or surveyor. The survey provides an invaluable level of detailed site information that includes contours, utilities, access, infrastructure, forest cover, and the exact locations of each.

BASE MAP

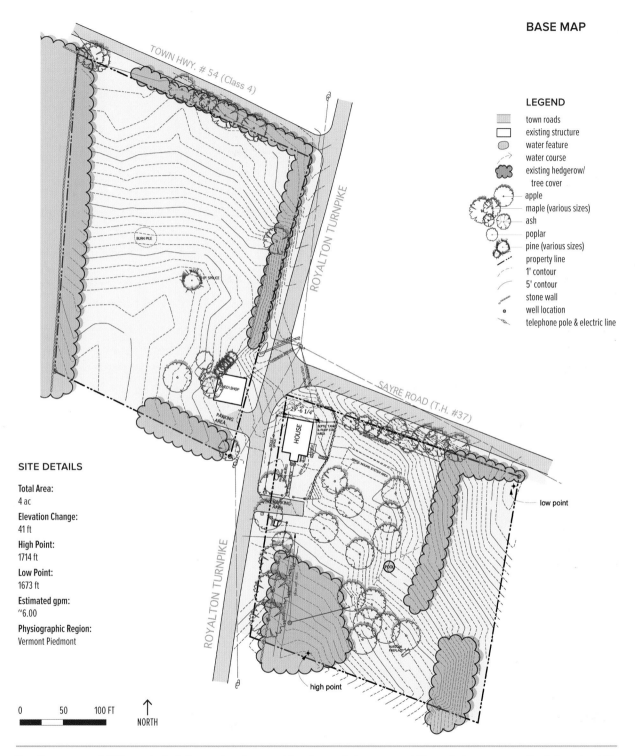

LEGEND

	town roads
	existing structure
	water feature
	water course
	existing hedgerow/ tree cover
	apple
	maple (various sizes)
	ash
	poplar
	pine (various sizes)
	property line
	1' contour
	5' contour
	stone wall
	well location
	telephone pole & electric line

SITE DETAILS

Total Area:
4 ac

Elevation Change:
41 ft

High Point:
1714 ft

Low Point:
1673 ft

Estimated gpm:
~6.00

Physiographic Region:
Vermont Piedmont

0   50   100 FT

NORTH

From the survey we distill the most salient site aspects and represent them in ways that are more useful to our design process, provide clarity, and assist in communicating with clients. This is our base layer of site design and planning, the base map. In lieu of a professional survey, there are other ways to create a base layer, such as using handmade surveying tools (for instance, an A-frame level) and triangulating locations using prominent site landmarks. Aerial photos and tax maps can also be helpful in creating this layer. Illustration courtesy of Whole Systems Design, LLC

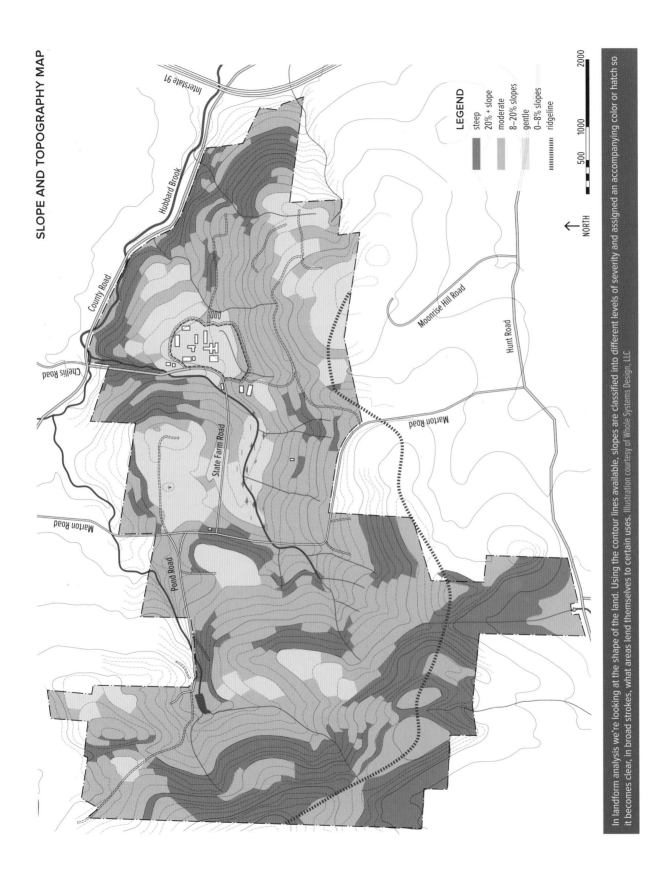

SLOPE AND TOPOGRAPHY MAP

LEGEND

steep
20% + slope

moderate
8–20% slopes

gentle
0–8% slopes

ridgeline

NORTH

500   1000   2000

In landform analysis we're looking at the shape of the land. Using the contour lines available, slopes are classified into different levels of severity and assigned an accompanying color or hatch so it becomes clear, in broad strokes, what areas lend themselves to certain uses. *Illustration courtesy of Whole Systems Design, LLC*

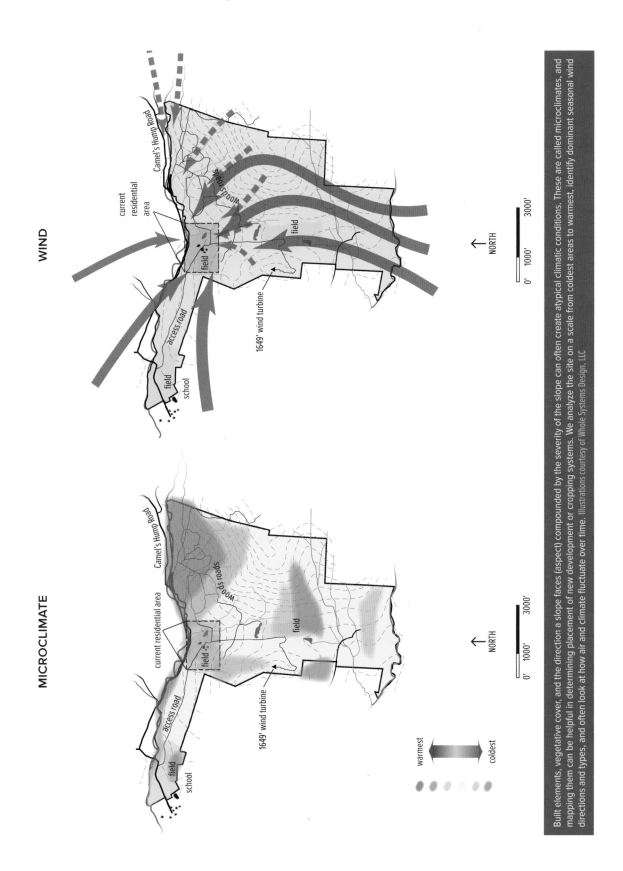

Built elements, vegetative cover, and the direction a slope faces (aspect) compounded by the severity of the slope can often create atypical climatic conditions. These are called microclimates, and mapping them can be helpful in determining placement of new development or cropping systems. We analyze the site on a scale from coldest areas to warmest, identify dominant seasonal wind directions and types, and often look at how air and climate fluctuate over time. Illustrations courtesy of Whole Systems Design, LLC

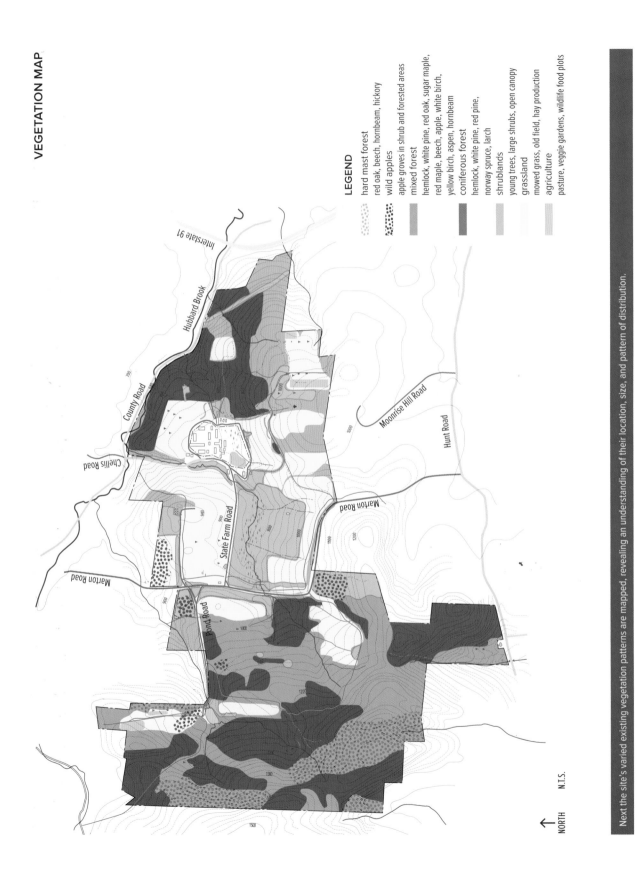

VEGETATION MAP

LEGEND

hard mast forest
red oak, beech, hornbeam, hickory
wild apples
apple groves in shrub and forested areas
mixed forest
hemlock, white pine, red oak, sugar maple, red maple, beech, apple, white birch, yellow birch, aspen, hornbeam
coniferous forest
hemlock, white pine, red pine, norway spruce, larch
shrublands
young trees, large shrubs, open canopy
grassland
mowed grass, old field, hay production
agriculture
pasture, veggie gardens, wildlife food plots

NORTH

N.T.S.

Next the site's varied existing vegetation patterns are mapped, revealing an understanding of their location, size, and pattern of distribution.

ACCESS AND CIRCULATION

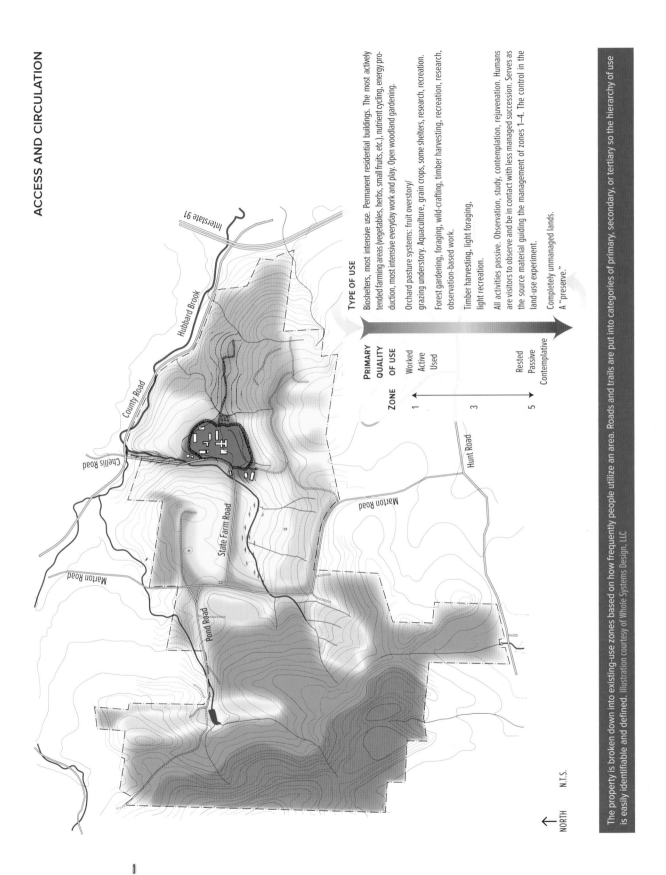

TYPE OF USE

Bioshelters, most intensive use. Permanent residential buildings. The most actively tended farming areas (vegetables, herbs, small fruits, etc.), nutrient cycling, energy production, most intensive everyday work and play. Open woodland gardening.

Orchard pasture systems: fruit overstory/grazing understory. Aquaculture, grain crops, some shelters, research, recreation.

Forest gardening, foraging, wild-crafting, timber harvesting, recreation, research, observation-based work.

Timber harvesting, light foraging, light recreation.

All activities passive. Observation, study, contemplation, rejuvenation. Humans are visitors to observe and be in contact with less managed succession. Serves as the source material guiding the management of zones 1–4. The control in the land-use experiment.

Completely unmanaged lands. A "preserve."

PRIMARY QUALITY	OF USE
Worked	
Active	
Used	
Rested	
Passive	
Contemplative	

ZONE	
1	
3	
5	

NORTH   N.T.S.

The property is broken down into existing-use zones based on how frequently people utilize an area. Roads and trails are put into categories of primary, secondary, or tertiary so the hierarchy of use is easily identifiable and defined. Illustration courtesy of Whole Systems Design, LLC

SOILS

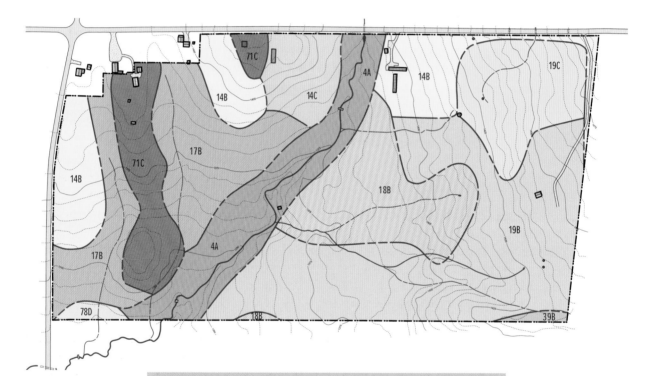

LEGEND

▯	building or structure
▨	demolished building or structure
⌒	25 ft contour
⌒	5 ft contour
⌒	stream
14B	soil types
▤	shallow depth to bedrock
▨	wet soils
▥	flood plain

↑
NORTH

N.T.S.

CLASSIFICATIONS

4A	Sunny silt-loam, 0–2% slopes
14B	Colonel fine sandy loam, 3–8% slopes
14C	Colonel fine sandy loam, 8–15% slopes
17B	Cabot silt loam, 3–8% slopes
18B	Cabot silt loam, 0–8% slopes, very stony
19B	Colonel fine sandy loam, 3–8% slopes, very stony
19C	Colonel fine sandy loam, 8–15% slopes, very stony
39B	Colonel fine sandy loam, 8–15% slopes, very stony
71C	Tunbridge-Lyman complex fine sandy loam, 3–15% slopes, rocky
78D	Peru gravely fine sandy loam, 15-35% slopes

SUMMARY

Soils on site are generally of a composition well-suited to very-well suited to traditional agriculture. The primary limiting factor to the site's open land is the pattern of generally wet soils. However, the soils mapped as "wet" on site need to be groundtruthed further as there are drier pockets within these areas that are well drained. While fertility of the soils can be altered, the soil's hydrology is more difficult to adjust. The wet-soil nature of the site in general calls for wet-hardy plantings to be cropped along with considerations for wetland and water-based farming systems such as aquaculture. These conditions also call for the driest areas of the site to be seen as special resources that are limited; places where crops requiring good drainage are prioritized carefully. Localized mapping of the drier soils with the wet areas as well as wet areas within drier zones will follow and adjust this map to allow optimal siting of all species and uses for cultivation.

There have been soil surveys done by the USGS for every state and region in the United States, and can typically be found online. These surveys classify each different soil type, depending on parent material, depth to bedrock, slope, composition, porosity, and many other criteria, and give a detailed description of each. These reports offer broad-stroke information on the different soil types suitable for a first phase of analysis. Soil samples should be taken and sent for testing before any detailed plans are developed. Illustration courtesy of Whole Systems Design, LLC

Design Criteria

The first part of this phase in the design process involves goal statements, which integrate both the goals ID and the site analysis—these can be called "design criteria." These statements serve as design guides—patterns that help massage solutions into alignment. They leverage the information gained during the site analysis and goals ID phase to form specific guiding statements that the next design phases can be derived from. Then when design directions do emerge onto paper or into words, they can be checked against these criteria to ensure that they make the most sense possible and fit within the parameters and patterns of both the client's goals and the site's character. We like to organize design criteria/goal statements according to aspect of the design so they can be most easily referenced later on. An example for one project—a large commercially oriented farm with an educational and research component—goes as follows:

VEGETATION

▶ Provide wind protection from the predominant cold or damaging wind directions.
▶ Be diverse and proven for multiple yields, including foods, medicines, habitat, biomass and fertility, thermal value, interest, and so on.
▶ Be established in patches and clusters, not homogeneously, and prioritize dynamic accumulators early on.
▶ Be self-maintainable wherever possible; for example, groups of cross-pollinating species, guilds for pest prevention, and fertility building.

INFRASTRUCTURE

▶ Capture, process, store, and distribute electricity and heat.
▶ Process farm products.
▶ Process nutrients generated within.
▶ Offer a diversity of microclimates in the landscape, and shelter outdoor spaces from wind, rain, snow, sun.
▶ Be composed of native materials.
▶ Be wheelchair accessible.
▶ Have the best possible indoor air quality.

▶ Be composed of and driven by biological systems as much as possible; for instance, living roofs, interior plantscaping, greywater gardens.
▶ Optimize views of the surrounding landscape, and promote direct sensory experience of the living world as much as possible.

SOCIAL SPACES (RESIDENCE, CIRCULATION, GATHERING)

▶ Create work centers
▶ Accommodate a range of inhabitants. These include guests staying a few hours or a few weeks and permanent residents.
▶ Offer a gradient of openness across the landscape, ranging from open pasture spaces and knoll-top lookouts to inward-facing gathering spaces within intensive gardens.
▶ Closely connect zones of intense use.
▶ Reduce energy expenditure in maintenance.
▶ Promote contact with living systems wherever possible.
▶ Offer outdoor use across the day and seasons.
▶ Provide a strong public-private gradient.
▶ Be legible in its design and construction, and help build awareness in the inhabitant.
▶ Encourage visitor interaction with gardening and farming systems.
▶ Be wheelchair accessible in certain areas.
▶ Preserve and highlight the site's land-use history.

Such design criteria can be thought of as quality control points and reminders in the design process. Have fun developing them, and try to appreciate them for the service they can provide, almost like checking your math in an equation. Criteria let us look back into the process to be sure some big amazing creative idea is not leading us away from actualizing a basic requirement of the design.

Imagination: Limiting Factor to Design

With the previous assessment and initial guiding work underway, the fun can begin. This is the time to pull out the pencils and tracing paper or drawing pad and

WARREN COMMONS
A POST-OIL COMMUNITY LAND PLAN

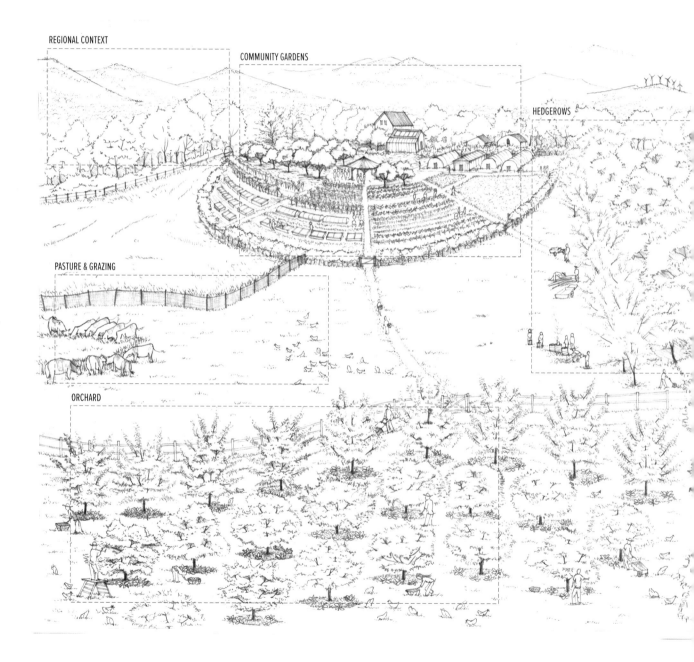

REGIONAL CONTEXT

COMMUNITY GARDENS

HEDGEROWS

PASTURE & GRAZING

ORCHARD

Clearly articulating the possibilities: a bird's-eye view of a fully developed master plan Illustration courtesy of Whole Systems Design, LLC

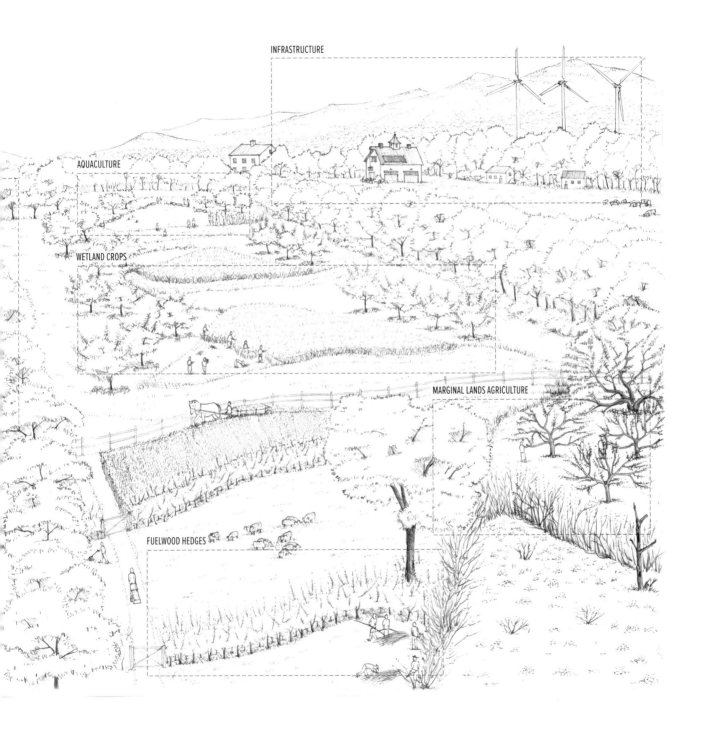

INFRASTRUCTURE

AQUACULTURE

WETLAND CROPS

MARGINAL LANDS AGRICULTURE

FUELWOOD HEDGES

Design is articulating a response to conditions that might be—but may not be yet; we must plan *as if*.

start sketching the possibilities for a space, an object, or a system. This is the time to think through the project in the horizontal and vertical planes and across time. During this process it is crucial to immerse yourself wholly in the project's possibilities.

About six years ago my colleagues and I were hired to plan a large new innovative farm development called Teal Farm in Huntington, Vermont. The objective was to design and implement a site that would model food systems truly fit for an adverse climate changed future with emerging resource constraints, human health, and toxicity challenges. It was crucial to reference historical patterns in such a job and to imagine carefully what future scenarios could actually be, to even begin to think about how to respond to them—and a good design needed to be *that* response. In this light we can see that design is articulating a response to conditions that might be—but may not be yet; we must plan *as if*.

To see what "as if" might actually involve, we need to immerse in the possible futures of a place and project in that place; we need to act it out, be inside of it, live it to the extent possible. If design is the process by which we anticipate and respond to changing conditions over time, it is fundamentally a phenomenological activity. Amid the year long initial master planning process I spent a handful of weeks immersed in a particularly deep way during the Teal Farm process. This experience was probably a lot like that of an actor who takes on their character in daily life to most fully get to know that person, their story, and how to communicate it. Conversations over dinner revolved around harvesting chestnuts in the understory of a two-hundred-foot-tall anthropogenic forest canopy. My friends on the design team spoke in detail about the stone barns our children would be playing in and what they would do with the pine nuts from the trees their great-grandparents had planted. We talked about what it would be like when you planted in garden soil that you could stick your

whole arm into with ease. We lived in this future world, intentionally ignoring the current state of affairs in order to be able to clearly envision other possibilities—higher possibilities. It helps to be single, to live alone and/or have other conditions that do not break this attention to the new reality which you are imagining. It requires unbroken attention and perhaps a certain obsession to get deep into a design. I found that the real breakthroughs came 8 to 16 hours into a work day and not usually inside around the drafting table but out skiing, walking, or climbing a nearby mountain.

At some point weeks into this design process, I ran into someone I knew in town, and he was surprised to see me; it had been a while. "Where have you been?" he said. "Haven't seen you as much as usual." I'd been around, not away, so it was strange he said this. But of course, I'd been spending a lot of time in Huntington, Vermont, in the year 2250. Our design for the Teal Farm could not have been possible without this mental time travel, and I would go so far as to say that any good design for other places involves such an approach during parts of the process. I think one of the primary reasons why most design around us today is so shallow is that it is too often born of someone doing it for a living, not as a calling, an obsession, something they actually love to do each day. The necessity of deep imaginative immersion here seems to apply to the creation of everything—be it a landscape, a business, or a lifestyle. Often, the results one seeks do not happen simply because the process required to activate them is not nearly as deep as is necessary. Design in this way is not a "day job."

The concepting phase is the time to immerse in this way, though at all phases of the design after the assessment phase, it is important to think "from within the design"—from within the place-to-be. This is rarely done enough, and I can honestly say that we do not do this as much as would be ideal in most projects—it takes a lot of time. Fortunately for the owner-builders, they can take this imagining time sitting on a rock in the garden, strolling the land, and in various other passive ways of hanging out on-site. For the professional designer this imagining time is highly limited for practical reasons—especially if they have a family or a "normal" life with a routine schedule. We still devote unbroken stints of obsession/immersion to projects but having an increasing family aspect to life limits it. We always encourage the client to spend time this way, as the possibilities for immersion are greatest for those living on site. They can then share the visions they have with their assisting designer—their site facilitator.

Schematic Design: Sorting Through Multiple Development Options

Once the process of goal clarification has begun, the site's context and characteristics have been explored significantly, and design-directing statements (design criteria) have been made, actual physical design plans can begin to be developed. The schematic design phase is what many think of when they hear the word "design," but as you can already see, it only represents a part of the entire process and is only as good as the analyses phases preceding it. In schematic design the designer puts to paper various versions of what seem to be the best solutions to the challenges thus far identified (new challenges will be identified in this process, so be on the lookout for them!).

The schematic design phase is the "take the best, leave the rest" part of the design process, where all sensible possibilities—drawn from the implications of the analysis and design criteria—are considered, mapped, studied, and tweaked. Schematic design is most often performed via plan-view drawings, sometimes accompanied by cross sections. Building schematics should be heavy on cross-sectional views, while landscape layout schematics would most sensibly be generally weighted toward plan views.

We have at times, when the budget allowed, performed 3-D perspective schematics, which are incredibly useful in helping think through the possibilities for space. Perspectives, even with computer-aided approaches, can often be time consuming and expensive. It is important to point out, however, that thinking in 3-D is crucial to developing a space in the most sensible ways. The old adage, "If you can't draw it, you can't build it," carries a ton of weight in the constructability of an object or space, but I would add another: "If you can't sketch it in 3-D, you can't think it and you can't think *about it* clearly."

And if you can't think about it clearly, it won't be a very high-quality space—you find that out later when

SCHEMATIC DESIGN   N.T.S.

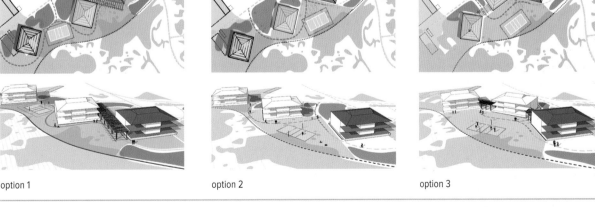

option 1   option 2   option 3

Different approaches explored during the schematic design phase lead to a working plan that highlights one of the options given or may combine the best elements of all three. These schematics were performed for a faculty housing addition to a school campus. Illustration courtesy of Whole Systems Design, LLC

SCHEMATIC

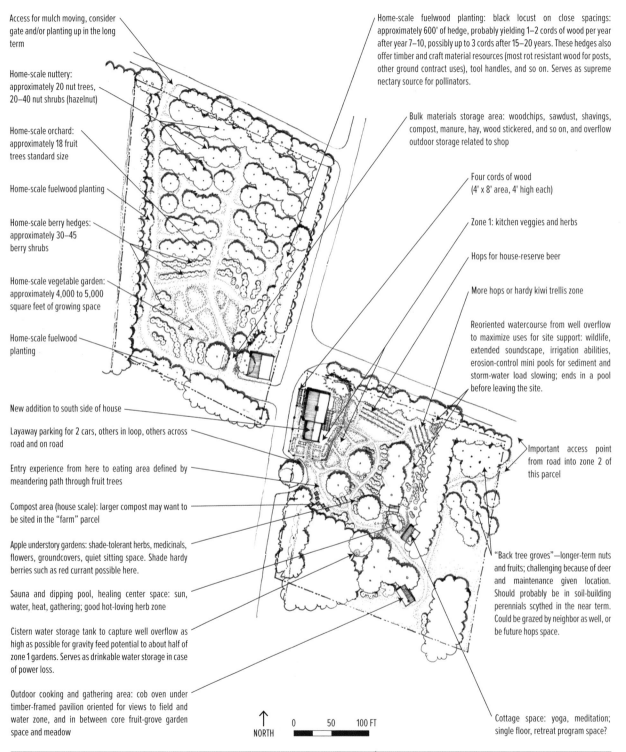

Access for mulch moving, consider gate and/or planting up in the long term

Home-scale nuttery: approximately 20 nut trees, 20–40 nut shrubs (hazelnut)

Home-scale orchard: approximately 18 fruit trees standard size

Home-scale fuelwood planting

Home-scale berry hedges: approximately 30–45 berry shrubs

Home-scale vegetable garden: approximately 4,000 to 5,000 square feet of growing space

Home-scale fuelwood planting

New addition to south side of house

Layaway parking for 2 cars, others in loop, others across road and on road

Entry experience from here to eating area defined by meandering path through fruit trees

Compost area (house scale): larger compost may want to be sited in the "farm" parcel

Apple understory gardens: shade-tolerant herbs, medicinals, flowers, groundcovers, quiet sitting space. Shade hardy berries such as red currant possible here.

Sauna and dipping pool, healing center space: sun, water, heat, gathering; good hot-loving herb zone

Cistern water storage tank to capture well overflow as high as possible for gravity feed potential to about half of zone 1 gardens. Serves as drinkable water storage in case of power loss.

Outdoor cooking and gathering area: cob oven under timber-framed pavilion oriented for views to field and water zone, and in between core fruit-grove garden space and meadow

Home-scale fuelwood planting: black locust on close spacings: approximately 600' of hedge, probably yielding 1–2 cords of wood per year after year 7–10, possibly up to 3 cords after 15–20 years. These hedges also offer timber and craft material resources (most rot resistant wood for posts, other ground contract uses), tool handles, and so on. Serves as supreme nectary source for pollinators.

Bulk materials storage area: woodchips, sawdust, shavings, compost, manure, hay, wood stickered, and so on, and overflow outdoor storage related to shop

Four cords of wood (4' x 8' area, 4' high each)

Zone 1: kitchen veggies and herbs

Hops for house-reserve beer

More hops or hardy kiwi trellis zone

Reoriented watercourse from well overflow to maximize uses for site support: wildlife, extended soundscape, irrigation abilities, erosion-control mini pools for sediment and storm-water load slowing; ends in a pool before leaving the site.

Important access point from road into zone 2 of this parcel

"Back tree groves"—longer-term nuts and fruits; challenging because of deer and maintenance given location. Should probably be in soil-building perennials scythed in the near term. Could be grazed by neighbor as well, or be future hops space.

Cottage space: yoga, meditation; single floor, retreat program space?

NORTH   0   50   100 FT

A schematic plan depicts how the different concepts will take physical form on the landscape. Illustration courtesy of Whole Systems Design, LLC

WORKING MASTER PLAN

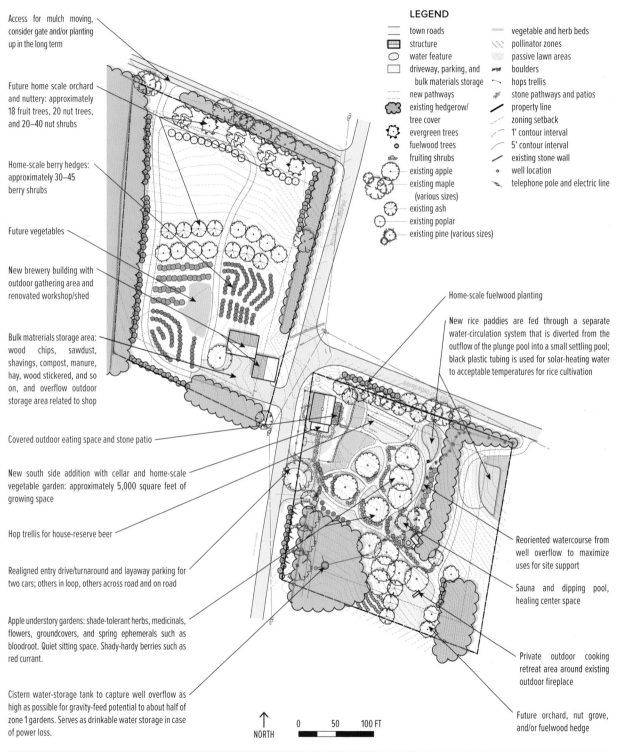

LEGEND

town roads		vegetable and herb beds	
structure		pollinator zones	
water feature		passive lawn areas	
driveway, parking, and bulk materials storage		boulders	
new pathways		hops trellis	
existing hedgerow/ tree cover		stone pathways and patios	
evergreen trees		property line	
fuelwood trees		zoning setback	
fruiting shrubs		1' contour interval	
existing apple		5' contour interval	
existing maple (various sizes)		existing stone wall	
existing ash		well location	
existing poplar		telephone pole and electric line	
existing pine (various sizes)			

Access for mulch moving, consider gate and/or planting up in the long term

Future home scale orchard and nuttery: approximately 18 fruit trees, 20 nut trees, and 20–40 nut shrubs

Home-scale berry hedges: approximately 30–45 berry shrubs

Future vegetables

New brewery building with outdoor gathering area and renovated workshop/shed

Bulk matrerials storage area: wood chips, sawdust, shavings, compost, manure, hay, wood stickered, and so on, and overflow outdoor storage area related to shop

Covered outdoor eating space and stone patio

New south side addition with cellar and home-scale vegetable garden: approximately 5,000 square feet of growing space

Hop trellis for house-reserve beer

Realigned entry drive/turnaround and layaway parking for two cars; others in loop, others across road and on road

Apple understory gardens: shade-tolerant herbs, medicinals, flowers, groundcovers, and spring ephemerals such as bloodroot. Quiet sitting space. Shady-hardy berries such as red currant.

Cistern water-storage tank to capture well overflow as high as possible for gravity-feed potential to about half of zone 1 gardens. Serves as drinkable water storage in case of power loss.

Home-scale fuelwood planting

New rice paddies are fed through a separate water-circulation system that is diverted from the outflow of the plunge pool into a small settling pool; black plastic tubing is used for solar-heating water to acceptable temperatures for rice cultivation

Reoriented watercourse from well overflow to maximize uses for site support

Sauna and dipping pool, healing center space

Private outdoor cooking retreat area around existing outdoor fireplace

Future orchard, nut grove, and/or fuelwood hedge

↑ NORTH

0   50   100 FT

The schematic plan is then refined and developed into a working master plan that will help guide development. Illustration courtesy of Whole Systems Design, LLC you inhabit places that aren't well thought through, with countless second-guessing: "Why didn't we put a switch here, steps there, a wall here, or a water hydrant there?" Anyone that would like to think through objects, spaces, and, indeed, places clearly should become fundamentally literate in drawing such that she can sketch with relative ease buildings and landscapes in perspective, cross section, and plan view. It's not that hard with some practice.

Drawing also has been shown to activate parts of the brain that are otherwise underutilized—those parts of the brain are likely crucial to wholly thinking about how to solve problems systemically and effectively in a place. Drawing is baseline empowerment in this regard. While drawing techniques are out of the scope of this book, I have highlighted several tips on developing visuals to aid your design that are particularly useful to the nonprofessional designer but earnest homesteader-farmer (and therefore designer by default!).

1. **Basic materials** to have on hand: rolls of tracing paper—12″ and 24″ by 50 yards; pens and mechanical pencils—0.7mm HB is the most versatile; an architects' and engineers' scale (ruler); graph paper with ¼″ grids; colored markers; a triangle or two, and a circle template. That's it—you don't need to spend much money on drawing supplies to do massive amounts of design. If you'd like to supplement your work digitally, I'd recommend learning Sketchup and Adobe Illustrator—both are accessible and can be self-taught by those who are computer inclined.
2. **Trace maps, plans, and drawings** you can find of things you like. Don't be bashful—cut out visuals from magazines, books, and other sources, and trace them. This gives you the feel of making plans, engenders patterns in plans, and starts to lend some fluency between the hand and the visual result, loosening one's creativity up a bit.
3. **Just draw**. The biggest reason that few homesteaders and farmers draw up enough graphic plans of their developments seems to stem from the simple discomfort associated with producing graphic work. While most of us grow up learning how to

communicate basic ideas via words, few of us receive any real training in conveying equally basic ideas in graphic form. This is actually a tragedy for society because it severely retards the creative potential of citizens at large. Drawing and being able to take a decent photograph is as important to imagining, conveying, and creating functional places as writing a decent paragraph about the place—probably a lot more, actually. So please, do yourself and society a favor—loosen up and start drawing!

4. **Photograph, print, and trace over**: Take an image of a space you would like to develop, print it, then trace over this photograph using lines to show the new changes on top of the existing scene. This is an immensely helpful way of teaching yourself how to draw in perspective. You can also project the image on a wall if you have a projector and use markers on tracing paper against the wall to do the same at larger scale. This works great for really getting into the possibilities of a larger site area. If you don't have tracing paper, you can stick paper onto a window and turn normal bond paper into tracing paper. A glass coffee table with a light underneath works even better (what designers call a "light table").
5. **Mock it up:** There's absolutely no better way to physically hint at and offer insight into the possible changes (and results) to a place than using large objects to lay out in a space for help in envisioning the changes. We use wood, tires, vehicles, people, barrels, potted plants, rope, chalk lines on the ground, and much more to do this. Mocking up a design is not always possible at larger scales, but for smaller spaces it is a fantastic tool.

Adapting Land to Rapid Change

Neither predominant agricultural models nor most housing and transportation systems are designed to withstand significant climate changes or resource supply changes. These systems currently depend on a constant and unbroken source of cheap energy and materials (read fertilizer, pesticides, shipment, parts, heat, electricity, and fuel) to operate. They also

DESIGN FOR CLIMATE CHANGE

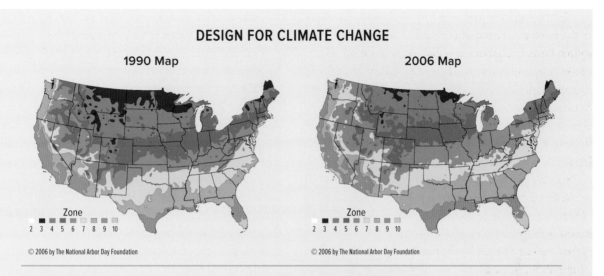

1990 Map

2006 Map

© 2006 by The National Arbor Day Foundation

© 2006 by The National Arbor Day Foundation

The USDA plant hardiness zone maps for 1990 and 2006—a clear indicator of the need to plan for change.

Earth orbits the sun at distances that vary by three million miles across the year. Volcanoes explode, ice fields melt, sea vents open and close, gases continually exchange between rock and plants, ocean and atmosphere. Human influence is only one factor in Earth's climate stability. Accurately engaging the issue of global climate change requires an understanding that Earth's climate has never done anything *but change*. With this in mind we move forward knowing that if the human project is to be successful on Planet Earth, it will be highly adaptive in the face of climate—and all other—forces of change.

Good design is design for change. Good design is structurally diverse and not dependent on any single element for its overall success. Good design harnesses the forces of evolution, leveraging both the built and the biological environment, and integrates them for maximum resilience. This chapter briefly overviews strategies for developing biologically adaptive, intentional ecosystems (permacultures), and climate-buffering landscapes (microclimates) in which humans can live more resilient lives if times become more difficult, or even if they don't.

Adapting to rapid changes entails developing resource systems (both built and biological) that will be functional across a wide range of conditions. This is true for changes in all systems, whether they be financial, cultural, or ecological. What specific challenges would we design for to be adaptive to changes in the global climate and resource costs/availabilities? Such changes are likely to include longer droughts, hotter summers,

> ## Good design is design for change.

possibly colder winters, more severe wind events, increased pest pressure, more acute precipitation events, earlier and later frosts, decreasing numbers of pollinators, rapid energy and material price increases, and other irregularities that have always tested humanity's ability to thrive and survive on this planet.

DIFFERENCES BETWEEN 1990 AND 2006 HARDINESS ZONES

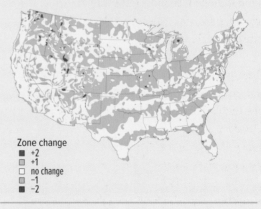

Zone change
- ■ +2
- ▨ +1
- □ no change
- ▨ −1
- ■ −2

A majority of the changes from 1990 to 2006 indicate a clear rise in average temperatures across the country.

depend on the climate's remaining largely the same as it has been for hundreds of years—the Midwest corn crop won't be harvested without both reliable and cheap energy *and* plentiful rainfall. Many homes and apartments (especially tall buildings) in the northern one-third of the country are not habitable in heat waves when the electric grid is down. Landscape-level developments that intentionally adapt to these changes employ the following components, among others:

▶ **Microclimate development,** including wind-breaks, snow-retaining hedgerows, thermal mass via water and stone, and sun-trapping vegetated or built arcs. These systems provide a buffer against regional climatic stresses by *localizing climate at the site level*.

▶ **High biodiversity** of crop species and crop systems, utilizing neighboring warmer and colder climate zone diversity (USDA hardiness zones +/– 2 zones) and the intelligence of complex ecosystems. Reviving the genetic diversity lost in the wake of global industrial agriculture is a prerequisite for adapting to rapid change. Since current challenges are so severe—from climate changes to persistent

biospheric contamination—we will likely need to not only revive past levels of diversity and health but evolve greater, unprecedented levels of biodiversity and ecological connectivity.

MICROCLIMATE DEVELOPMENT

A microclimate is any discrete area within a larger area of differing climate. Microclimates exist unintentionally in nature, but good design creates microclimates intentionally. Since cold is a limiting factor (along with light) in sustainably inhabiting the New England landscape, developing warm, protected microclimates is the top priority here. Cooling strategies, however, will likely become increasingly important, especially in southern New England, if conditions continue to warm. Optimized microclimates result in the following:

▶ Lower active energy needs for buildings (less fuel, less cost, less pollution). Example: passive solar house within a passive solar landscape.

▶ Longer growing seasons relative to the surrounding environment. Example: climate-designed garden spaces that stay frost-free for weeks longer in the spring and especially in fall than adjacent areas.

MICROCLIMATE CROSS SECTION

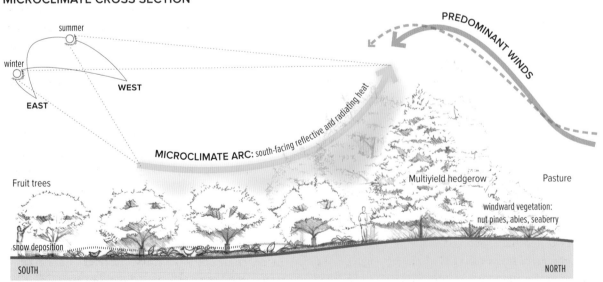

Larger and hardier tree species are placed to the north of high-value zones to both shield prevailing cold winds and create a south-facing space that holds and disseminates heat. Illustration courtesy of Whole Systems Design, LLC

DESIGN OF WARM MICROCLIMATES CHECKLIST

1. Face: southerly, and avoid cold-air drainages and dams
   • South–southwest = warmest
2. Slope: vertical-space harvesting
   • The farther poleward, the steeper the slope should be to capture the most solar energy
3. Bowl: solar arc/sun trap
   • Utilize energy-harvesting forms
4. Minimize radiative losses: provide cover
   • Nighttime losses of heat are the most difficult to avoid.
5. Wind-shelter
6. Buffer and deflect, create eddies, preserve and enhance hedgerows. Examples of microclimate-creating features include hills, fields, trees, cliffs/stone, gullies, ridges, groundwater, ponds, lakes, roads, walls, lawns, roofs, and courtyards. Employing such features in the development of climate-protected spaces is more effective than attempting to create new microclimates from scratch.
7. High mass
   • Stone and water are the primary heat-retaining materials.

8. High absorption (low albedo)
   • Utilize color effectively.
9. Time your microclimate.
   • Design for a particular time of day and year, usually whenever limiting factors are most present. If, for example, you're making a greenhouse to extend your gardening season, it does not need to be sited or designed for all-day sun each day of the year. In this particular instance a tree that shades the greenhouse at noon on the winter solstice doesn't need to be cut down if your goal is season extension, not winter production. There are many examples of how timing your microclimate can enable opportunities that would not exist if the microclimate was not thought of in such a nuanced and precise way—aim to find these optimizations through timing in all of your microclimate development work, especially when it comes to built systems and solar design.

► Higher yields from plants and animals, via better growing conditions. Examples: warmer environment for heat-loving crops; cool-shaded spaces for domestic animals in the hot summer; wind-sheltered spaces for plants, animals, and buildings.

► More enjoyable, lower stress, and healthier human habitats. Longer outdoor living season; more fresh air; more contact with water, plants, living systems; and greater physical activity and mental stimulation. Example: outdoor living spaces comfortable in the summer, warm in the winter.

It is relatively easy to adjust the climate of the spaces we inhabit, whether they be horticultural or for human enjoyment, and the results are stunning. Fortunately good examples of microclimate design abound in the living world around us and in vernacular design, from beehives to termite mounds, to deer wintering areas, to traditional farm layouts and building configurations. It's only in the most recent era of cheap energy that humans have been able to forget about harnessing innate patterns in the local climate to our advantage.

MICROCLIMATE DEVELOPMENT STRATEGIES

The first step in crafting beneficial microclimates is proper site selection, as some landscape features cannot be changed at all or only to a small extent. The second step in localizing your climate is site design. Once a site has been chosen and a handful of strategies planned for and implemented carefully, you can optimize the existing climate of the site to more fully meet the needs of its inhabitants. Examples of microclimate-creating features are hills, fields, trees, cliffs/stone, gullies, ridges, groundwater, ponds, lakes, roads, walls, lawns, roofs, and courtyards. Employing such features in the development of climate-protected spaces is more effective than attempting to create new microclimates from scratch.

Diversity and Connectivity

Of primary importance for increased food security and regional resilience is developing diverse and interconnected food-crop systems. The following strategies highlight the benefits of high-biodiversity, polycultural food systems.

MANY CROPS

Early and late frosts; intensifying drought, heat and cold; and other stresses select against certain crops. A broad range of species with different flowering cues and hardiness capabilities is insurance against poor fruit sets, pollination failure, and other problems due to capricious weather. Such an approach is exemplified in a planting scheme that includes apples and plums (early flowering) with elderberries or kiwi (very late flowering)—one of these flowering periods is likely to be okay each year but possibly not both. A wide array of dry and wet-hardy crops is a hedge against a season of drought or inundation. We mix pears on quince rootstock with pears on pear rootstock and plant the same tree in both dry (high) and wet (low) situations in the landscape to hedge against the possibilities of drought or inundation in each growing season; it's often one or the other, after all. Intentional genetic diversity in species and variety is fundamental to any resilient ecological system.

NEW CROPS

Developing innovative new crossbreeds also helps to ensure resiliency of food systems. For example, crossing a sweet cherry (*Prunus avium*) with a Nanking cherry (*Prunus tomentosa*) can create a next-generation cross that flowers like the Nanking (late, thus avoiding the killing late-spring frosts) but has the larger, sweeter, and more marketable cherry. Hybrid vigor is crucial to develop across plant and animal families. Adapting our food system to ever-changing conditions entails continually increasing the fitness of each component in the system, from human to plant to animal. That is good breeding. Breeding never ends but continually adjusts to fit changing conditions with each successive generation.

WARMER AND COLDER HARDY CROPS

Rapid warming and cooling trends will probably outpace the agility of current agricultural systems. Durable farming systems should be designed to adapt to changes of 10 to 15°F warmer *or* colder within the span of a few decades. This is possible to achieve through highly diverse crops, highly connected agricultural-ecosystems, microclimate buffering, keylining,

> **What if our task was fitness within, not attempted control of, the ever-changing conditions of our existence?**

mycelium webs, and other biological resiliency strategies but also by developing crops that can extend into warmer or colder temperatures. If zone 4 became just 10 to 15 degrees warmer (an average low of −10°F), a diversity of bamboos and palms could be grown. Some apples can withstand −50°F or colder—a real plus if the global ocean conveyor belt stops or changes direction.

EVOLUTION AS ADAPTATION TO, NOT CONTROLLER OF, CHANGE

The question is not *if* or *how much* things are changing; changes in Earth's climate, in human society, and in every other planetary system are *guaranteed* by the full faith and backing of the Milky Way Galaxy. The question is whether or not we will view such changes as an adaptive challenge—as a challenge to cultivate biodiversity, ecological resilience, and an increasing, not dulling, sensitivity to the possibilities around us. What if our task was fitness within, not attempted control of, the ever-changing conditions of our existence?

Working Plans and Implementation Documents

Once you've worked out various scenarios for developments, it is time to develop a plan of action. These are your plans for development and implementation and range from overall plans for the site over a span of time—a "master plan"—to specific drawings and documents that detail how to actually build, plant, install, or otherwise make the components of the development—the construction or implementation documents. These plans and textual specifications are often drawn to scale and include both plan view and cross-sectional drawings. Master plans are not solid, set-in-stone documents—although everyone wants them to be. Heck, I am hired many times largely because people want a plan that's solid, unwavering, and something they can

A student in our permaculture design course working on a design

follow now and in ten years. Sorry—they don't exist. Most plans are iterative. And despite the authoritative sounding name, master plans are no exception. A good "master" plan is a "working" plan—in other words, it's the latest version of good approaches. It will change; that much is guaranteed. The important part to

> A good "master" plan is a "working" plan—in other words, it's the latest, greatest version of approaches, but it should change over time.

HYDROLOGY AND GRADING PLAN

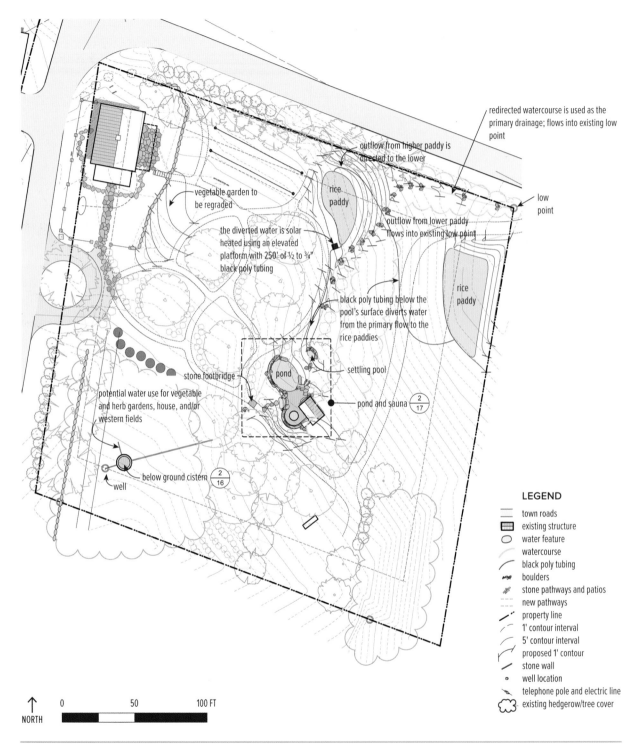

redirected watercourse is used as the primary drainage; flows into existing low point

outflow from higher paddy is directed to the lower

rice paddy

vegetable garden to be regraded

the diverted water is solar heated using an elevated platform with 250' of ½ to ¾" black poly tubing

outflow from lower paddy flows into existing low point

low point

rice paddy

black poly tubing below the pool's surface diverts water from the primary flow to the rice paddies

stone footbridge

pond

settling pool

potential water use for vegetable and herb gardens, house, and/or western fields

pond and sauna   2/17

below ground cistern   2/16

well

LEGEND

	town roads
	existing structure
	water feature
	watercourse
	black poly tubing
	boulders
	stone pathways and patios
	new pathways
	property line
	1' contour interval
	5' contour interval
	proposed 1' contour
	stone wall
	well location
	telephone pole and electric line
	existing hedgerow/tree cover

NORTH   0   50   100 FT

A detailed drawing of a proposed grading and hydrological system that takes the working master plan a step further. These plans are used to estimate materials and cost, think through the system in greater detail to identify challenges, and as a reference throughout the building process. Illustration courtesy of Whole Systems Design, LLC

HYDROLOGY AND GRADING DETAILS

HYDROLOGY SYSTEM CROSS SECTION

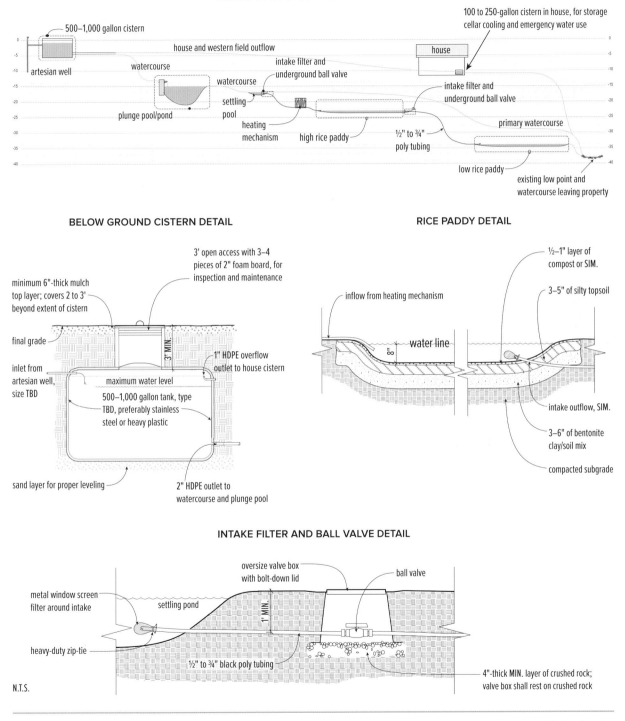

Cross-section and other detailed drawings accompany the implementation plan to give detailed information about specific features and design direction.

Illustration courtesy of Whole Systems Design, LLC

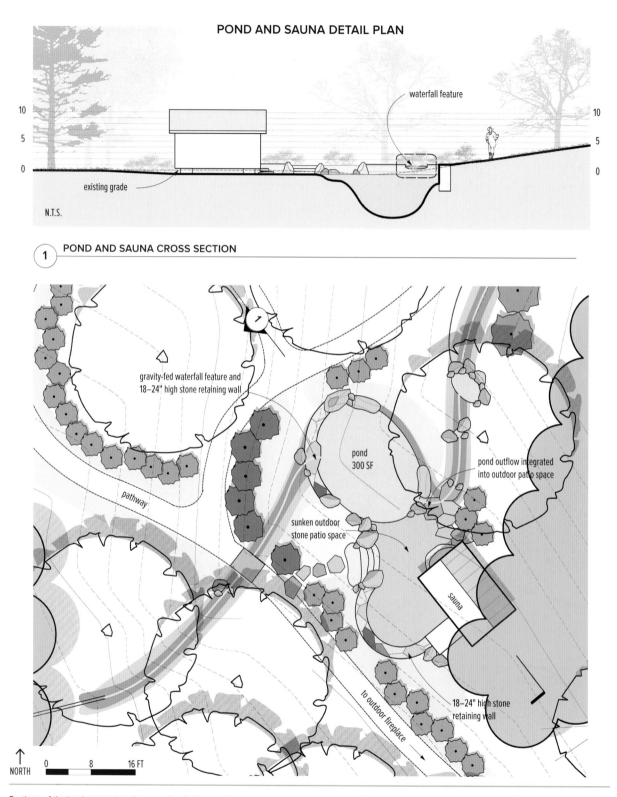

POND AND SAUNA DETAIL PLAN

POND AND SAUNA CROSS SECTION

waterfall feature

10

5

0

existing grade

N.T.S.

1

gravity-fed waterfall feature and
18–24" high stone retaining wall

pathway

pond
300 SF

pond outflow integrated
into outdoor patio space

sunken outdoor
stone patio space

sauna

to outdoor fireplace

18–24" high stone
retaining wall

NORTH   0   8   16 FT

Portions of the implementation plans are detailed enough to warrant a more thorough level of documentation. Features such as ponds and small outbuildings are often high-priority and costly items that require a more in-depth study so that expensive mistakes are avoided. Illustration courtesy of Whole Systems Design, LLC remember is that it's a guide for next decisions, not an ultimate life map or site oracle. Land and the lives unfolding for them are far too complex, unpredictable, and mysterious for any vision of a "way" to hold up year after year.

And they have one more primary purpose: to avoid huge mistakes—for instance, not putting the house in the wrong place or putting the orchard where a road for the eventual barn will need to go. Such plans are "master" only in that they locate elements that are thought to be inevitable in locations such that other actions can be made down the line. The paralysis that dominates a place when such a plan doesn't exist or, conversely, the repeated mistakes made when such a plan is not in effect are spectacular. In this way a master or "working" plan is essential. But don't abuse it—remember, it's a living document. It *must* change to remain valid.

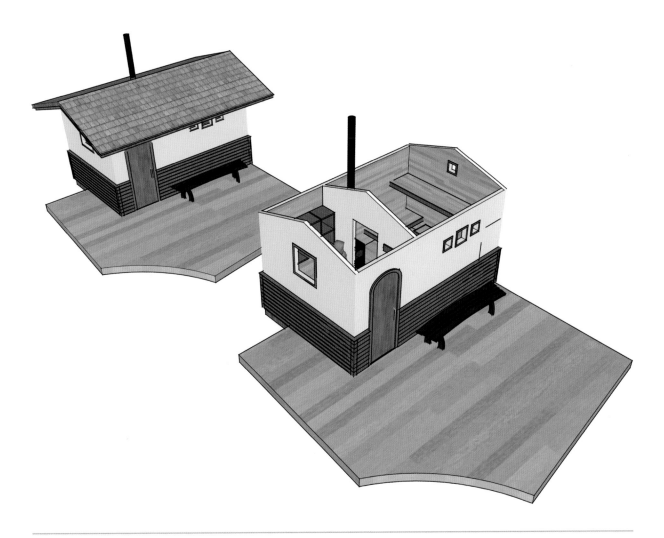

A three-dimensional model facilitates better communication and understanding between designer, client, and builder. The detailed model increases both the efficiency of the construction process and the quality of the end product. Illustration courtesy of Whole Systems Design, LLC

A wood-fired sauna and sunning deck with dipping pool built by Whole Systems Design, LLC, set in an outdoor room within an old apple grove. The dipping pool captures an artesian well overflow that was formerly running immediately off-site into a road ditch. Photograph courtesy of Whole Systems Design, LLC

Water is life. The saying has become a cliché, but its profundity cannot be overstated. Think galactically for a moment: Earth is the rare and tiny blue ball teeming with life amid a solar system—at least—of rocky gaseous orbs. It is Earth and only Earth—so far as we know—that is three-quarters water. Planet Water would be a much more accurate name than Earth. It should therefore be unsurprising that nearly all life systems have evolved to harvest, utilize, and cycle water—and the nutrients that are soluble in water. A resilient homesteader and designer, then, must be a *water process facilitator*. Through an awareness of how water affects living systems, we must orchestrate the interplay of systems in a manner that is vitalized rather than limited by its presence.

Nothing defines the nature of a place more than water.

Nothing defines the nature of a place more than water. The quantity, qualities, forms, distribution, and intensity of its entry into a landscape determine nearly everything else that happens ecologically in a place. Though there are many physical aspects of a place, including the type of bedrock, soils, and climate, it is the play of water that most directly determines an ecosystem's behavior and capacity for production or

regeneration. Thus, the most optimal design within which to fit human activities in a place start and end with water. First, to design for and with water, we must understand how to read the tendencies and behaviors of a place as they relate to its hydrology. This is not as simple as the sheer volume of water received; a site with 120″ of rain per year can be less resilient (more brittle) than a location with 50″ a year if the former location receives most of those 120″ in a short period of time via heavy storms (monsoonal) and the latter location receives a third of that volume across the entire year via less intense storms. Think Southeast Asia versus Great Britain.

Brittleness and the Quest for Resiliency

The most useful way I have found for understanding water's impact on a place and consequently how to work with it is through the scale of "brittleness." In a world of well-designed water systems, deserts grow into forests. It is important to remember that many droughts and certainly many of their acute impacts are human created. I first heard a place characterized as being brittle or nonbrittle from Allan Savory, the founder of Holistic Resource Management. Having done work in some of the most difficult areas of Africa, and actually having learned how to reverse desertification, Mr. Savory is intimately aware of the interplay

A brittle landscape is not simply a dry one but one in which regeneration naturally occurs more slowly.

between water, climate, and place. His work has resulted in significant regeneration across a range of climates, particularly in harsh desertifying areas of eastern Africa.

A brittle landscape is not simply a dry one but one in which regeneration naturally occurs more slowly. In simple terms this means: You clear a forest, you get a desert. This is true for the vast majority of acreage between the Mississippi River and the Coastal Mountain Range in the United States. In a nonbrittle (resilient) landscape, the rate at which the ecosystem can rebuild biomass and biodiversity is comparably high. Think of New England or the Pacific Northwest. There, you cut down a forest and a new one grows up in its place even with no replanting or seeding. These are places where moss grows on north- and east-facing slopes, where the growing season is short, and the dormant (read: rotting and soil accumulation) season is short. These are places where precipitation usually exceeds evaporation for much of the year, places where life has been lent a big helping hand by the soft hand of climate and water availability.

Here in Vermont—a hyper-nonbrittle region—it's not uncommon to see large forest trees reaching fifty feet or more into the canopy growing wholly on bedrock. For that kind of phenomenon to occur, you need a place where the climate promotes seed growth on all levels, from water availability to nutrient accumulation. Just think of a seed of a hemlock or yellow birch actually taking root on bare rock—first, a small amount of soil must accumulate (from leaf fall); second, the seedling must survive dry periods in that thin soil-moisture reservoir. Nonbrittle climates are truly miraculous in the helping hand they lend to life. In a brittle place you clear a forest, and even with replanting, seeding, and mulching, the job of redeveloping biomass and biodiversity is a difficult and long one—for the biological mantle/forest was the primary way moisture was captured and stored

(and even made in the first place).* In brittle landscapes any action that reduces overall water availability or concentrates it into smaller periods of time, rather than spreading it out, usually creates positive feedback loops, reinforcing the brittle tendencies further. We see this happening nearly everywhere on the planet, ranging from deforestation, ditching, and road construction to poor grazing—not necessarily "over"-grazing.

A primary directive of all regenerative, high-functioning land-use systems is the evening out of water availability in the landscape, distributing precipitation and soil moisture availability across time. Such "humidity distribution" can take many forms, and this chapter will focus on each of them individually.

When faced with the question of where to direct water and how, the designer must always be asking the same question: How can I slow it, spread it, and sink it?

The challenge and opportunity represented by water in the landscape can be grasped most easily if you think of water as fertilizer. If you had a rivulet, creek, or river of valuable compost running through your landscape, what would you do? You'd slow it down, spread it out, and distribute it across the site so the plants received its benefits. All water systems on a homestead or farm should be thought of in these same terms. The resilient farmer and homesteader needs to be aware of how her site developments affect the movement and storage of water on site. When faced with the question of where to direct water and how, the designer must always be asking the same question: How can I slow it, spread it, and sink it? Additionally, the following questions should always be kept in mind:

---

* Forests actually produce their own moisture, including rainfall and cloudy humidity—all of which can be harvested by plants, through evapotranspiration and the "wringing out" effect they have on the atmosphere. This is one of the reasons deforestation often leads to desertification, not just through soil loss but through actual reduction in moisture: Forests make rain, and removing them creates drought on multiple levels—in the climate and in the soil.

Table 3.1: Brittleness Spectrum

	Very Nonbrittle		Nonbrittle		Semibrittle		Brittle	Very Brittle	
	1	2	3	4	5	6	7	8	9	10
High Annual Production	Tropical rainforests			Subtropical and temperate tall-grass prairies				High-rainfall tropical savannas		
	Temperate rainforests							Mild-rainfall tropical savannas		
Medium Annual Production	Mild-temperate forests			Mild- and cold-temperate midgrass prairies				Low-rainfall tropical savannas		
	Cold-temperate forests (WSRF)							Mild- and cold-temperate steppe and grasslands and shrublands		
Low Annual Production	Subarctic coniferous forests			Tundra and alpine grasslands				True deserts: tropical, mild, cold, arctic		

A standard brittleness spectrum overview showing the Whole Systems Research Farm as the nonbrittle zone it is, relative to other zones of the world. It's likely, however, that the cold-temperate region of Vermont is actually far less brittle than tropical rainforests, which when clear-cut take longer to reorganize than the northern hardwood forests of the world.

This once relatively lush and productive area of north-central New Zealand shows many signs of desertification in the making as rampant deforestation leads to eroding hillsides, slowing production of rangeland, and a drying landscape. The creek bed in this image drains thousands of acres. Although it is spring at the time of this photograph, the drainage is bone dry and pasture grass growth has essentially stopped.

ASSESSING YOUR WATER BUDGET VIA A FIVE-GALLON BUCKET

Leaving a bucket outside for a year tells you a lot about the brittleness of your location. Here in the northern hardwood forest of New England a bucket usually fills up within four to five months and stays full for most of the year, dropping three to ten inches or so in most growing seasons. This can vary significantly, however: In the summer of 2012, a full bucket evaporated by about 50 percent by the end of the summer.

Here's an easy water-assessment strategy that anyone can afford to do and that some of us may have even done without meaning to. Leave a five-gallon bucket or similar water-holding container outdoors across the year. Make sure it is in an open area that receives the sunshine and precipitation affecting the site. Now neglect it. Make sure it doesn't fall over, and observe what happens over the course of a year. Here in central Vermont our buckets fill up within two to three months, then stay full, dropping an inch or few or so every now and then in a dry July or August, then overflowing again before becoming a block of ice sometime in November or December.

This simple test speaks volumes about your region's climate and your site's unique microclimate. If the bucket fills and stays full or even half full, you're in a water-rich part of the world. If the bucket remains empty and fills only part way before emptying again, you are in an arid zone. Everyone else falls somewhere in between. Pay particular attention to the times when the bucket is empty or full. That means you have a drought or overflow situation. The bucket tells you what is happening in the water table, but remember that the table itself is much more delayed in its response than the bucket. During those times of the year when the bucket is empty, the table will be dropping, and the opposite will be true when it's full. On the surface of the ground you want to ensure that as much water is captured as possible throughout the year to create a bucket-full situation: This way plants are not limited by water scarcity, and it's available for all other uses (which we'll go into in a moment). Consider yourself lucky if the bucket is half full or more most of the year—that puts you in the slim minority of an increasingly water-stressed, drought-prone world.

► How can I increase the time in which it stays on this site?
► How can I move water from the wettest areas of the site to the driest?
  • How can I water the ridges from the valleys?
► How can I use water to move nutrients from where they are abundant to where they are scarce (forcing additional questions concerning nutrient sources and sinks)?
► How can I lay out the site especially in terms of animals/humans such that excess nutrients can be distributed via gravity downhill? "Animals above plants" is usually the best layout.
► Where can I capture water from, especially as high on the site as possible, and where on the site can I store it as high as possible?

► How will I access this water and distribute it?
► How will these systems behave in freeze-thaw situations and in intense flood or drought events?
► How will these systems be managed over time, and how will they be changed when the climate gets wetter or drier, hotter or colder?
► How can these systems function if parts from global trade are not available? How can the system be as low tech as possible?
► How can I avoid pumping water in favor of letting gravity do the work?
► Where might toxins be entering the water system on site, and how can I mitigate that? (Think old septic system, driveway with cars, neighbor's land uphill, and so on.)
► How will freezing affect the water system on-site?

Gravity-Feed Systems

For both agricultural and domestic purposes a gravity-driven system, as opposed to a pump-driven system, is superior for obvious reasons. Gravity is free, never sleeps, and doesn't break. Pumps, at best, require maintenance and energy input. To gravity feed water we must activate the "store water high in the landscape" principle of permaculture; you can only move water downhill. So the primary directive here is to locate water storages as high as possible but low enough that significant amounts of water are able to be captured.

What's "significant"? you might ask. That depends on what you are trying to do with the water. For domestic uses a tiny spring can be dug and stored above one's home that will harvest $1/30$th of a gallon a minute (GPM) and stores two hundred gallons. For certain agricultural uses, such as watering twenty cows, that would not suffice, and a lower location on the property that can harvest, say, five GPM and house a storage of two thousand gallons might be needed. The point remains the same: the higher the better, so long as you can capture enough water in the time frame (period) in which the cycle of deplete and replenish occurs.

IN THE LANDSCAPE

At the Whole Systems Research Farm, our primary high water storage serves agricultural purposes (not in-house domestic—though it could in a water emergency) and is composed of the first pond we constructed. Please refer to the map of the WSRF for pond locations. (We will be developing water-table-accessing spring boxes in the future for domestic use.) This pond holds about fifteen thousand gallons of water and was built with a mini excavator in about three days. It is clay lined from material on site and has a surface area of approximately twenty by twenty-four feet.

What makes this pond somewhat unique is not its size but its water table relationship. The pond has no continuous supply, such as a spring-fed or stream-fed pond. Rather, it's mostly an exposure of the water table, with large amounts of intermittent surface flow harvested from the landscape immediately uphill of the pond (with the help of ditches aimed toward the pond).

ECOSYSTEM BIOMIMIC: WETLAND

Ponds are one of the best examples of how human impact can be a positive influence on an ecosystem. In areas without direct human intervention, landscapes comprising hilly and mountainous terrain (where the climate allows) often develop a gradient of wetlands, from those adjacent to river channels in the bottomlands to midelevation wetlands to high-elevation depressions near the ridges of the watershed. These wetlands serve crucial functions, ranging from animal and plant habitat to groundwater recharge and storm water runoff/flood reduction. To the detriment of ecosystems, in many regions—even where the climate allows for ponds and wetlands to form—they are scarce.

Ponds and wetlands need depressions to develop, but glacial action and the influences of mass wasting make their appearance relatively rare in most landscapes. This represents an enormous example of underutilized ecosystems; biological systems that are not optimized in terms of biodiversity or biomass because of the structural deficiencies present within them. It matters not whether an ecosystem has significant amounts of human disturbance or not—the effects of lacking depressions in a landscape are the same: low relative habitat opportunities, lack of storm water infiltration, and a consequent high tendency of these watersheds to experience flash flooding.

Humans have a capacity to be conscious ecosystem change agents. With this potent ability comes a responsibility to identify limiting factors to ecosystem health and to stimulate the emergence of systems that reduce such limits so that biodiversity and biomass increase in the system as a whole. All regenerative work is part of this process. Ponds are one of the most regenerative anthropogenic systems because of the potent impact they provide related to these two functions. On our research farm site, both habitat and groundwater recharge/storm water pulse absorption can be easily seen, and both of these functions manifested immediately after construction.

As such, the pond rises and falls throughout the year in accord with the table's variations. When significant surface-flow events occur—anything above a quarter inch of rain or so—the pond captures hundreds to thousands of gallons of water, depending on the saturation level of the soils before the storm.

Often, this capture will be immediately infiltrated back into the local aquifer. For example, we have witnessed multiple-inch rain events here when this and the other ponds' levels increase from before-event levels of one to four feet below capacity to multiple inches and sometimes feet above those levels, only to fall back down close to their prestorm level within a few days after the event. This shows the spectacular ability of ponds to serve as water table buffers: to capture the water shunted into them by swales (when they overflow) and ditches, then to retain that water without letting it leave the site into the river. Quickly, this held water begins to seep back into the water table—essentially representing a "reverse spring" effect. This reverse spring is the most regenerative aquifer-recharge action that can occur in a landscape and is "naturally" found in mid- to high-elevation wetlands.

DOMESTIC USES

Similar to water for use in the landscape, domestic (in-house potable water) is best gravity fed from a reliable source found and developed high in the landscape. The classic vernacular strategies for spring locating, developing, and utilizing are the most resilient approaches to this age-old problem of human habitat. There are several books on the market covering spring development in detail, so the scope of this section will not include that, but it will cover the basics of locating, developing, distributing, and utilizing gravity-fed water from the landscape in the home.

Locating Water

For the purposes of creating a highly resilient habitat, I always aim to locate surface or near-surface (shallow well) water resources on any property under consideration for development. This may seem to fly in the face of the current ease with which we can drill deep wells using machinery. However, because such deeply bored wells require not only the initial high cost of drilling and casing but also a perpetual cost/energy expenditure of pumping water from depths of twenty to a thousand feet or more, these deep wells should be seen as highly vulnerable to systems failure. Such failures range from power outages (a deep well pump needs a lot of power), to the pump breaking, to parts for the pump being unavailable for small to large periods of time, to time and expertise required to pull up the pump and fix or replace it. When you consider the fact that your water supply is second only to your heat supply and more important even than food supply, you realize how shockingly vulnerable we are when we depend on deep wells for our water. Drilled wells are, at best, a compromise.

Finding potable water on a site is among the most primary challenges in human habitat developments, to be grouped with only the most pressing and foundational needs, including site access and location, solar access, and slope stability. Even slope angle and soil quality pale in comparison to the necessity of having good-quality water in significant enough quantity to provision crucial needs. Potable water access, therefore, is near the starting point of my site evaluation and development consultations here in New England. This should not be surprising, for it was the same way for pioneer settlers to this region: Having no ability to drill deep wells or pump water, they only considered a site worth inhabiting for any length of time if there was existing spring access or the possibility of developing one. What I mean by "spring access" is clear enough: an existing seep or flow of continuous water in potable quality emerging from the ground. How to tap into that water and channel it to a location is covered below.

Spring development is another matter that is understandably much less clear to anyone who has not engaged in the millennia-old task of digging at a wet spot of ground to get water. Finding water involves two approaches that we like to combine where possible. The first is ecological detective work—reading the landscape for signs of water—and the second is dowsing, which is discussed briefly below. Physical locating of water involves noticing various factors about the landscape, including:

WHAT TO DO WHEN YOUR WELL PUMP FAILS

Picture the following scenario: It's January, and the first arctic blast of air is upon you and your family. You've been in your current house for five years, and the water supply has always been a nonissue. You had the water tested when you moved in, and it was good. The realtor told you the well was two hundred feet deep and had a capacity of five GPM. It's always provided good water in abundance. So naturally, you are baffled when you quickly step into the shower ready for the hot water to take the chill off this 5°F morning and there's no water. You hop out of the shower and get dressed, totally confused.

Your confusion only deepens when you realize that the power is still on in the house. At this point most people are able to do nothing but begin looking through the phone book under "water wells," preparing themselves mentally for a really big bill and no water for a while. You, however, are a clever modern rural homesteader and handy around the house. "Aha," you think, "it's probably the fuse for the pump." Upon inspecting the circuit breaker box, you find no breakers tripped and everything in order. Now you're starting to get anxious. You know the chickens and sheep in the barn are not going to stand for any excuses for going without water, and their supply is the same as yours. You start to think less about that interrupted shower and more about cooking dinner. Your hierarchy of needs quickly begins to take shape. Stop now, and think about it. What would you do next?

There are only two reasonably likely possibilities: Either the pump has failed or the distribution (waterlines) connecting the well to the house has failed. Your work of being a frost-intrusion detective and electrician begins. How do you determine where in the 150-foot run of buried well line between the house and the well the line has become frozen or broken? Or how do you diagnose the pump to see if it's failed? Here's a hint: Start with the pump first. The first order of business involves putting your ear next to the top of the casing and listening to see if you can hear the pump running. The "choose your own adventure" starts with that.

But this book's scope is not to help you diagnose the many problems that can come up with technological systems like this. It is, however, to point out how fragile they are and how to get by comfortably without them. The water for tonight's meal and the sheep and chickens will likely need to come from a source other than your well—chances are that no lines or fittings broke but that the pump died or the lines froze (usually at the well casing or where it enters the house). If you have a pond you can access by cutting through the ice, a cistern in the basement or attic, or a spring box well protected from frost, any one of these will get you by for days or week with some physical exertion. Melting ice or snow on the woodstove could serve as a last resort as well. Depending on electricity to have water would be like needing a car to have food—a vulnerability no one desires. Fortunately, there are more passive and reliable ways to secure this basic need.

- ► **Slope size and shape:** Springs and areas of high water table are most often found at the toe of a slope—the larger the slope, the more plentiful the springs, seeps, and high water table areas usually are and the larger the quantity of water found there. Water pools underground the same way it does on the surface: Micro valleys and depressions are more likely to have water underground in the same way they hold it above ground.
- ► **Geological features:** Water tables tend to surface or at least rise just above and below cliffs and exposed ledge (bedrock).
- ► **Vegetation composition:** Sedges and other wet-loving plants are clear indicators of water present close to the surface. In this part of the world, the following plant composition often represents the gradient from dry to wet found on most sites: grasses and other nonwetland plants—plantain and white clover—sedges—rushes—cattails. Very wet ground is usually easy enough to identify; it's the seasonally high water table land that's most challenging and often crucial to locate because its implications for water locating and planting are enormous.

THE ART OF DOWSING

Dowsing or water witching is the art—some might say a science—of locating water presence without physical surface features. Thought to originate in ancient Europe—though probably an art used by many peoples for millennia—dowsing today in this part of the world involves walking across land holding a tree branch (often of willow, alder, or other wet-loving species) or metal rods. As the dowser walks over or near the presence of water, the branch or rods move in such a way as to indicate water presence. I've witnessed dowsers' branches bending sharply and abruptly downward and metal rods quickly snapping toward one another as the dowser approached a specific spot. They then will release this movement as the dowser continues walking, then perform the same motion again when encountering the same location.

Many a modern human upon hearing about dowsing will dismiss it out of hand, as I once started to. However, witnessing the act of dowsing and its results has a tendency to make one a believer. I for one consider myself open to the possibilities of dowsing but not an adherent. I have hired dowsers multiple times on both my own site and for clients who needed to find well locations or a spring site. Do the dowsers produce good results? Often, yes. I look at it this way: If you're going to spend between $3,000 and $12,000 drilling a well (the cost in most American locations, based on typical depths of one hundred to six hundred feet), spending $50 to $100 dollars on someone who's well acquainted with the need to find water could be an enormous gain, while the potential for loss is minor. If you don't have access to a dowser and simply need to develop a minor spring or seep, ecological identification of water resources may likely be all that's needed.

Slowing and Infiltrating Water

Regenerative influences on a landscape—and thereby influences that promote resiliency in the human habitat that the land supports—must always involve providing water to plants (and soil biology) in as even a manner as possible across the year. Since no climate is so kind as to reliably provide even moisture provision (though certain maritime hyper-nonbrittle places, such as parts of the UK, come close), the regenerative farmer and homesteader must configure her landscape

POINT SOURCE NUTRIENT DISTRIBUTION
Curtain fertigation with concentrated nutrient sources

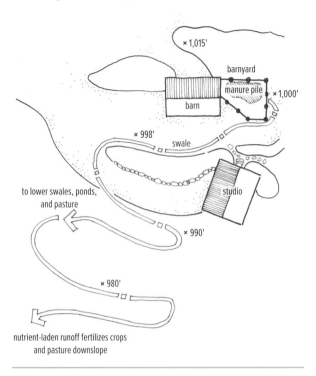

nutrient-laden runoff fertilizes crops
and pasture downslope

Dispersal of nutrient-concentrated runoff through cropping areas to reduce runoff from site into local watershed and fertilize food plants at WSRF

to (1) capture as much water as is reasonably possible, (2) store that water for arid periods, and (3) distribute that water when necessary across the site.

This can be accomplished in myriad ways. The most common method many of us are familiar with is also the most destructive and most vulnerable to failure: pumping groundwater from deep aquifers and irrigating with it—for example, on a two-thousand-acre midwestern farm. This approach not only uses a slowly replenishing resource

PRIMARY MECHANISMS FOR SLOWING, SPREADING, AND SINKING WATER

► Vegetation and soils
► Swales and mounds
► Terraces and paddies
  (see chapter five for details on paddies)
► Ponds

(deep groundwater deposits), it's energy intensive and expensive and has many failure points, from pump to irrigation lines. It's also wasteful of the water resource itself unless it involves drip irrigation. Conversely, we aim to "irrigate" using the freely provided service of rain and snow from the sky. This means we must adapt the land and our systems to capture, store, and distribute the water that falls from the sky (a renewable resource, not fossil water) as effectively as possible. This involves using the soil of the site itself as the primary storage mechanism and ponds/paddies as secondary storages.

Activating this involves various aspects of the land-infrastructure system. First, we must select deep-rooted vegetation and arrange these plantings in such a way as to slow, spread, and sink water. These plantings should represent enough of the site as needed to harness moisture in the landscape—especially crucial in arid regions. Second, we must promote organic matter–rich topsoil at all possible points because that is the best location in which to store water—it's the largest storage resource available on most sites, and the benefits of upping organic matter have huge gains in other aspects of site performance. Third, we must shape the land in such a way as to slow-spread and sink water. It is this third aspect we call "earthworks," and although it is only one of three primary approaches to fundamental site resiliency, it is the one that must be performed before much work should happen on the other aspects, because of the disruptive nature of changing land shape.

WATER-HOLDING EFFECTS OF VEGETATION AND SOILS

Varieties and layouts of vegetation contribute significantly to the overall water balance. All things being equal, from a strictly water management approach, the ideal situation involves deep-rooting perennial plants covering the entire surface of a site—think an oak forest. In practical terms perennial plants won't cover the entire site but at a minimum will be planted along the contour in a regular pattern that slows the flow of water as it moves across the slope. When water hits the planted hedgerow (and the mound of detritus and root-lifted soil beneath it), much—sometimes all—of it slows and sinks into the ground.

Soil improvement in three years: Sample at left was taken from halfway up a swale mound, while the one on the right is from the bottom of the swale. The original subsoil material from which these are derived is the exact same.

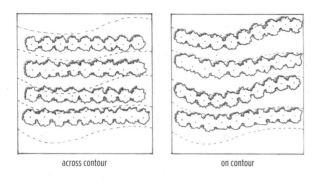

across contour   on contour

Planting on contour versus off the contour: Following a line of equal elevation offers many more benefits when laying out plants, many earthworks, and, usually, fencing and roads.

Naturally, the deeper, lighter, richer in organic matter a soil is, the more readily it can absorb water and the more deeply it can infiltrate that water. All aspects of biological health and productivity are increased as soil depth and health increases, and the effects of water systems on resiliency are no different. By building deep, high organic-matter soils, water flow across the surface of the land (what we're trying to avoid) is greatly reduced.

An Agriculture as Diverse as the Landscape

Only about 10 percent of the state in which I live, Vermont, is composed of "agricultural" land, while the vast majority of the state is too wet, dry, steep, shallow soiled, or infertile to reliably support conventional field-based crop production, though it's been tried before. Vermonters once farmed much of the state's non–"ag' land, clearing about three-quarters of the state by the mid-1800s, mostly for pasture. Devastating soil erosion resulted, along with rapidly decreasing yields.

As we enter the twenty-first century, land that is still clear of forest represents Vermont's most forgiving landscape—generally, low-angle slopes with deep, well-drained soils supporting (usually with constant inputs) pasture and annual row crops such as corn and grass for hay. Currently, nearly all of Vermont's food production is derived from one-tenth of its land base, and this land's capacity shrinks in both area and output each year. "Prime soil" lands, having been abused for nearly two centuries, continue to lose significant production capacity each year as mechanized, tillage-based farming compacts soil structure, exposes the soil to erosion, and damages soil health through continual inputs of liquid fertilizer. The actual acreage of "prime soil" land is also shrinking under the influence of suburban sprawl and transportation developments.

As the need to establish a more resilient, sustainable, local, and secure resource base becomes increasingly clear, we are confronted with the need to produce a reliable supply of food and fuel from the vast majority of our landscape that we have not yet managed to utilize productively without incurring significant damage. In a future of diminishing resources and increasing stressors such as climate change, sociopolitical instability, and economic insolvency, we will need to generate value sustainably on the majority of our landscape without depending upon one unit of production's sustaining nine units of consumption.

How do we produce lasting value on challenging landscapes with poorly drained, droughty, or degraded, infertile soil? Fortunately, this has already been done to a large extent in other parts of the world. Both degraded and inherently challenging landscapes can be regenerated and maintained as highly productive, low-input, no-till, perennial agricultural systems offering yields of fruits, nuts, fiber, fuel, meat, milk, and even perennial grains and vegetables.

In America, however, we have few examples of such systems and need to look elsewhere to find truly sustainable cold-climate agricultural systems. Permaculture, with its emphasis on low-input, self-fertilizing, diverse crop arrangements (otherwise known as "guilds") and no-till approach, is particularly suited to producing food and fuel crops on degraded and sensitive landscapes (which is most of America) that reliably fail under large-scale, mechanized, input-dependent, soil-exposing, tillage agriculture. Land design needs to continually adapt to America's hill lands, cold climate, and abused soils.

Successful versions of "agriculture for the hills" from elsewhere—such as the oak, walnut, and chestnut pasture agroforestry systems of the Mediterranean—are

Holding water where it belongs—on the hillsides (in ponds and in the soil) so it can infiltrate slowly, fertilize, and not contribute to acute flooding in the river floodplains below

not likely to succeed here by simply attempting to replicate them. Establishing reliable, sustaining, and regionalized food systems is an innovative process requiring researching and developing techniques that function across the majority of our landscape. Here in Vermont that means a "new-old" hybrid agriculture for rocky, thin, infertile, seasonally inundated land. This involves at least three primary strategies:

1. Identifying and breeding new plant and animal varieties (and reviving formerly used heirloom varieties) that are optimized for the diverse conditions of the cold climate landscape
2. Developing cultivation techniques such as contour swale-mound planting that help buffer both droughty and inundated land conditions to allow production of a much wider array of plants than would otherwise be possible in the same location
3. Changing the scale and mechanics of production systems from large to small, and mechanized to human and animal-powered, and making other adaptations in the ways we can produce on the vast variety of land types

Relocalization in the cold-climate regions of the world will involve the skillful use of the incredible diversity that our landscape contains, from the acidic conditions of a pine plantation to the anoxic clay soils of a wet, abandoned field to the thin, dry, dead soils of an abandoned steep pasture.

Ponds, paddies, swales, and mounds—all land forms that allow productive use of both seasonally inundated and droughty landscapes. Here, rice paddies and a terrace planted in squash mounds are situated between ponds and swales at the Whole Systems Research Farm.

Utilizing "marginal" lands requires significantly more skill and care than "prime" agricultural lands with erosion, infertility, or simply lack of production easily resulting from their mistreatment. "Marginal" lands also represent some of the most important and sensitive ecosystems on the planet, while containing possibilities for some of the highest crop yields possible anywhere—the largest food staple in the world is rice grown in poorly drained wet soils. Use of these landscapes must be undertaken with careful planning and great understanding of the existing opportunities and challenges of the site.

Fortunately, gleaning yields from these ecosystems can be done in ways that not only promote the health of the natural ecosystem but offer human yields as well. Site specific by necessity, agriculture for "marginal" lands must be highly diverse—given the astounding variation in landscape conditions present beyond the typical large, flat agricultural field. Farming landscapes other than typical "ag" land not only requires it but benefits from humans working in synergy with the local ecosystem as beneficial members of the site's living community to support ongoing fertility development and long-term yields. I'll now outline approaches particular to several commonly found growing conditions.

Droughty and Rocky Land

Land that is dry and sloped presents an interesting challenge for agriculture in the cold climates of the world and elsewhere. Overall strategies for dealing successfully with these conditions involve the following:

▶ Species selections for plants that can not only handle but actually thrive in dry, poor soil and improve the soil for different future plants. Rocky soils, in particular, are most suited to a perennial-focused agriculture. Example species include sea buckthorn (*Hippophae rhamnoides*), black locust (*Robinia pseudoacacia*), buffaloberry (*Shepherdia argentea*), and various other berry and nut shrubs and trees. All three of these species are nitrogen-fixing land-healing plants.

▶ Earthworks such as on-contour swaling, in which ditches are dug along contour to slow and trap water

as it travels across the slope. This allows water infiltration into the soil horizon, where it irrigates deepening plant roots and delivers oxygen and nutrients. Both the swales/ditches and the mounds below are planted with nitrogen-fixing plants and dynamic accumulators, helping to build soil structure and soil biology, creating conditions that eventually support a larger array of plants to thrive in the same location. After an initial establishment period with "heavy giving" plants (as opposed to "heavy feeding" plants), species that would otherwise not be supported on such sites can thrive. These include more sensitive fruit trees, berries, and other multiuse food and fuel trees and shrubs.

▶ Mulching with fungi-inoculated wood chips helps keep soil moisture optimal, build healthy soil biology, and suppress weeds.

▶ Drip irrigation systems that allow a very small amount of water and energy to be applied precisely across a landscape at timely intervals allow the establishment of plants that would otherwise be unable to survive.

Seasonally Inundated Land

An enormous amount of Earth's landscape is underutilized because of perched water tables and low-angle slopes underlain by poorly drained clay soils. Useful responses to such conditions involve similar approaches: for one, selecting species that are well suited to perennially or seasonally wet conditions and inundated conditions. Species particularly well adapted to wet conditions include currants and gooseberries (*Ribes* spp.), elderberries (*Sambucus canadensis*), cranberries (*Vaccinium* spp.) and highbush cranberries (*Viburnum* spp.), Chokecherry (*Aronia* spp.), willow (*Salix* spp.) and alder (*Alnus* spp.) for fuelwood and craft wood, and many others. Other useful strategies include grafting nonwet tolerant species onto wet-tolerant rootstock, such as pears onto hawthorn or quince.

In addition, on-contour swales and island mounds (at various scales) simultaneously lower the water table in the immediate area of a crop plant while raising up the plant itself. Systems in Europe have practiced

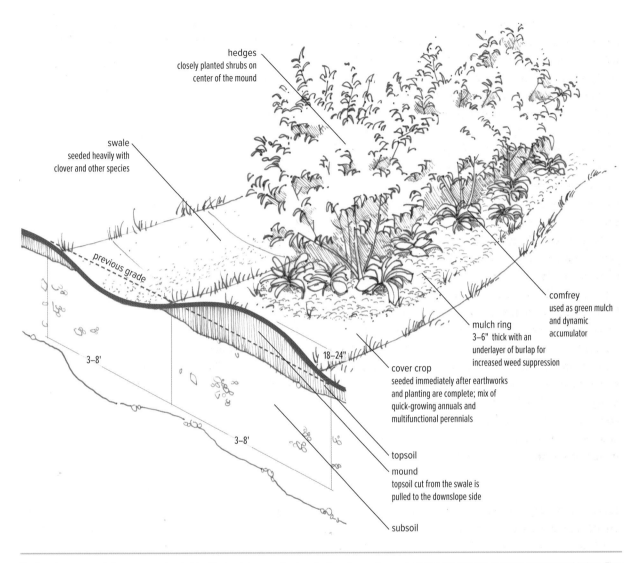

hedges
closely planted shrubs on
center of the mound

swale
seeded heavily with
clover and other species

previous grade

3–8'

18–24"

3–8'

comfrey
used as green mulch
and dynamic
accumulator

mulch ring
3–6" thick with an
underlayer of burlap for
increased weed suppression

cover crop
seeded immediately after earthworks
and planting are complete; mix of
quick-growing annuals and
multifunctional perennials

topsoil

mound
topsoil cut from the swale is
pulled to the downslope side

subsoil

Swales at our research farm are constructed by digging into a swath of earth along the contour and placing the material along an even mound below. That mound is then planted at its height with berry and tree crops.

tree-based agriculture in wetlands for thousands of years under the name *hugelkulture*, in which they utilize woody debris to help form the raised planting mounds. Gradually, the woody material breaks down into soil, feeding the plant over time while catching leaves and other nutrient-rich debris that circulate via wind currents in the area.

SWALES AND MOUNDS

We call the wave pattern of mounds and ditches running with the contour "swales." They can be made of woody debris (a hugelkulture strategy), earth, or some combination of the two. The effects they have on water movement down a slope are desirable and the same: They check water's movement as it descends and forces it to stop or slow, allowing it time to infiltrate into the ground below. At the Whole Systems Research Farm, we make swales with the native earth on-site by pulling earth from uphill downhill, forming a mound. We then use these high surface area drier locations for cropping. A swale "waters" the area immediately below the mounded location and, depending on soil type and

LAND TRANSFORMATION WITH SWALE BUILDING

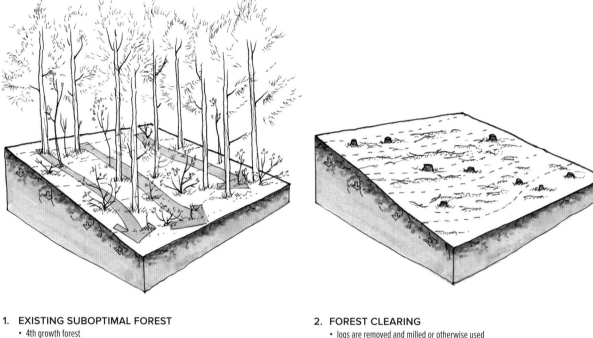

1. EXISTING SUBOPTIMAL FOREST
- 4th growth forest
- much surface water moves across slope
- soil is built very slowly ~1" every 100 years

2. FOREST CLEARING
- logs are removed and milled or otherwise used
- slash and stump are left on-site
- stumps may be ground flush but not ripped out

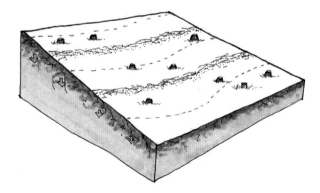

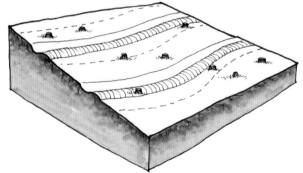

3. WOODY DEBRIS ORGANIZED
- swales are sited and laid out along contour
- slash and small debris laid out along swale line or just below

4. SWALES CUT AND SHAPED
- existing ground cut and filled, then roughly graded on top of and slightly above debris grade
- swales are shaped using hand tools
- bare soil is seeded and heavily mulched

Forest or field to perennial crop succession is accelerated through the use of small earthworks, planting, and grazing during both implementation and establishment phases.

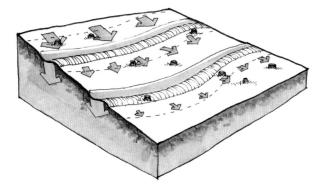

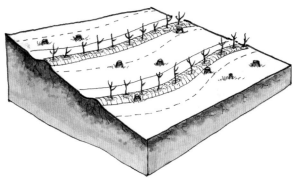

5. SWALE INFILTRATION
- water collects in swales and slowly saturates subsoil layers
- soil formation begins to speed up

6. PLANTING
- young bare-root plants are placed on top of mounds; spacing varied according to species composition and cropping intention
- soil formation continues to speed up as plant roots reach deeper subsoil layers

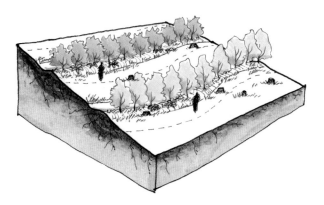

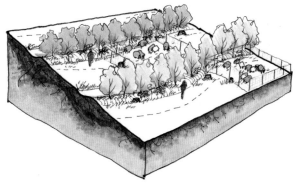

7. PERENNIAL PLANT SUCCESSION
- plants become established and additional understory perennial guilds are formed
- soil formation increases more rapidly as guilds work to bring subsoil minerals and nutrients to surface layers

8. GRAZING AND PERENNIAL PLANT SUCCESSION
- rotational intensive grazing is integrated between swales
- rapid topsoil formation occurs from continual root dieback and regrowth below surface while animal manure is spread above surface
- coppicing/pollarding/pruning begins for animal fodder and to increase root dieback and regrowth and vigor of woody plants

rainfall amount, can disperse water that would have run off the surface into the soil well below the swale—five, ten, even twenty feet downhill. This "capture, store, and even out" of moisture is one of the reasons swales are such a soil- and plant-regeneration tool.

The productivity of swales and mounds is astounding. We have noticed that all species of plants respond positively to being on a mound, and the increase in growth and health seems to vary from moderate to extreme. On average it can be said that a given plant at the research farm will grow at least half again as fast as the same plant rooted in flat, unmounded, and unswaled earth nearby. Often, we've seen plants respond with twice the growth rate, including species such as black locust, goumi, elderberry, currant, gooseberry, cherry, peach, and apple. The pattern seems to exist across all species except truly wet-loving plants such as sedges—and we don't really want to grow sedges.

Why do plants prefer a swale or mounded location? The answer seems to come in two parts. The first has to do with our wet climate and high water table; in desert climates a mounded planting would be much more drought stressed because it's "high and dry"—the land on a mound is more exposed to the drying effects of the atmosphere (wind and sun). Such plants in a dry climate would often do more poorly. In this climate we have found that getting above the high water table and periodic inundation caused by rainy periods and snowmelt is of prime value.

The second reason is more universal: A flat piece of ground has less surface area than a wave-shaped piece of ground. Biological activity and soil health is concentrated most heavily in the upper layers of the soil at the land-atmosphere interface—this is where organisms have the highest capacity to metabolize, where roots are most perfuse, and where organic matter is

The author constructing swales with an 8,500-pound compact excavator—the most useful small earthworking tool for swales and paddies at this scale

A SWALE OR A DITCH?

Swale construction on a relatively steep (approximately 35 percent) grade in early spring at the WSRF. Note that swales are seeded and planted immediately.

Hand-working swales to smooth the grade out after machine forming them

There are many variations of the terms "swale" and "ditch" within the permaculture world. Some refer to them synonymously, while others consider a ditch to have a pitch (not be on a true contour) and a swale to always be on true contour. I prefer the latter definition; that is, a ditch moves water (conveyance) and a swale stops water flow downhill and holds it, still, on the slope. Swales, therefore, are more applicable for the pure need to check the flow of water and infiltrate it. Ditches of very low angles, however, are very important and employed throughout our site to convey water from the valleys to the ridges. These low-angle conveyances—1 to 4 percent grade—allow us to get water from where it is abundant in the valleys to the droughty ridges. Our soils, being high in clay content, allow a low-angle ditch to convey water while infiltrating significant amounts simultaneously. The sandier the soil and faster the infiltration rate, the steeper a ditch needs to be to move water effectively.

highest as a result. So this is where most plant feeding occurs. When we contour a piece of ground and turn a flat patch of ground into one wave or a mound shape, we instantly add surface area and soil-area interface. This relates back to a primary permaculture design strategy: *The edge is where the action is.* Swales and mounds create edge—highly productive edge, and we get more land from making them: Literally, our acreage is increased when we contour the land. This last reason is profound, and the results we've witnessed are surprising.* We now only wish we had contoured nearly the entire farm in the early years. Hindsight is always 20/20, to be sure.

Water Ridges with Valleys: Keyline Agriculture

The concept of keyline agriculture in modern form emerged from the drylands of Australia. Farmers there, most notably P. A. Yeomans, discovered a simple and glaring truth: that aridity limited land productivity over large acreages while, simultaneously, certain areas of the same region were literally swamped with water. While keyline agriculture contains many concepts, its most fundamental is this: Spread the abundance of water from where it is concentrated in wet areas to those areas that are consistently too dry. Keyline is based on the understanding that water is often the most severe limiting factor to plant (thus animal) productivity.

This is such a basic fact that it's often overlooked. Think of your nearest pasture or yard. In it you will find microvalleys and microridges—areas of consistent wetness and consistent drought. Depending on your location you will most often find that the ridges are poorest in productivity. The drier your location, the more productive the valleys will be. In many climates there is a sweet spot just uphill of the valleys that is

most productive. The goal of a resilient agriculture from a keyline perspective is to make as much of your land be like that sweet spot, hydrologically, as possible. That's why we spread the water from the valleys toward the ridges.

It is not the scope of this book to get into the details of how keyline is laid out, designed, or performed, as it is well covered by other texts.† However, the foundation of keyline agriculture is crucial to understand for all resiliency agents. From the water perspective it is to water the ridges with valleys and to make deep soil-watering events occur as much as possible. This second strategy is covered in the rapid topsoil formation part of chapter four.

Ponds

The rain woke me up this morning, again. Falling now in sheets across my ponds, fruit trees, and vegetable beds, drenching the sheep and the ducks alike, is the seventeenth or eighteenth inch of rain that's fallen this spring, and we still have more than a month to go. "Normally," we get four to six inches of rain in May *total*; it's now May 19, and we've already received eight or nine inches on this site in central Vermont since the month began, and it's still raining hard. The forecast calls for another inch or two to finish this week alone.

What do ten-plus inches of rain in a month mean for us? For me as a homesteader and small farmer, it means some washed-out vegetables—my cabbages are looking somewhat poor in their low-lying bed—and slow starts to other vegetables; luckily, many are in a raised cold frame, but my beans may now be rotting in the soil after sitting there for the better part of a week with no sun to warm the soil for sprouting. Yet my rice paddies have just started overflowing, the ponds are brimming, the ducks are finding slugs and snails wherever they look. This rain is very good for the perch and bluegill in my ponds, for the ducks that make eggs and meat, and for the now fast-growing rice crop that thrives in

---

* While these are two primary reasons of swale value, they are only relative to water capture. The fact that swales and mounds are composed of loosened, aerated material probably also accounts for the high productivity they offer. Sepp Holzer actually calls his mounds "raised beds" to indicate this effect. In this country raised beds usually refer to garden beds with soil retained via constructed materials such as timber.

† *Water for Every Farm* and other works by P. A. Yeomans are original resources for learning more about the revolutionary approaches keyline agriculture offers.

The Whole Systems Research Farm bottom ponds in early spring

water-logged conditions. Another foot of rain doesn't hurt a crop that's already flooded and liking it.

My pasture also looks great—with every inch of rain, it seems to grow two to four inches this month, and the sword of clover, vetch, and rye is thicker every day. The sheep seem to tolerate the cool rain, thankful for the bounty of fresh grass it delivers. Pasture growth in May is normally about three times faster than July growth largely because of moisture. If it keeps raining the pasture will keep on growing rapidly. The tree and berry crops look fantastic as well. We've earthworked the landscape of this research farm so that our perennial plantings are on top of mounds running sideways across the slope "on contour." These plants have all the access to water they could want in the bottoms of these swales but are free from inundation, being planted up high on each mound.

Since rain pulls down significant quantities of nitrogen from the atmosphere as it falls (washed from the air at greater rates than any other time in the historical record), it stimulates plant growth; it's literally liquid fertilizer. Accordingly, rainforests, the rainiest environments in the world, have the fastest biomass production. So plant crops that can avoid inundation because of their growing situation, along with those that don't mind the lack of sunshine and heat, are thriving. Along with the fish, ducks, and pasture, this includes the perennial crops: apple, pear, plum, cherry, quince, peach, walnut, hickory, chestnut, oak, blueberry, aronia berry, seaberry, honeyberry, gooseberry, currant, and a score of other permanent producers. Some aspects of this farm system are actually greatly benefiting from this cool wet weather, while conventional fields of corn

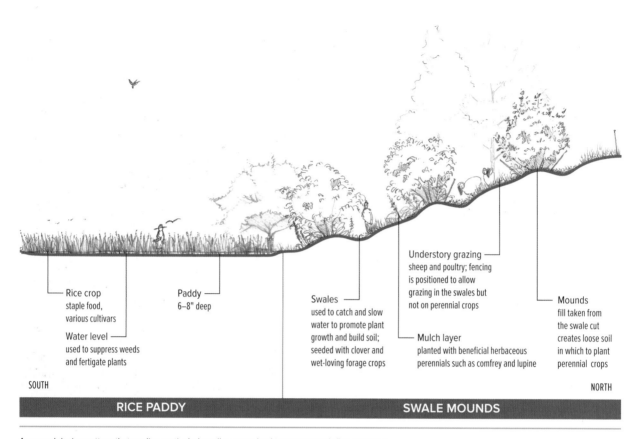

Rice crop
staple food,
various cultivars

Water level
used to suppress weeds
and fertigate plants

Paddy
6–8" deep

Swales
used to catch and slow
water to promote plant
growth and build soil;
seeded with clover and
wet-loving forage crops

Understory grazing
sheep and poultry; fencing
is positioned to allow
grazing in the swales but
not on perennial crops

Mulch layer
planted with beneficial herbaceous
perennials such as comfrey and lupine

Mounds
fill taken from
the swale cut
creates loose soil
in which to plant
perennial crops

SOUTH

NORTH

RICE PADDY

SWALE MOUNDS

A general design pattern that applies particularly well to steep land interacting with flat, wet land

and other fragile bare-soil annuals sit mired along the river bottoms, now too soft for machinery to deal with.

Ponds, swales and paddies have been a part of the working landscape since agriculture emerged, especially on sloped lands. Since water is the basis of productive biological systems, retaining and distributing this storehouse of fertility and life within a landscape is key to the success of any living landscape. The climate, topography, and soils, along with the ease of access to machinery and cheap energy (for now), in the northeastern United States offers a particularly timely opportunity to capture, store, and distribute water via ponds on farms.

Ponds, paddies, and swales in this climate can be cropped for a variety of outputs, most established of which are fish, rice, and berries, respectively. Shallow-water systems such as paddies have the unique ability to be fertigated easily (nutrient-rich water delivery), which allows rapid growth of heavy-feeding plants in

an otherwise poor fertility situation. These systems can also be perpetually productive on account of water being the nutrient delivery mechanism: Witness paddies that have produced a staple rice crop over centuries upon centuries in sloping landscapes. It is likely that other cropping possibilities beyond rice will emerge with continued innovation of fertigation in both paddies and swales in the coming decades.

Ponds, especially, have many uses beyond what can actually be produced *inside* them, and it is these uses that make them an especially attractive working landscape feature. These include:

▶ **Microclimate enhancement:** Water bodies capture and store solar energy and release this heat slowly, especially in the autumn, to the adjacent area. In our testing on the Whole Systems Design Research Farm, this effect varies from year to year with the severity of the fall's first frost: Our three ponds

Some of the rice paddies at the Whole Systems Research Farm in late spring just after planting

will often not buffer against frost if the first freeze is about 27°F or less yet will extend the growing season by weeks if the first frost in the fall is a mild one, which is usually the case.

► **Wildlife:** There's perhaps nothing we can do to more quickly enhance the biodiversity of species in our landscapes than by creating water bodies. In addition, amphibians are in need of particularly strong support, given the decline in health of their populations in recent years. Ponds with large wetland edges are ideal—and often rarely found—habitats in many areas of the Northeast. Each time we've built a pond, at least three species of frog and two species of salamander arrive on-site within weeks. The values ponds offer for beneficial insects, birds, and mammals can also be observed in short order.

► **Storage for Distribution:** Large water storage is invaluable for fire control and irrigation, as well as for drought-proofing a landscape over time. It takes two to three days or fewer to make a pond that can hold a hundred thousand gallons or more, making it the most economical means of storing large quantities of water. Farms with a need for irrigation often recognize the opportunity to gravity feed such water via a supply located high in the landscape such that its water can be fed to the entire farm without pumps or electricity.

Capturing surface water from as many acres as possible is important for farms wishing to be adaptive to shifting climate conditions and the adverse effects of drought punctuated by intense rain events. A well-sited and properly integrated pond can be the most crucial "shock absorber" farms have for large precipitation fluctuations. Ponds in this capacity serve like batteries, storing excess energy (water) when it is abundant so it can be distributed slowly over a long period of time (drought).

The center of the Whole Systems Research Farm looking west in early summer. Note solar orientation of the design studio and gardens. Photograph courtesy of Whole Systems Design, LLC

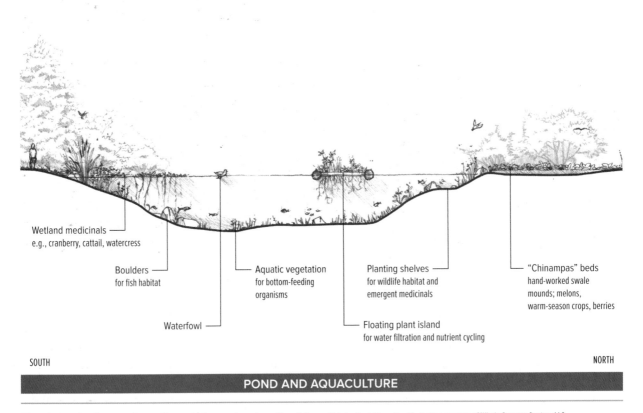

Wetland medicinals
e.g., cranberry, cattail, watercress

Boulders
for fish habitat

Waterfowl

Aquatic vegetation
for bottom-feeding
organisms

Planting shelves
for wildlife habitat and
emergent medicinals

Floating plant island
for water filtration and nutrient cycling

"Chinampas" beds
hand-worked swale
mounds; melons,
warm-season crops, berries

SOUTH   NORTH

POND AND AQUACULTURE

When designed and managed correctly, pond edges are teeming with activity and biological diversity. *Illustration courtesy of Whole Systems Design, LLC*

► **Other:** Recreation, food storage, and increasing radiative light for crops and building interiors are several other important side benefits of well-integrated ponds, which demand more space to discuss than is possible here but are worth mentioning.

Management over time is more involved than the scope of this chapter permits, but the following guidelines are basic ground rules for ecologically enhancing multipurpose ponds:

► Don't mow to the water's edge. That's the best way to wreck the most abundant wildlife habitat a pond offers. If you must make access via a mower, do so in limited areas along its perimeter.
► Seed any bare areas that are not greened up every spring through early summer until there are none

left—this can take two to three years or more, depending on vigilance and weather.
► Keep a watchful eye on overflow spillways (recommended) and drainage fixtures/pipes (not recommended) to prevent clogging and the waterline rising to a dangerous point.

Ponds, swales and paddies are some of the most important features we can install today to ensure a more productive, multifunctional, and resilient landscape tomorrow. Well-designed and constructed water-retaining and distribution systems such as these can help homes and farms become more fit for a future that is likely to bring with it adverse conditions, including drought, flood, increased pest pressures, increased costs of inputs, and other stresses that only highly resilient, low-input systems will handle successfully.

Chapter Four
Fertility Harvesting and Cycling

The nation that destroys its soil destroys itself.
—**FRANKLIN D. ROOSEVELT**, 1937

The State of Vermont soil maps say that I live on land composed of six to twelve inches of silty loam underlain by gravelly clay subsoil—not prime farmland but plenty good enough for growing fruit and nut trees and agreeable to raising vegetables as well. Yet for the past eight years, I've been gardening and planting trees across this site and have found only pockets of silty loam soil a few times; it's almost totally just clay, some sandy soil, boulders, gravels, and more clay. Where's all the topsoil—are the maps wrong?

Local elders in their 80s who have tended farm animals on this hillside have helped me complete a picture that should not be surprising, for it's the story of this corner of the world's landscape—and similar to much of the world's. It begins with land clearing for timber

> **Topsoil is washing off the exposed heartland of America at a rate of about one billion dump-truck loads per year. Only a comet or large asteroid collision with Earth has ever destroyed so much biological capital so quickly.**

extraction during colonial times, followed by potash production with the remnant forest combined with a sheep craze (two million-plus sheep in my tiny state of Vermont). Add to this a ravenous diet of ten to twenty-plus cords of wood per house for heat, followed by hardscrabble grazing in the early half of the twentieth century to send the rest of any remaining topsoil into the rivers and lakes. Overgrazing sparsely vegetated sloping land yields predictable results: massive transport of topsoil from the hillsides into this region's great storm-water detention basin—Lake Champlain.

And Vermont's soil story is no different from the rest of the United States, with the Gulf of Mexico catching the topsoil washing off the exposed heartland of America at a rate of about one billion dump-truck loads per year. Only a comet or large asteroid collision with Earth has ever destroyed so much biological capital so quickly.

The Great Soil Erosion of the nineteenth and twentieth centuries represents the most massive transport of material on Earth since the last ice age. My small ten acres on which, say, about eight inches of silt was lost, amounts to roughly twelve thousand tons of topsoil, or about 350 dump-truck loads. Standing at the sunset of the cheap-energy era, we now have to build a renewable society starting with about two hundred million fewer dump-truck loads of soil than the first European settlers to this region had.

Amazingly, most farming seems to go on as usual. We've managed to keep a small group of crops producing by trucking our fertility in from afar after extracting it from the ground, burning up ancient energy and fertility at once. As we begin to transition out of the cheap-energy era, the reality of Earth's missing topsoil will be felt more deeply than we might imagine. Healthy soil is, of course, the foundation of any agriculture and culture; food can only be wrestled from the land via fossil fuel fertilizers and pesticides temporarily, at best. Societies have long existed without highways and electrical grids; it's when the soil and water give out—or the climate shifts quickly—that civilizations collapse. This collapse may continue to be delayed for as long as fossil fuels can be substituted for soil, climate change is tolerable, or until potable water becomes scarce.

It's becoming clear that one of the most direct ways humanity can simultaneously triage the soil–climate–water–human health emergency is through rapidly building topsoil. As we begin to digest the news about the role of topsoil as linchpin in ecological health and human resource sustainability, we are waking up to a world of new possibilities, including global carbon negativity, agricultural yield improvements (while simultaneously *reducing* inputs), flood mitigation, and biodiversity rehabilitation. Only topsoil formation does all of these things with enough amplitude to matter. Building topsoil is a deep solution that doesn't create a multitude of new problems while attempting to solve the first one; in contrast, it actually solves many problems synergistically.

Disenchanted with the failure of each silver-bullet techno-fix, humanity is beginning to realize that the resource-generating system we need most has already been invented: Photosynthesis is the production, and soil is the storage. If being the "toolmakers" sustained humanity through the last epoch, evolving into soil makers and water purifiers might just get us through the next.

The more we learn about the living matrix underfoot, the more we understand it as a vast, synergistic composite of ingredients and processes. Although soil is composed of known substances such as minerals and particles from the underlying bedrock "parent material," organic matter from plant and animal tissues, and water, we have just begun to understand the almost

> **If being the "toolmakers" sustained humanity through the last epoch, evolving into soil makers and water purifiers might just get us through the next.**

magical existence of living soil born from nonliving matter. Despite its many mysteries, we do know that soil is several things:

► It is the principal in our trust fund with Earth (the assemblage of species and water are also part of that inheritance).
► It is generative. Along with water, it is the living medium from which life stems (with the influx of sunshine).
► Its quantity and quality set Earth's thermostat. Soil is where most of the carbon is: Two percent organic matter (carbon) in the top foot of soil represents more carbon than has been produced on the planet since the Industrial Revolution began. This amount of organic matter can be built in one growing season, easily, with sound land practices.

Solar photovoltaics, wind power, hydrogen fuel cells, smart grids, nanotech, clean coal—do these methods sound exciting? Maybe, but each of these creates a host of its own new problems. What if the best news in humanity's prospects for a more livable future is not these and other new technologies manufactured from factories but ages-old living material manufactured by water, fungi, wind, and plants? How will soil (and biological systems in general) again become our baseline resource generator and storehouse? How can we enhance the soil system to sustain humanity's resource needs while at the same time sopping up the excess carbon we've left in the atmosphere during the Great Fossil Fuel Party?

A century ago we began producing our resources with oil instead of soil. Now, we're beginning to realize just how bad a deal this was; we needed the soil not just to produce our resources renewably but to temper our climate, sustain biodiversity, deal with drought, and maintain our health. But how can we possibly rebuild, say, three billion tons or a hundred million dump-truck loads of soil in the state of Vermont alone, maybe one trillion (yes, a trillion) dump-truck loads in the heartland of the United States? It takes a "natural" system hundreds of years to make just one inch of topsoil, so we need a way to make soil that's a thousand, maybe ten thousand times faster than the historical rate of soil formation.

Is this possible? There's only one way to find out, but the evidence is that it is indeed. Because of the sheer volume of matter needing to be converted into topsoil, any system that builds soil rapidly will utilize the most abundant and potent resources at hand, including:

▶ **Subsoil** (mineral source)
▶ **Atmosphere** (carbon dioxide/carbon and nitrogen source)
▶ **Water** (oxygen and nitrogen source and nutrient delivery)
▶ **Sunshine** (energy source for converting plant matter into soil organic matter)
▶ **Nutrients** ("wastes"): manures, urine, crop residues, woody biomass, food scraps, rock minerals, sand, and other available soil components (nutrients and organic matter)
▶ **Tools (and human beings)** to optimally utilize the above resources, measure soil formation, and utilize continual feedback for improved soil formation over time

Strategies are emerging for combining these ingredients to make fertile topsoil with great speed and to first capture and cycle all existing nutrients on the site that are available. These include compost, urine, and humanure; biochar; fungi; remineralization; cover cropping; intensive, tall-grass grazing; subsoil plowing and keyline agriculture; and cultivating deep-rooting perennials. Most if not all of these strategies can and should be combined. Some are suitable only on the farm scale, while others are more suitable at the home scale, and strategies vary according to the type of landscape in which soil formation is applied. The following is an overview of each approach and how we are utilizing it on the Whole Systems Research Farm.

Compost, Urine, Humanure, and Biochar

Composting is a no-brainer. No families or societies that maintained a direct and durable connection to their food and other basic resource systems over time (unless they were successful nomadic peoples) trashed their excess organic nutrients—including food scraps, urine, and feces. These "human effluents" are primary resources to keep the fertility cycle going, and landfilling such valuable assets would be as sensible as raking up the leaves falling onto the forest floor and hauling them off to the dump. Insane. Yet that's what most industrial humans do today.

So if this is the *opposite* of resiliency and regeneration—the antithesis of what we're after—what is the desired goal? A good friend and design colleague, Buzz Ferver, uses the following way to describe an integrated human–land fertility relationship that sees human beings as simply one step in the flow of water and nutrients: *"You are a solar powered biological being, symbiotically co-evolved with plants, fungus and bacteria, primarily functioning to process and concentrate plant soluble nutrients in exchange for fruits, nuts, grains and other foods."* The primary strategies we've used for harvesting and cycling fertility—the basis of a productive and vitalizing ecosystem—are overviewed in this section.

COMPOST

All cycling of organic materials can be termed composting, but we use the term here as is commonly the case—to mean recycling of kitchen-, garden-, and barnyard-generated food scraps and other materials. Humanure use is also composting but will be treated separately in this section. Of all the topics in this book besides vegetable gardening, there is probably more written about composting than any other area. The literature is vast and complete, so I only wish to share

It's challenging to make truly "correct" balanced soil.

the specifics of what we do on the research farm, or what we have learned about composting, where it is somewhat novel or departs from typical composting as most homesteaders know it. If you are new to composting, I would recommend you seek out some of the numerous resources available on it. Composting is a crucial transition skill and is as basic to any gardener, homesteader, or farmer as using a tape measure is to a carpenter. It's also easy to do moderately well but difficult to do with true expertise—it's challenging to make truly "correct" balanced soil.

We have tried various approaches to turning our kitchen, garden, and barnyard nutrients into soil over the years—from piles contained in pallets to piles contained by mesh screen/fencing to the simplest layered piles right on the ground. As with basically everything on this homestead, the best approach has proven to be the simplest. One caveat here: We don't make perfect soil on this farm. We make pretty good garden soil and do so with minimal effort. For challenging applications such as seed germination, our soil is not ideal—it works, but damping off and slow seedling growth are common. We have a long way to go in this regard, and I plan to introduce larger quantities of leaf mold into the process in the hope that our germination mix improves.

For composting kitchen scraps the method we currently like the best is layering browns and greens in a three-sided container made with scrap pallets. We choose the pallets carefully for ones that don't have very large gaps and don't have treated lumber, which is rare around here anyway. We wire or nail or screw them together, with whatever materials are handy at the moment. We put a very tight-gapped pallet on the bottom first—that's important, as so much good soil and nutrients tend to be lost out of a compost pile in this wet climate.

In the past two years, I've also experimented with using metal mesh to hold piles and think it could work better because there's more aeration on the sides. I use 2″–4″ galvanized metal fencing material that is excess from planting jobs (we use the same material for tree protection). I like the three-foot height, however, rather than the four-foot we use for tree protection. I simply make a cylinder with this material and layer up compostable material inside—usually a four- to five-foot diameter seems to work well.

I am experimenting this year with growing squash in the compost piles over the summer—they seem to volunteer there anyway, so we'll see if they can yield a good crop. I just pulled the compost I'd need for the summer from the piles first and stored it in buckets. We layer up the pile over a month or three as it's made, adding in kitchen scraps and garden weeds (if it's gone to seed, we pile it in an area we call the "farm" compost for a very long-term breaking down). We use a large input of yard scythings from the most fertile area, especially the leach field, once or twice a year to really bulk up the piles, adding in those grasses while green. We scavenge leaves in the woods in the fall to add to the mix as well. During the growing season we also take some walks to gather comfrey, which is extremely valuable as a compost accelerator and mineralizer of the soil being made (because of its ability tap into deep layers of the soil horizon).

At times, when there is an excess of nutrients being made, we gather them up and add them to the compost piles to get more nitrogen and organic matter especially—as kitchen scraps don't add up to much soil. Those bulkers and potentizers include barnyard manure and bedding mixed together and chicken manure–laden bedding from our movable chicken coop. The chicken manure especially helps to accelerate and heat up the compost pile, quickening its breakdown. Once a pile is done, we cover it with a few scraps of burlap to keep the bulk of rain from washing through the piles, thus diluting their value. We only turn piles on occasion, mostly letting time do the work. However, when we see a need for more soil quickly or need to consolidate piles, we will turn one pile from a container into the neighboring container, thus flipping over and aerating the contents well and making new space for a fresh pile.

We have never watered a compost pile here and don't consider it better or worse to have the compost located in the sun or shade—though if I could choose, shade is probably better because of the stable temperatures and lack of drying. Locating the piles in the sun does give the added advantage of being able to grow in the piles themselves, utilizing the nutrients just sitting there in half-finished piles. Instead of sun/shade criteria, however, we locate the piles for complete ease of access and for consideration of what we want to fertilize downhill from the pile, as runoff is unavoidable unless you compost inside a shed or similar location.

URINE

Human urine is a near-perfect plant food, and a hydrated, healthy human being urinates a half dozen times a day or more. That's hundreds of easy opportunities each season to feed back into the system that feeds you. Using urine as fertilizer on the homestead can only seem strange in a relationship between people and plants with incomplete cycles and distance, rather than connection. Raising plants without offering them back your excess nutrients is like being given a gift by someone repeatedly without returning the favor.

Because of failing septic systems and especially urban waste treatment systems, industrial humans are literally "pissing on" fish in the rivers downstream from their sewage treatment plants and the creatures of the sea into which that river flows. This is not proper nutrient cycling—animals don't benefit from human urine, *plants* do. Concentrating human feces and urine into massive centralized systems not only deprives the land of the fertility from which these nutrients were derived, it loads the oceans with excess nutrients and toxins. This lineal flow is the opposite of the arrangement between land and sea that living communities rely on. Fortunately, if you live in a rural area, placing yourself in beneficial relationship with the living world around you is as easy as growing food plants, composting, and walking around the site to simultaneously water and feed the plants most in need of nutrients.

It is said that each human being excretes enough plant nutrients to grow enough plants to sustain him- or herself. This cyclic concept should not be surprising, as humans and plants have evolved with relationships between each other for millennia. Think of the synergy:

What if it so happened that urine was toxic to plants, or that it simply didn't contain nutrients plants need? Of course, just the opposite is true; all our excess bodily nitrogen goes into our urine—the same nitrogen that is often the limiting factor to plant growth. Coincidence? Doubtfully, but either way, the imperative is simple: Cycle value in the system—transform a waste from one element into food for another, always.

Urine is one of the most valuable resources generated on the homestead, and no human habitat firing on all of his cylinders would waste much of it. We have found it relatively easy to use urine during the growing season but hard to make optimal use of it during the long dormant season. When plants are growing—generally from about April to October—I urinate either directly on the base or near the base of trees and shrubs that need more fertility. I look out for the plants that are growing the slowest and fertilize those as a priority. I aim to grow twelve to twenty-four inches of new shoot per year on most fruit trees, a lot less on new nut trees, slightly more on older nut trees, and maybe six to twelve inches on most shrubs, aside from elderberry, which is super vigorous and can grow two feet per year easily for the first few years.

Not surprisingly, these are found more often the farther I walk from my zone 1 (kitchen, office, workshop, bedroom). When you find a plant in need of nitrogen, urinate at the base of it—the younger the plant, the closer to the base the better, because of its limited root development. As plants get older I like to fertilize them farther out from the base. Avoid depositing urine in the same spot over and over again or repeating use on the same plant. When you do fertilize with urine, get the liquid gold in deep, where the plant roots can access it. During or before a real rainstorm, you can spread more broadly on the surface.

Using human urine on plants entails walking around the landscape to a larger extent than we might otherwise. And while it's easy to cycle your fertility back into the landscape during the growing season, the dormant season presents another challenge entirely. In my cold climate, for six to seven months of the year, storing urine presents several challenges. First, although sterile when it leaves your body, urine becomes highly active and odoriferous quickly—read, overnight. Odor is a challenge, if you're storing in buckets or jars. Second, urine tends to lose its nitrogen quickly into the air. So come spring the amount of nitrogen actually available for plants would be questionable.

I have come to the following approach as my best current method for solving the winter fertility-extension challenge. First, I urinate outside at the base of wood-chipped plants when possible in the winter months. I do not know for sure how much of the nutrients will be available for the plant come spring, but it seems to do some good and is a nonpolluting way of releasing the nutrients.

Another great use of winter urine is inoculating biochar, which is made in my woodstove. This is one way of achieving the necessary step of activating and "charging" the biochar and serves an equally important function of nutrient storage. It seems, from what little research on the subject I've done, that some of the nitrogen will vaporize by spring, but some will remain locked up by the potent nitrogen-absorbing biochar (pure carbon). Then come spring the biochar is mixed into garden soil, where the winter's fertility can be extended across the growing season.

I have enough mixed results with biochar to feel as though I am still a number of growing seasons away from feeling confident in the above strategy of urine-inoculated char as a soil amendment in beds directly. I would think this material as a compost amendment would be very good but have reservations about direct-bed additions, as I've seen biochar do weird things to vegetables in garden beds—mainly in the form of slowing plant growth immensely in the early stages of transplanting and growing from seed.

Another more recent experiment we are doing with urine on-site is to fertilize our just-built Jean Pain wood chip water-heating mound. The Pain mound offers a fantastic opportunity to deposit high-nitrogen urine in a location that can actually absorb it for the long winter and utilize it, since these mounds are an enormous carbon-rich store of material: They need nitrogen to fully compost in short order. While the goal of the Pain mound is not compost primarily (it is heat)—though it's certainly a secondary goal—the mound offers an

opportunity, like a "bedded pack" becoming popular with innovative graziers in recent years. For more information, see the "Compost Hot Water Heating System" sidebar in chapter six on page 219.

A bedded pack is simply a thick, high-carbon layer of organic matter—whether it be brown hay, wood chips, sawdust, shavings, or the like. Such a carbon diaper is more able to absorb a massive influx of nitrogen than anything else. And like the Jean Pain mound, the bedded pack heats up because of the microbial composting action, offering animals (typically cows in this application) a warm spot to rest in the winter. For us fertilizing the Pain mound is about as simple as it gets: Deposit urine in a small container if it's too cold to go outside in the middle of the night, and pour it into the mound during the day. You can also urinate directly on the mound itself. We put the nitrogen-rich duck water from the barn into the mound as well.

We'll know a lot more about how this approach works in the coming year, but for now we can say that it's probably going to work very well at least for a composting and nutrient-capture approach. Signs that the nitrogen is helping the heat production process are good—the mound is only three weeks old, and already it's at about 140°F.

HUMANURE

We started cycling solid human effluent only a year ago here because of reliance on an already in-place septic system for such. (This is one reason to build your homestead from scratch, rather than having to retrofit and, in the meantime, be forced to incorporate a senseless system.). Our humanure system in its early stages consists of a composting toilet (an old Clivus), which is large enough to collect a year or more of manure. The human feces and sawdust/planer shavings migrate and are pulled downhill with a wooden raker, so we can keep them from getting freshened by new material. The breakdown of the material continues in this bottom area before that material is scooped out of the hatch to be composted further, for at least another year, mixed with grass clippings, leaves, comfrey, and the like, in the outdoor pallet or metal mesh piles I described above. I add biochar to the bottom of the compost toilet receptacle to sop up the extra carbon- and nutrient-laden water/urine/funk, which percolates through the pile and to the bottom.

I have not used the humanure yet but will report back with results in the next edition of this book. I am sure it will be rich, high-organic-matter compost that should be fantastic. We will treat this material separately from the other compost and make sure it is more thoroughly finished before using. We'll also tend to focus its use away from root crops, most likely. And we'll do all we can to encourage visitors who might be on pharmaceutical drugs to use the conventional septic. Cycling your "waste" into food certainly makes you think about what you put into your body.

BIOCHAR

New research is being done on biochar, an ancient soil ingredient that has the potential to substantially increase agricultural production while helping to reverse climate change via long-term soil carbon storage. This ingredient is often referred to as terra preta, char, and agrichar. Biochar is made through a process called pyrolysis: creating charcoal through the burning of dried plant material under controlled low-oxygen, low-temperature conditions. Roughly half of the organic carbon found in crops can be returned to the harvest location by turning the crop residue into biochar and integrating it into the soil.

Biochar increases soil moisture retention, facilitates beneficial microorganisms, and radically reduces nutrient leaching by binding nutrients securely to persistent carbon molecules. Unlike compost, biochar is incredibly persistent in the soil and is often found in soil deposits from fires that occurred thousands of years earlier. Compost, on the other hand, is metabolized by organisms in the soil and lasts a few years or fewer, with much of the original carbon deposition released back into the atmosphere and lost from the soil. Charcoal stays in soil for millennia, allowing us to make long-lasting deposits into the soil carbon bank. When we inoculate the char with biological activity and nutrients (*bio*char), we also add a nutrient-laden sponge into the soil horizon off which plants can feed for long periods of time.

Think of biochar as a doubly whammy of goodness for the soil and a major climate benefit. We've made biochar very simply via outdoor brush pile fires and in our woodstoves. The latter results in a more optimal char, but the former is a yield from a farm activity that wants to happen regardless of the char yield. When we clear land we make slash piles. We use some of the slash, the common term in rural New England for treetops left behind or used for other purposes aside from milling during logging operations, as modified hugelkulture beds where we install the woody debris on the downhill side of berms below swales—or simply willy-nilly incorporated into the swale berm.

However, there is always more slash than we can integrate in this way. This slash gets piled around the landscape, usually dried for a few months or more, then sometimes burned, but not mostly to ash. We ready ourselves before the fire with buckets and sometimes hoses, ideally before a serious rainstorm. Since the pile wants to burn right down to ash (not a very useful or permanent soil amendment), we need to ensure that the burn stops before that point and leaves us with a large amount of char. In practice much ash is made, but if we have plenty of hands and water, we can usually douse the fire about three-quarters of the way into its normal burn cycle and end up with large amounts of charred wood. That charcoal is then very useful in swale-berm incorporation and in more pure biochar making.

I have started to call this hybridization between biochar and hugelkulture that has emerged over the years as a matter of course on my farm *hugelchar*. We pick out the best, most thoroughly charred pieces and crush them in buckets using sapling mashers—bringing that material into the vegetable garden soil flow, rather than leaving it in the rough less-perfect-soil demands of the perennial system. We've also put these large pieces of charred wood on a tarp and driven over them with a tractor. In both cases we end up with nice, small aggregate and powder—ideal char from which to inoculate and make biochar. The woodstove approach simply involves pulling embers from the firebox of the stove while using the stove. We then quench the embers with water or urine to explode the structure of the char (increasing its surface area) and let the material cool. We powder it by hand as in the above example and urinate in that material or pour compost tea into it. That material can then be used in garden beds, when tree planting, in compost piles, and so on after it has absorbed the nutrients for a couple of days.

A note of caution when using char is important to mention: Char or poorly made biochar can stunt your plants in a spectacular manner. The ability of char to absorb nutrients and water is widely accepted, and after poorly inoculating char the first time I used it, I can attest to this, unfortunately, from a negative affect. Excited about biochar, I hastily applied some to garden beds near my house after soaking the char in some urine water for a night. This was in the fall before putting the garden to bed for the winter. We then transplanted kale and other hardy early-season vegetables into this bed in the early spring. For a month these plants did not grow a millimeter. I've never seen anything like it— complete state of suspended animation. My girlfriend at the time, and primary garden grower, almost had my head—she had been skeptical of my haphazard garden soil experiments since day one.

Moral of the story: make sure your char is fully charged with nutrients and water before applying to the soil, and if you want to be very careful, consider mixing the biochar with compost first, then aging for months or even a year or two to ensure that the soil is "corrected" before growing vegetables in it. This soil "correctness" is not a specific chemical or physical composition, although you know many soils are "incorrect" when you see them—partially decomposed woody debris, leaves, or food scraps are classic signs. Soil that is truly correct and ready for vegetables is a somewhat mysterious thing that I am still learning much about.

For instance, after almost fifteen years of making compost, I still cannot make a germination/potting mix that performs very well with little to no transplant shock of starts and little to no damping off or that encourages very rapid growth. I have used Vermont Compost Company's famous compost and Fort V mix for this use for years, although not consistently, always wanting to use my own carefully made compost. Finally, this spring I was sick of poor seedling performance and

was chatting with a friend, a very talented grower, about it. He said, "I gave up on making my own potting mix years ago; Karl Hammer knows how to make it, I don't!"

It all clicked at that moment. Making soil is hard. My friend is a fantastic grower—his gardens are stunning with vigorous, lush plants and little disease or pests. And even he has trouble making his own germination mix. While I felt better about myself after speaking with him, it does leave me somewhat disconcerted. Seedling mix is a fundamental food-production material; needing to buy it each year is not a model of resiliency. Since Vermont Compost Company is close and I have a good stockpile of their soil on hand (it's not particularly perishable), I am not terribly concerned. My own compost also does germinate viable plants, just poorly relative to what's possible. However, the entire gardening season is made much more sluggish than necessary by having suboptimal potting mix. This weak link in my own home resiliency system is high on my list to address.

Fungi: Quiet Ally to the Whole System

Fungi process and feed on woody debris much like plants feed on soil. And in the cold-temperate climates of the world, which contain some of the world's great forests, woody debris is often much more abundant than good soil. Fungi play a crucial role in breaking down this woody material, cycling it into soil and other

Fungal roots—mycelium—feeding on one of many wood chips at the farm

A student in a Whole Systems Skills workshop practices one of the steps in shiitake mushroom cultivation: drilling holes for the fungal mycelium to be inserted into. This process is immensely aided by using a hyper-speed drill bit for an angle grinder rather than a conventional drill and boring bit.

Morel mushrooms growing six months after we prepared the site with a burn and inoculation via sawdust spawn

components that in turn plants can utilize. Some of these plants are trees, which in turn make more woody debris, thereby feeding the cycle endlessly.

During the last decade or so, people are putting together a picture of the crucial and immense role fungi play in ecosystem health and resilience. Researchers such as Paul Stamets have found evidence that mycelium—the roots of fungi—can cover vast distances in the soil and actually seem to have the ability to transfer nutrients from one area where they may be abundant to areas that are lacking. It's as if they can

> **Harness the fungal force—it's powerful, works to make systems more resilient, and offers food and medicine yields as a bonus.**

balance out entire areas of an ecosystem by intermediating the nutrient and energy flows so the system as a whole can thrive most optimally.

Paul Stamets calls mycelium "nature's Internet," noting this incredible capacity for fungal networks to actually serve as a web of fundamental connections in an ecosystem. Healthy forest systems have been observed to contain immensely large and intricate webs of mycelium, with some studies concluding that a mile of mycelium can be found in one cubic inch of healthy forest soil.[*] Whatever the emerging details of the science of fungi may be, the implications for the regenerative designer-doer are the same: Harness the fungal force—it's powerful, works to make systems more resilient, and offers food and medicine yields as a bonus.

---

[*] More on this can be found here: http://agroforestry.net/overstory/overstory86.html.

Jackie Pitts and Kristen Getler wax the inoculum-filled holes along with the log ends to retain moisture and prevent competition from other decomposers.

We have facilitated the growth of fungi and its roots, mycelium, in various ways on the farm—some have been more persistent in their positive results over time than others. Taken as a whole, wine cap stropharia was the most spectacular, easiest, and fastest fungi to work with but seemed to stall out after two to three years, while shiitake takes more work but seems to be more resilient and reliable over time. It's also easier to be a source for your own logs (used as substrate for shiitake) than for wood chips (used as substrate for wine cap), in that a chain saw is fairly essential for many tasks, while a wood chipper is a pain and expensive to own at best.

Our wine cap (*Stropharia rugosa annulata*, or the Garden Giant) facilitation started when we began to plant this site about seven years ago. We used bare-root trees, which we dipped in a specially prepared root dip containing humic acid, beneficial bacteria, liquid kelp, fish emulsion, and various mycorrhizal and other fungal organisms. We planted all the bare-root trees by first dipping them in buckets of water and this root dip, followed by a heavy mulching of wood chips. We didn't realize at the time that every tree we planted was also a biological inoculation of fungi across the site. This much became clear about three months after planting, when in midsummer droves of mushrooms began sprouting up all over the place. Many of these reached large sizes—up to that of a dinner plate.

We marveled at the growth of these but didn't know they were edible until a friend of mine and a soil-fungi aficionado, Buzz Ferver, was walking around with me and asked, "Why aren't you harvesting all of that stropharia!?" We began to and found ourselves awash in wine caps for the following two years. At many times throughout the summer, when the rains were just right, we would be able to harvest a couple of shopping bags full at least once a week just going around the mulch

Wine cap stropharia mushroom (*Stropharia rugosa annulata*): We harvest from vegetable garden soil in large quantities.

rings in zone 1. The flushes were so abundant for the first couple of years that many of these mushrooms formed and went by before we could even get to them.

Wine caps, however, don't keep well, and you have a very small window of harvest to actually get them in good condition. They are best while still burgundy but after they've opened enough that they are at least silver dollar size (usually), often much bigger. As soon as they begin to go from a red-wine color to something more pale, they have started to go by. You will notice they release massive amounts of spores from their underside (gills) at this stage as well—harvest just before that happens. These mushrooms keep for a few days or fewer in the fridge and for only hours unrefrigerated—growing maggots within a day.

Paul Stamets has recommended using these mushrooms for fish production because of the immense number of flies one could grow on them, which is a

Spores released from the rapidly reproducing wine cap mushroom captured on a piece of bond paper

Golden oyster mushroom that we grow on totem pole–style production using otherwise low-value poplar wood

great idea we'd love to try sometime. As we began to see the power of this fungi and its tendency toward this site, we started to "run" with the species, promoting it by offering it food in ever-expanding areas of the site, paving the way for its expansion with wood chip mulches.

The epic wine cap days of walking around each morning to massive flushes everywhere are no longer, however. I am not sure exactly why but surmise that these flushes no longer occur because of a combination between two factors—reduction in feeding the organism and complete/sufficient colonization of the site's soils such that the fungi does not need to fruit (read reproduce) anymore. The first few years of site establishment, we kept up with mulching so the feedstock for the wine cap was continuous. We also began putting down large amounts of horse manure, in which wine caps seem particularly prone toward producing.

Remineralization

The land system, like the human body or any other system, is often limited in health and vigor by missing or insufficient nutrients, minerals, energy flows, water, and other components. We call such components "limiting factors" and are always trying to identify which are emerging, understand which have influenced the system in the past, and attempt to anticipate them before they arise in the future. Indeed, that's much of the game of productive land management. For a long period of time, from the early biodynamicists to recent soil biologists and savvy farmers/gardeners, those working with land have realized that limiting factors often come in the form of missing soil minerals. Elements such as calcium and boron, along with other micronutrients, have been found to acutely affect plant growth—both negatively and positively depending on their balance in the system.

Remineralization is the movement and action to identify and ameliorate the limiting minerals in the soil ecosystem. This work is often best done through comprehensive soil tests to establish a baseline of data, followed by additions of key minerals that have been identified as lacking or present in imbalanced proportions. Most soils happen not to contain the optimal balance of soil minerals (and other nutrients) that should be identified as early as possible in site development. These minerals are heavily determined by your soil and parent material (bedrock) type. The USDA Natural Resources Conservation Service (NRCS) soil maps available for most of the United States have significant information regarding the underlying materials in each region, from which you can obtain many clues about what may be limiting factors in your soils.

Usually, experienced gardens and farmers in your area will be able to elaborate on what these limitations may be—in our area calcium is usually lacking. "Weed" and other plant identification is also a great way to understand what nutrients may be in abundance or lacking in your soil—check out the literature available on that subject, as reading the weeds is an important aspect of general ecological literacy and will greatly aid your land development and maintenance work for a lifetime.

Once you've established what components might be most limiting to your soils, it is time to set about amending them. One caveat: Some amending is useful no matter what your soils may be, and lacking a soil test, you can still rely on certain soil modifications to be helpful. These include organic matter additions/ compost, nitrogen additions (often but not always), mineralogical additions (not all, but many minerals). Fans of remineralization often take the view that soils in general, due to a long legacy of abuse, are limited in many minerals and a general addition of broad-spectrum minerals is key to growing healthy plants and animals.

Whether this makes sense to you or you'd like to take a more calculated approach, the amendments are similar, with the following being favored. We've used

many of them, and it's too soon to say what kind of effect they've had on the system—we have also not run side-by-side trials to know. In general I subscribe to the remineralists' view that our abused soils are lacking minerals and from even one-time additions can greatly gain in health for generations into the future. If you think of it in this way, you can imagine a soil's being "released" to manifest its full potential, given the addition of something (mineral- or nutrientwise) that was limiting it. Favored amendments in remineralization include Azomite, Greensand, fish and kelp-based products, sea salt, and bone-based materials such as char. I highly recommend looking into these materials to find out more about the variations of their uses and other specifics.

Cover Cropping and Winter Cover

Cover cropping during the growing season, as part of good crop rotation, will probably be considered standard practice in life after the cheap-energy economy. Our age of peak oil holds a future in which imported growing media and nutrients in the form of soil, manure, and other fertilizers will be increasingly expensive or unavailable. Cycling fertility on site perpetually will be crucial in a postpeak future and will determine, in large part, the economic success of a farm and the viability of a homestead.

Since nitrogen is the most often relied upon off-site input, producing (or harvesting) nitrogen on-site via fertility farming will be central. Indeed, home and farm landscapes in general should be thought of as nitrogen-capturing nets, sopping up atmospheric nitrogen and scavenging excess soil nitrogen for plant growth and soil organic matter production. Nitrogen-fixing plants such as clover, vetch, pea, and myriad perennials are components of this system, as are animals; land committed to these resources represents a significant portion of the productive acreage in a regenerative landscape. This is why permaculturists are fond of planting at least one nitrogen-fixing plant for every "feeder" plant such as a fruit tree or berry bush. They serve as the fertility factory in the landscape, reducing or eliminating the need for fertility importing from off-site.

Pollinator heaven white clover amid an overstory of rye on a pond berm, less than three months after this was bare ground

It's surprising how many organic home gardeners, and even organic farmers, tend their crops carefully throughout the season only to clean out the garden or field at the end of the year and leave exposed earth, spreading not a single pound of cover-crop seed, leaves, straw, hay, or other mulch. This simple task protects the soil from six months of rain, snow, and wind until the next gardening season. Without such cover, bare soil loses its finest and best material to the percussion of rain drops, the leaching effect of snow melt, and erosion by wind. In addition, a dormant-season planting of cover crops fixes nitrogen and sends organic matter (via roots) into the soil, boosting fertility and soil health for little cost and effort.* Some covers also aid in optimizing phosphorous, micronutrients, and other chemical levels in the garden. All cover crops contribute organic matter to the soil, the foundation of good soil and plant health. At WSRF we spread cover crops the day the garden is cleaned out at the end of the season, rake it in if necessary (depending on species), water if needed, and turn in the cover as soon as possible in the spring if it does not suffer winterkill. Buckwheat is one of my favorite covers (though it doesn't fix nitrogen), because in my climate it winterkills reliably.

While we have not dialed in a nearly perfect cover-cropping strategy, we have found that the following principles, when followed, result in consistent winter cover of our beds. Truly, the challenge is dealing with the cover (turning it in/grazing it down) in the spring.

► **Sow a cover as soon as possible after cleaning the bed**—crucial to get ahead of any weeds. We like to harvest, graze the bed very hard with chickens, then pull any remaining weeds with their roots and rake in a cover crop—especially oats or buckwheat.

---

* It takes time for a nitrogen-fixing plant to actually convert atmospheric N into soil N, so turning an N-fixing cover crop in before this process happens is a common mistake.

A cover crop of forage peas coming in strong with snow still occurring in early spring. A combination of raised bed, salvaged glass window for the cold frame, and a hardy plant allow a four- to eight-week jump on the normal growing season. The peas were used as a cover crop and turned in, as well as cut for salads multiple times first and regrown until tough.

▶ **Use whatever is at hand for cover.** We cover garden beds with tarps, membranes from pond work (EPDM works great), burlap, or simply mulch beds heavily in the fall to get a solid winter cover. This covering—but not cover cropping—has the advantage of protecting the bed from erosion while leaving you with a nice bare planting surface or seed bed in the spring but with the distinct disadvantage of no C and N capture and organic matter building as a result.

We are not zealots about using only straw for mulch. I've seen as much "weed" come up from straw as from hay, and the idea that hay isn't clean but straw is is simply not true where I live. Straw is also not available locally here, as almost no one grows grain. I now make my own straw via the rice production, but that is not enough to mulch many beds heavily, though I am using it along with rice hulls as mulch and having great results.

▶ **Turn the cover in early**, or if you have enough materials, sheet mulch it. By "early" I do not mean necessarily when snow melts, but it should be done as early as is needed to kill the cover crop before you need to seed or plant the bed. When establishing new beds we will devote the resources needed to actually mulch each spring for one or two springs. This takes a lot of material but results in rapid bed formation. This looks like the following:

• Sheet mulch an area (usually grass) to kill existing vegetation using cardboard, burlap, straw, hay (we avoid newspaper due to the dyes and other likely chemicals).
• Add compost on top.
• Plant in that in year one—ideally, beans or other fast soil-builders and light feeders.
• Cover crop it heavily in the late summer or fall.

Daikon radish or tillage radish, sown via broadcasting lightly and often through the growing season, especially immediately after heavy grazing and earthwork disturbance. We never have to devote garden space to radish or turnip with this approach, and most rot in the fields, forming deep soil deposits.

- Sheet mulch that cover crop the following spring, and repeat compost application.

I should note that this is one area of farm/homestead work in which we find that chickens have a distinct advantage over every other form of livestock we've tried (the other is in disturbing groundcovers to create seedbeds in pasture—essentially the same effect). We use sheep to graze down a really weedy bed and the ducks to keep snails and slugs out, but with their immense scratching effect, only chickens will truly turn over the bed, wreck all or most of the weeds, create a perfect seedbed, and fertilize the bed simultaneously. Turkeys would likely have a similarly positive disturbance, but we have not tried them yet.

PERMANENT COVERS IN VEGETABLE BEDS

An additional aspect of cover cropping that we have found some success with (but that seems to have even greater promise over time) is successional planting of vegetables through a cover crop—à la Masanobu Fukuoka* style. This approach involves seeding a crop such as squash, cucumber, tomato, kale, or most anything else that tends to leave a lot of open bed space or bare soil between plants for a large period of time during its establishment. These crops create major weed and soil-damage problems because they encourage so much bare soil for such long durations.

To avoid this we have tried planting buckwheat a few weeks after the squash or other vegetable gets established. Giving the squash a head start allows it, in theory, to get above the cover crop, before the cover then comes in behind it to fill out the bare soil. We have had this work to a functional but not nearly optimal degree. The cover crop (especially buckwheat) tends to get too high and shade the main crop too much. Ideally, we would use a very low-growing cover that can be broadcasted, is easy and cheap to make or acquire, is fast establishing, and is nitrogen fixing. Ideally, the seed could take a bit of shade as well.

This ideal seed would be something like white clover but much faster establishing and not perennial. White clover is a perfect plant in many respects, but it's slow to establish. I'd like to try to maintain garden beds of permanent white clover, like Fukuoka did in his rice paddies, and also grow vegetables through the clover cover. It would likely require planting somewhat large and vigorous seedlings into the clovered bed, but that seems doable. The challenge in this approach would be the constant need to seed clover, as other plants come in constantly, and keeping the other taller grasses and other "weeds" out of the bed. In our climate I am not sure that the perfect groundcover for permanent cover in vegetable beds exists, but I will keep looking for one.

---

* Masanobu Fukuoka was a renowned farm innovator living in Japan. His experiments with "natural farming" and a particularly Taoist approach of letting nature do as much as possible—a lot like permaculture—received international acclaim over the years as his success in growing high yields with minimal inputs was astounding. I regard him as a particularly important reference because of the nuanced way in which he practiced successional plantings.

The need for a successional cover to bolster crops is crucial to protecting and building soil while working an annual system. Much work needs to be done to improve this area of home and farm soil maintenance and enhancement.

There are many good resources for cover crops available online and in print form. Seek out the most appropriate covers for your phase of land development, your soils and climate, your particular cropping cycle, and other relevant aspects of the site's system, such as animal forage needs.

Tall-Grass Grazing

Picture a ten-thousand-head mob of buffalo or wildebeests thundering across the plains of North America or the Serengeti. Is healthy topsoil what comes to mind? For most of us, steeped in conventional ecology and environmental science, the answer is, "Hell no!" Instead, we imagine ecological destruction from overgrazing and desertification. And in truth the association between grazing animals and land abuse is not unfounded: Poor grazing practices (not necessarily "*over*grazing," just poor grazing management) contribute greatly to the desertification of large areas of the earth, wrecking soil and water and leaving a high climate-change bill. But that's different from the quick movement of densely packed animals through a landscape (otherwise known in agriculture as mob or stock grazing), as in the American West, Africa, and other places where deep-soil prairie lands and massive herds of animals coevolved.

Low-labor, industrial grazing is typified by low-density animal stocking occupying the same land area over long periods of time. (Picture your typical pasture dotted with a few animals here and there.) This, while seemingly idyllic, is the opposite approach to regenerative grazing, which builds soil and grows the healthiest animals. Innovative graziers have recently been realizing that high-density, very short rotations (the cycles naturally performed by grazing herds for millennia) is a far more productive approach to managing

Sheep in a deep sward—mob stocking grazing animals for rapid soil formation and herd health

INTENSIVE ROTATIONAL GRAZING AND KEYLINE CULTIVATION

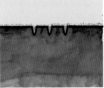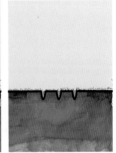

STEPS 1–4

1 PLOWING
subsoil loosening allows water, nutrients, and oxygen access to the soil horizon.

2 SOIL AMENDING
small amounts of fertility-promoting nutrients such as rock powders, biochar, and beneficial bacteria access the soil via subsoil rips.

3 IRRIGATION
The plowed field is flooded briefly to deliver water, nutrients, and oxygen deep into the soil.

4 INFILTRATION
Water sinks into the soil, moistening and aerating the soil, allowing optimal root development and plant growth.

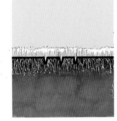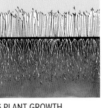

STEPS 5–8

5 PLANT GROWTH
Pasture grasses utilize enhanced soil by deepening root penetration into the subsoil.

6 PLANT GROWTH
Roots dig ever deeper into the soil horizon, further loosening subsoil layers and transforming subsoil into rich topsoil.

7 INTENSIVE GRAZING
Dense grazing action quickly manures the landscape while harvesting grasses. Root dieback adds carbon to deep soil layers.

8 MICROLIVESTOCK
Microlivestock such as fowl utilize remaining nutrients on the surface and add additional nitrogen while reducing herd disease.

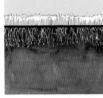

STEPS 9–12

9 PLANT REGROWTH
As plant roots regrow, the soil-building process repeats itself.

10 PLANT REGROWTH
New roots grow to ever-increasing depths as they move the newly available subsoil.

11 INTENSIVE GRAZING
Continued root dieback creates multiple biological climaxes, yielding new, stable, high-organic-matter topsoil from former subsoil.

12 MICROLIVESTOCK
Note the difference between step 1 and step 12. The dark brown color (topsoil) has seeped deeper into the tan color (subsoil). New topsoil becomes subsoil from grazer and plant-root mechanisms.

INTENSIVE ROTATIONAL GRAZING AND KEYLINE CULTIVATION

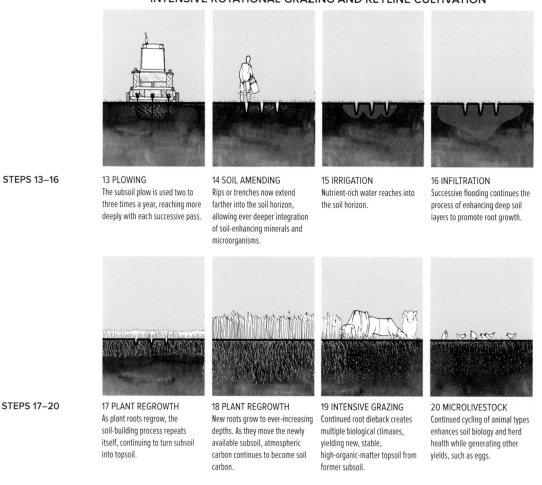

STEPS 13–16

13 PLOWING
The subsoil plow is used two to three times a year, reaching more deeply with each successive pass.

14 SOIL AMENDING
Rips or trenches now extend farther into the soil horizon, allowing ever deeper integration of soil-enhancing minerals and microorganisms.

15 IRRIGATION
Nutrient-rich water reaches into the soil horizon.

16 INFILTRATION
Successive flooding continues the process of enhancing deep soil layers to promote root growth.

STEPS 17–20

17 PLANT REGROWTH
As plant roots regrow, the soil-building process repeats itself, continuing to turn subsoil into topsoil.

18 PLANT REGROWTH
New roots grow to ever-increasing depths. As they move the newly available subsoil, atmospheric carbon continues to become soil carbon.

19 INTENSIVE GRAZING
Continued root dieback creates multiple biological climaxes, yielding new, stable, high-organic-matter topsoil from former subsoil.

20 MICROLIVESTOCK
Continued cycling of animal types enhances soil biology and herd health while generating other yields, such as eggs.

The process by which intensive rotational grazing and subsoil loosening encourage rapid and deep topsoil formation via the transformation of subsoil into organic matter–rich living topsoil Illustrations courtesy of Whole Systems Design, LLC

pastureland. Enter *intensive tall-grass grazing, mob stocking,* or whatever your preferred term may be for regenerating the landscape through the application of grazing animals. When a large number of densely packed, heavy animals moves through a landscape quickly, occupying that landscape just once or twice a season, the following soil-building events tend to occur:

1. Tall grasses, with correspondingly deep roots, are grazed down to within a foot of the ground *but not completely down to the ground*; grazing to the putting green level damages the plant's ability to rebound.

Plant roots die back as a response to this pruning, leaving organic matter (carbon) in the soil strata. The deeper the roots have penetrated, the deeper into the soil this organic deposition occurs. And the taller the grass was before grazing, *the deeper the roots were able to grow*. This is the organic matter/carbon-pumping stage in the system, where atmospheric carbon is transported into the soil by plants. Think of tall-grass mob-stocked grazing as a potent carbon-negative conveyor belt reclaiming atmospheric carbon and putting it back into the earth from which it came.

The research farm's herd and flock hard at work on the bottom pond berm and lower fuelwood hedges, turning comfrey, clover, grasses, and black locust leaves into fertile soil, eggs, meat, and wool

2. Densely packed animals provide nitrogen in the form of urine manure as they graze. They also turn up clods of sod, allowing access for rainwater to bring the newly deposited nitrogen and biological activity (microbes in the manure) into the soil. Think of grazing animals as an enormous living rotovator spewing soil-enhancing nutrients behind them; that's the action of a massive animal herd if allowed to move through, not loaf upon, a patch of ground. Rains wash the fertility and biological inoculants into the newly broken-up soil, where it can penetrate deeply and not run off the landscape, as it would more readily were the soil surface unbroken. This is the fertilizing and soil biology–enrichment stage of the process.

3. Grasses left standing six to twelve inches by the quickly moving herd rebound rapidly and are allowed to grow to hip height or taller before the herd is brought back again. This is the resting and regrowth/root-penetration stage of the system.

These three steps are the primary reason vast areas of land have been improved and sustained, not desertified, by the presence of massive animal herds. Modern "mob stockers" such as Joel Salatin and Abe Collins are applying this understanding to ecologically ("biomimetically") manage their animal herds for the multifunctional production of meat, milk, soil fertility, drought resistance, greenhouse effect reversal, and the many other benefits of healthier, deeper soils. The take-home points here for the modern homesteader or restoration farmer are the following:

Let it grow! You only build soil as deeply as you can get plant roots to penetrate (what comes up must go down), so the taller you let your yard or pasture grow before it's cut or grazed, the more soil you're making (and CO_2 you're sequestering). Think of any areas in grass as pasture or vegetable gardens-to-be—areas where you want good soil. An upshot here is that mowing, *if* you mow, wants to happen three to six times per year, max. This begets the need for white clover and other low-growing groundcovers, unless you mind the prairie look in your front yard.

And to the extent that you can, manage your animals for short grazing periods in tall grass. Plan the grazing rotations carefully when your landscape allows.

Pasture Reclamation: Why Not to Let Your Field "Go"

The ten-acre beat-up old hillside farm I inhabit has presented us with various restoration challenges in the past nine years of site development. Of all these challenges, from poor soils to high water table, from slope to ledge, none has proven more intractable so far than redeveloping a good sward of grass (forage) in the former fields. When I first arrived in September of 2003, an acre or so around the house was nice thick grass, having been a mown lawn. Another two acres or so farther out of zone 1 was patchy grass but still mostly grass.

The remaining three to four acres of open land was a brush-hogged field that contained a lot of grasses, along with much fern and only a small number of woodies (since it was cut every year). Upon moving in I promptly decided that it was crazy to use fossil fuel to mow down the field every year, and I let it go. I figured at that point that "letting nature repair itself" was the best way to a healthier and more productive field in those five or so acres. I also nearly completely ceased mowing the area around the house.

Aside from ending up with a messier looking property as a whole, the results were surprising. Instead of increasing in diversity, the field as a whole became mostly goldenrod, poplar, willow, alder, and birch, with some pine and lots of brambles. Fern also came to dominate the "overstory" of the now fully abandoned field. The grass around the house became taller but patchier. After a few years I realized that I needed to at least weed whack the yard, as I did not have a scythe or the skills to use one back then. Within four to six years of "rest" (read abandonment), the lawn looked pretty sorry, with large patches of bare soil interspersed throughout, and the field looked even worse. What was at one time grassy pasture (a long time ago) and later a poor but grassy field was now becoming head high with brambles and pioneering tree species with a fern understory. Grasses became almost totally absent, and sedges (nonedible to grazing animals) moved in to occupy large areas of the field in the wetter locations.

By the time I realized that these four to six acres could be a useful part of the emerging small farm, the job of turning this area into a productive and more diverse pasture/tree crop system was much, much larger than it would have been if I had not let the field go to woodies, brambles, sedges, and ferns.

The movement to a less diverse and less productive state of the field was becoming clear to me in 2007 and

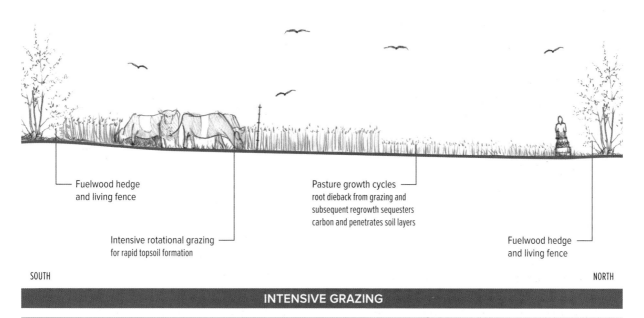

Fuelwood hedge and living fence

Intensive rotational grazing for rapid topsoil formation

Pasture growth cycles root dieback from grazing and subsequent regrowth sequesters carbon and penetrates soil layers

Fuelwood hedge and living fence

SOUTH   NORTH

INTENSIVE GRAZING

Intensive rotational grazing integrated with other productive land-use systems

Forest to perennial polyculture (field) conversion: much of this frame was the dark understory of a white-pine monoculture, comprised of fern and pine needles one year before this photo was taken.

2008, but I did not yet have the time or motivation to graze. I knew, however, that some use of the field was needed to keep the woodies from overtaking the whole field. So I began scything the field in 2007, figuring that this mowing action would not only knock back woodies, fern, and goldenrod but would also promote some grass. After a year or two of this action, I began to realize that this knocking down of the "sward" (if you can call it that) was only serving to mulch the field and actually was counterproductive, as it was effectively suppressing new growth of the seed I was beginning to distribute. This mulch was also keeping new seed from even reaching the soil surface.

I realized after a year or so that knocking down hip-high goldenrod is a great way to keep everything but goldenrod from growing! I could have, and should

have, been scything the field when plants were very young early in the season, but scything is tricky when plants are not a foot tall or higher. All in all the efforts at scything the field to get it into a better condition were unsuccessful and where the material was scythed when young, positive change was still painfully slow. Scything, I realize now, is not the same as mowing.

I began grazing in 2009 with four sheep, and because of such poor-quality forage, and so little of it, I had to move them at least four times a week, and sometimes more than once a day, for the first two years. At this time I also began cutting saplings and hand-thinning brambles. In some areas I spread a small amount of lime, and over most of the field, I seeded massive amounts of seed at least a dozen times during the early and middle part of the growing season; species included white,

Aretha, one of our best Icelandic sheep, amid a tall stand of comfrey—one of her favorite and nutrient-rich foods at the homestead

red, ladino, and sweet blossom clover; hairy and crown vetch; purple top turnip; daikon radish; orchard grass; sheep fescue; winter and annual rye; field pea; alfalfa; and probably a few others at times. This seed was spread mostly by hand-broadcasting using lime, Greensand, Azomite, bonechar, and other materials as a carrier. I inoculated all nitrogen-fixing seed most of the time.

The seed was spread often before, during, and after grazing under the assumption that the animals could trample the seed into the ground, offering good soil-seed contact. Any areas then left bare after the animals were done would get seed spread onto them as well. Two to three applications of seed per heavy grazing area seemed as though it should revolutionize the field. I also even resorted to feeding seed to the sheep in small amounts of grain used as a nutritional supplement—later being told that sheep's digestive abilities are so thorough that they would not dispense with any viable seed. The sheep were moved in electro-netting, and stocking density was high because I was aiming to create enough disturbance (both mechanically through eating and hoof action and biologically through manuring) that a new succession of plants could be initiated.

In the second summer I began to see a tiny amount of grasses emerging over most of the now-grazed area, and in a few small locations, the sward even looked half decent, with some clovers, vetch, and orchard grass emerging. But despite the grazing, heavy seeding, and light soil amending, most of the acreage had changed only a slight amount. At this rate it was going to take a decade to get to decent pasture. I persisted, however, and kept on with grazing, slowly seeding less and less.

Fire, one of the most rapid and successful land-renovation tools we've used at the farm in the journey from beat-up and abandoned thin field to lush pasture

After a second full season of grazing (with fewer animals, losing one in the pond and one to old age), the pasture emerged in the third spring with surprisingly little change from the year before. Larger areas contained some grasses, and the areas that had some grass density emerge earlier now looked like four sheep could feed for maybe a day in one 164-foot length of electro-netting.

But for the most part the third grazing season emerged in late April with a shockingly unimpressive change to the field. Ferns and goldenrod still dominated, despite the woodies being knocked back (mainly by hand-cutting and even some uprooting in the winter with a mini excavator). Seeing such little successional change in late April and early May of the third growing season, I realized that the *ecosystem inertia* of the old field was simply more powerful than the disturbance that we were applying. A greater degree of, and likely other types of, disturbances was needed to shift the species and their arrangement in this field from bramble, fern, sedge, and woodies to ones containing a much larger degree of grazeable grasses (along with the couple of dozen species of perennial food crops we were planting in the field).

So in early spring of the third grazing season, we started prescribed burns of the field when the weather allowed. Two such weather windows emerged, and we carried out what has so far seemed to be a successful burn of about one-third of the field. During this process we made firebreaks by raking dry herbaceous material into rows, which we burned first. Burning these lines then gave us a nonflammable break for the rest of the fire to hit and stop because of lack of fuel. The fires were initiated with a propane-powered flame weeder, and the weather allowed just enough air movement to promote a slow but steady burn of the one-hour fuels (dead fern, grass, sedge, and goldenrod, mostly) to move across the field.

At the time of writing this, it appears as if most of the dead material that was in effect mulching the field, and suppressing new growth, was burned off. Immediately after each burn, we seeded the area with a mix of forages described previously. We were lucky to have a wet spring period after the first burn, and within two weeks the burned zone was greening up very thickly.

The Benefits of Mowing

After growing up in suburbia, mowing has long been a bane of my existence. Naturally, mowing is a severely overused and abused land-management tool in the United States, and in no area is it more manifest in all of its spectacular wasteful and noisome glory than suburbia. So I had learned to hate mowers, mowing, and all things mown. And that's why it has taken me almost years to realize that mowing actually has a wonderful disturbance power to it and that, when applied with the correct timing and severity, it can help transform certain successional phases into more productive and biodiverse communities quickly.

Here's why mowing is so special and can do things that scything cannot and that grazing (often initially) has a tough time doing well: It reliably increases stem density, and it promotes grass. Stem density is the number of actual grass blades (or vetch, clover, or other stems) in a field per unit area. Just as with organic matter in soil, you want as much stem density as possible, within reason. When a field-growing plant, especially grass, is cut, it tends to shoot up other stems in an attempt to grow where it could not—the same way a pruned woody perennial plant tends to sucker or grow multiple shoots from a point where one stem was pruned.

Most fields are actually very low in stem density, even when the sward might grow tall and look lush. Given that you want as much fodder as possible and organic matter production (root growth and dieback) as possible, you want stem density to be high with as little patchiness in the sward as possible. The ability of mowing to increase stem density and promote grasses while discouraging woodies, sedges, and ferns shouldn't be surprising—just look at many poor soil fields, sometimes even very wet ones: If they are mowed very often, the stem density is high, and it's almost all grass and clover. That's a perfect jumping-off point to start grazing.

Scything is problematic before stem density is high because it tends to mulch out and suppress new growth and the opportunity for it. If you harvest the cutting, it won't do that, but then you're pulling fertility from the field—not something you want to do when establishing new pasture on beat-up soils. Grazing is great for fertility, and it's solar powered, but it's unselective in general, with animals leaving the plants they don't like—even in an intensive rotational setting, this happens significantly. Mowing is nonselective: You can roll in with a mower or walk in with a weed whacker, and for a small amount of fossil fuel (or human power if you use a push-powered mulching mower), you can cut down and chop up everything in the area. The chopping-up aspect is crucial and is what allows mowing to not suppress new growth, in contrast to scything. You can also mow a very low sward, which you cannot do with a scythe, and you can mow many times per year—far more than grazing unless you are okay with compromising animal health; you can mow a field a half dozen times or more in a season but grazing it that many times is terrible for parasite loading and nutrition.

And cutting back plants many times a season is crucial, in some areas like my field, to promoting the filling in of patches by stems from neighboring plants. This third year of pasture reclamation saw us grazing as intensively as in the last two years, with continued seeding and mowing behind the animals after they've manured and left behind a portion of the sward. The mowing in year three did significant good work for the pasture, and I am very glad that we took the time to do it. Without this mowing it would have been impossible to alter the course of succession in such an optimal way, and the momentum of goldenrod and

fern in particular—above and beyond all other plants here—would have continued undisturbed enough to dominate the pasture.

Once stem density is high and the field composition is almost totally highly valuable fodder crops, the need to mow goes way down, truly ending up with little to no need: The animals will eat nearly everything, and what they don't can be scythed and harvested or left—as the biological activity in the field will digest the dry matter remaining, and it will be such a small proportion of the field that the suppressive mulching effect won't happen much, if at all.

The lessons I have learned during this pasture reclamation process are still occurring, but so far they include these:

► Abandoned poor-soil fields have a stubborn inertia—it is difficult to transition out of an assemblage of moss, fern, woody plants, and brambles. Doing so requires careful timing and significant disturbance force. The WSRF pasture reclamation project has been slower than necessary because I did not realize until recently the amount of force required (animal density, soil disturbance). I did realize early on the importance of timing after seeing May and June seeding take hold while any pasture we applied grazing and seeding from July on changed very slowly from one year to the next. April, May, and June are the windows of opportunity for serious pasture transition, given the coolness and wetness with which to establish new plants. The length of growing season remaining also helps with seeds broadcast early in the spring or summer.

► Manage the water early on: We've installed ponds, paddies, and swales in our field that change the hydrology, usually for the better. A high-water-table field is a tough condition, but we have managed to drop the table in certain areas while raising it in others. Areas where grass is desired should not have a water table near the surface because sedge proliferation is difficult to quell—and no one grazes sedge as far as I know. Consider ditching around a field to dry it out a little, then capturing the water lower down. In general this is not a recommended approach holistically, but it can help get grass going. A water table can be higher and grazing more sporadic once the sward is established with fewer ill effects.

► Seed early (for above reasons).

► Seed lightly and often: Weather for the week to two weeks just after seeding seems to be the largest determinant, along with soil germination surface condition, as to how successful a seeding will be. We have found that seeding a dozen or more times early in the year ensures some likelihood of hitting the right window with minimal seed waste and expense. Better to go light and have some of the seed take than go heavy and run a much higher risk of no seed taking because of weather.

► Seed immediately after disturbance: In keeping with a basic principle (see chapter two), we always are most successful when filling niches intentionally, and that requires filling them quickly after they are opened by a disturbance force. Seed requires soil contact to germinate, along with moisture and light to grow. The ideal scenario we are shooting for is (1) get seed-surface contact, (2) get moisture to seed, (3) get light to seed, and (4) get nutrition to seed. Seeding right after animals leave a paddock is great as it is during the rotation, so they press the seed into the soil (ruminants, not birds!).

► Mow early and often to get stem density high: This means early in each growing season but also early in site development if you are transitioning an old field back to pasture. You can do more with a field in April, May, and June than in July and beyond.

► Consider permanent fencing right away, and invest in it if you can (unless you have an endless labor supply). If we had done this early on, it would have already paid for itself many times over in the amount of labor we've put into moving electric netting.

► Consider feeding on pasture: a great way of getting nutrients into the field and ensuring bacterial proliferation over fungal presence, which is key for grass growth.

► Don't let your field go! The wisdom of old Vermonters, and likely rural people everywhere, to never let a field go fallow for more than a year or so likely comes from a visceral place—from past experience

that has seen how difficult and slow it is to get back to a pasture condition once a woodland succession begins. It's also likely that such people knew very well the massive amount of work involved in transforming the land from forest to field originally.

Scything: The Most Resilient (Mechanical) Biomass Harvesting Method

Of all the hand-powered land development and maintenance tools I have used, the scythe is probably the most effective in terms of amount of work yielded per amount and quality of time spent performing the work. Splitting wood with a good ax and pruning small trees probably come in at a tight second and third place, respectively, in this hypothetical, but useful, contest. When I say "scythe" I am not referring to the hardware-store-variety heavy-handled tool—the American scythe—or a laborious chopping-at-vegetation activity. I am referring to the Austrian scythe—a slender instrument that when wielded in the correct sweeping motion results in an enjoyable, devastatingly effective means of mowing light brush and grass.

I have been using an Austrian scythe for about seven years, starting with the tool available from Scythe Supply in Maine, later adding a higher quality scythe from Scythe Works out of New Brunswick and British Columbia, Canada. These tools, except the blade, can also be made without enormous difficulty if you have good woodworking skills, but the process does require steam bending of the snath (shaft).

With a proper scythe, good technique, and a little conditioning, one can mow an acre or two of grass in a handful of hours or so. If the land is brushy, double that estimate. While a fuel-driven machine can certainly mow more land, it cannot do so well over highly varied and rocky terrain, and doing so is less beneficial for the body and mind than the Zen-like practice of scything. A scythe also costs a fraction of the cost of a mechanical mower and will outlast it a hundred times over if maintained well. It can also be completely maintained in-house with a few basic tools. The scythe, however, requires far greater skill than the mowing machine. Such is the general pattern with hand tools compared to power tools; the elegant, often slower, but long-term healthier solution requires more experience and skill than the easier, short-term, faster approach.

Proper scything equipment consists of a snath (shaft), handles, blade, and hardware attaching the blade to the snath. The handles should be fitted custom to the user; as with all fine tools and finely performed craft, the fit between user and tool is crucial. Sharpening equipment is equally essential, as the scythe only cuts well with a nearly razor-sharp blade. Lack of blade sharpness is certainly the most common error among new mowers, since sharpening a blade is actually quite difficult.

Any athletic person with good coordination can learn the scything motion well within a season of mowing, but getting a blade very sharp is something that often will take a number of seasons. I am still learning to get a decent edge a handful of years into scything, and I had a bit of varied blade-sharpening experience before beginning to scythe. The blade on a scythe is sharpened every five to ten minutes, depending on the hardness of what is being mowed, using a curved narrow whetstone that is carried submerged in water on a belt-mounted holder. While this sounds excessive, it's actually the most efficient way to work, since sharpening only takes ten to thirty seconds, and a honed blade slices through the material with far less strain on user and tool.

After a dozen or two dozen hours of mowing, again depending on the quality of material being mowed and on the skill of the sharpener, the blade must be peened (thinned). This involves a hammer and small anvil-like tool or simply a curved-head hammer and the pounding of the blade's edge very specifically so that metal is drawn out and thinned right at the edge of the entire length of the blade. Peening is not easy at first and is unique to scything for most of us. I don't peen very often, but when I do it usually requires about five minutes per blade. I find that a jig supplied by Scythe Supply does makes peening easy because it simplifies the process. Thicker blades require a little more work—brush blades being the thickest, toughest blades compared to longer, thinner grass blades. The

rougher your land, the thicker and shorter a blade you want, while the more tame and succulent your land, the finer the blade (and easier the mowing!).

The Scythe Book by David Tresemer is an invaluable resource for anyone interested in scything tools and proper technique. There are also a number of very helpful videos on YouTube showing highly developed mowers in action, particularly a video made by Peter Vido, founder of Scythe Works, of his daughter and master mower.

Scything involves a low sweeping motion, with the blade carried just above the ground in a broad arc, usually about the width of the height of the mower. The wider an arc one can mow, while still maintaining a balanced and efficient stance, the better. Mowing is accomplished more in width than in forward motion as the mower only steps forward short steps after each sweep of the scythe. With proper technique involving a strong but slow twist of the torso and application of core strength, a mower can mow without tiring for a few hours after some practice.

Traditionally—and in parts of the high Swiss Alps and other regions of Europe especially, mowing was always done from dawn, or even the predawn hours, into the midmorning, with all work finished before the grass dried thoroughly, when work became slow and dusty. The scythe operates much more efficiently on wet grass, and the cool of the morning makes for more effective and enjoyable mowing. We mow in the morning hours here and during or just after a rain.

Mowing in the rain is actually quite enjoyable, and one can do so safely with bare feet soaking up the health and goodness in the mown grass and damp earth. The health benefits of the scything movement and the contact with living systems that scything facilitates are immeasurable. The scythe shows us clearly how the enjoyability and health-influencing aspects of a task on the homestead and farm are often more important than the speed with which results are achieved. Is the job that slowly degrades one's health but only takes an hour a day more effectively done than one taking three hours a day but that maintains the vigor of the person performing the task? Since the work on a farm and homestead is never truly complete, the imperative seems clear enough: We must enjoy and be invigorated by the bulk of the work we perform in life—no destination, just a journey.

Ducks, Chickens, Dogs, and Sheep

In the development of the WSRF, we have thus far used three species of animal consistently: two birds—ducks and chickens—and one grazer—sheep. We tried goats and pigs but decided that the quantity and type of food they require is not a good match for the resource flows of this farm. Animals well fitted to a farm ecosystem must utilize an excess of a resource and transform that into a resource area that is lacking. For us that so far has been two things: (1) browse and forage (leaves and some grass) into soil and more grass, and (2) slugs, snails, and bugs into eggs and soil. This need for the transformation of one resource into others will always change over time and at some point here will go from seeking soil and more grass alone into seeking other yields such as meat, milk, and fiber. It is the system-establishment phase and the fact this land is an abused

and abandoned farm that requires that as a foundation we establish healthier soils and a better sward of grass from which to raise future animals.

As a whole, when evaluating animal suitability for your systems, keep in mind that the most sensible animals in a homestead geared to be adaptable to a rapidly changing world should be chosen based on the criteria below. And keep in mind that trying various animals is often the only way to find hidden synergies and constraints in a specific animal's interaction with your unique system. Each site's conditions are different enough that no solution found on another site will be wholly adaptable to your own. Find the closest examples, and learn from them, then try, tweak, and try some more. In all likelihood it will take a number of years to establish a synergistic animal aspect to your system. Criteria and considerations for selecting animals in a functioning permaculture include:

▶ **Input-output ratio:** The most outputs, in both quality and quantity, relative to inputs should be a primary determinant of an animal's suitability. This

Some of the first ducks—Indian runners—at the homestead, taking in the view at sunset from a lush area of late summer pasture

aspect includes time, often forgotten as a crucial input (see below). This aspect is contextual and requires an understanding of how the farm/homestead fits into its surroundings. An output such as meat or fiber, for instance, may have a huge value if your neighbors want it, even if you do not. Or your local community may not want or need any animal products from you, and their outputs are only valuable if they can be used on-site.

► **Likeability:** What animals do you get along with the best, pay most attention to, are naturally inclined to observe and relate to? Those, all other aspects being equal, will always do better on your site than those you feel no connection with—simply for utilitarian reasons: You can't care for someone you aren't attentive to as well as someone you are. Domestic animals, like people, thrive based on their connections and the degree to which they are cared for. Care means something different to each animal as well. Care for a beef cow is good grass, lots of room, and good water but does not involve tons of human contact. Care for a milking sheep involves more human contact as they run into more problems healthwise that require human care.

► **Infrastructure needs:** These range from a dry space for the toughest grazers, which can spend all winter out in deep snow, to goats, which do best with some cover from even mild, warm rainstorms. Pairing your infrastructure with the needs of the animal is key.

► **Soil needs:** Are you starting with good-quality agricultural soil or a beat-up subsoil slope?

► **Vegetation needs:** Do you need to grow the vegetation you already have on site (e.g., good pasture forages), or do you need to change the composition of plants radically (abandoned field or young forest)? The more you need to change composition, the greater animal and human impact you'll need, the greater the work and time frame involved.

► **Health needs:** This aspect should be considered under "Input-output ratio" above but is so crucial and oft-missed that I've listed it separately. I am amazed how many people endeavoring to carry out a self-reliant homestead and farm (even those doing

Kosher King meat birds (a.k.a. "the meaty ones"), along with Ancona ducks, enjoying the newly terraced area beneath a hemlock

One of the many "happy accidents" on the farm: The discovery that chickens guard sheep against fly infestation, made by grazing them together. This sheep was found with fly strike two weeks after being separated from these chickens after an entire summer of fly avoidance while cohabitating with the poultry.

full grass-fed and refusing to use grain) think little of the medicinal and veterinary needs of their animals. The need for wormer, vaccines, birthing aid, disease management, and other specialized or time-consuming medical needs of an animal vary enormously by species. This is a primary reason I view sheep as transitional for my farm and not viable at this scale or even remotely close to this scale—they need too much health maintenance inputs (simply in time alone). This plays especially into the next aspect. . . .

▶ **Time needs:** This is the most often overlooked selection consideration I run into. How much time is the animal going to need daily, yearly, and in special (or likely) circumstances? Sheep, for instance, don't need much maintenance if nothing goes wrong, but they are parasite prone, and often things *do* go wrong from a parasite standpoint. Then the time suck of such an animal really starts to hit home. Time is

your most valuable asset in a functional home/farm system, and it's limited, so choose to apply it wisely. Nothing in the system short of another human being or infrastructure emergency can suck up the kind of time that a sick, injured, or otherwise problem animal can—not a fruit tree, or a berry bush, or a vegetable bed. Animals are a big commitment, and when they have problems, the devotion needed for that part of the system goes through the roof.

Thus there is a certain social robustness needed in the human management of a human ecosystem before animals should be introduced to it. You need a reservoir of time from which to draw when an animal has a problem, and the larger the animal, the more of them you have, the kind of animal, and the health of their home all determine the consequences of such an occurrence. We have experienced a spectacular variation of this as we've kept ducks, chickens, and sheep—and their attendant needs in that order.

THE ANIMAL GENERALIZATION MYTH

It's essential to point out at the outset of this section that making generalizations about animals is about as accurate as it is about people. It's pervasive, and you read or hear such nonsense as the following:

- ► "Goats eat everything!" (Actually, goats are one of the most selective grazers in the world and more picky than most creatures when given the choice.)
- ► "That heirloom chicken breed is great for pest control."
- ► "Chickens eat fallen fruit—put them under your fruit trees."
- ► "That variety is such good foragers."

Why are such statements nonsense? Because they treat an entire species or variety as though they all act the same. Excuse my "French," but when you actually work with such animals, you see immediately that such ideas are complete bullshit. Animals are individuals, just as you and I are. Let's get that out of the way immediately because it really retards the conversation about animals and only comes about from too much reading and not enough doing.

The point here is to remember that animals act based upon not only their instinct (breeding results) but out of their training, environment, stimulus, what they've learned, and many other factors. So we need to think in as nuanced a way about animal behavior as we do about people behavior. The accurate way to think of it is "this individual duck does this" or "this particular sheep does that." And also, like people they change from year to year. Our ducks never ate mature vegetation during the growing season, just during the winter, for three years. Then in year four they attacked my large cabbage plants.

Why? Maybe because it was very dry and the slug population plummeted. Maybe. The point is just because an animal or group of animals tend to have acted in certain ways in the past is no reason to think they will always act that way. They respond to conditions just the way people do, actually probably more.

And they learn, too. Birds didn't touch our rice crop for three years, then in year four they decimated it. When asked during tours of the property, "What do you do about birds?" I'd respond, "They don't eat the rice." Then they did. That's happened a dozen times here in all animal aspects. Take our first chickens a few years ago. We put them under the orchard in June just like a good permaculturist is told to do. "They'll eat the fallen fruit!" Well, ours didn't. Why? No idea, but they didn't, and it wasn't because they weren't hungry, because they were—subsisting on almost no grain.

Here's another: "Sheep don't eat bark, only goats will." Nope. Ours followed this rule for two years and in the third year took out our oldest pear, a peach, and some other trees. They learned that bark was good. They broke the rule. Given enough time, most animals seem to make similar decisions. Our most recent meat birds, Kosher Kings—were "great for rotational grazing!" I was told. Yeah, well, not ours. They never stayed in the poultry netting. Why? No idea. They didn't fly out; they just found their way under and through the fence. The birds we had the year before did stay—Cornish giants—and those are supposed to be poor choices for ranging. A clear example of the recommended approach not working at all. That happens a lot, so you will have to experiment countless times with countless approaches to find out *what is true for you in your site.*

Ducks:
The Water-Loving, (More)
Vegetable-Friendly Chicken

Our best success on the animal front has been with ducks, bar none. Our flock has consisted of from four to fifteen birds at any given time, with nearly all females and up to two drakes (males) of four different breeds. Early on in site development, we started growing vegetables. Given our wet ground and abundant clover, that led to massive slug populations, which in turn led us to ducks. As is always the case in permaculture, when some species is a "problem," we ask, "What eats it?" Ducks eat slugs, and fast. With the introduction of ducks our slug problem disappeared (we also reduced some of the heavy garden mulching we were doing), and we began to get eggs—a nutrient-dense and nearly free yield.

The input-output balance for ducks on this piece of land in its current state is phenomenal. We raise between four and fifteen ducks a year on two to four bags of grain, with all other inputs derived from their

The gang of runner ducks moving through the farm with typical camaraderie.

free-range foraging. The grain is only used to get them through the winter, and we hope to experiment with fodder crops for the ducks that we can raise ourselves, which might include corn, amaranth, or another grain such as rice. Carol Deppe's *The Resilient Gardener* covers fodder crop raising thoroughly, and I'd recommend checking it out if this is of interest to you.

We started our duck flock three years ago in the spring by ordering the one-day-old ducklings from Murray McMurray Hatchery in the Midwestern United States. All the ducks arrived in good condition, and we have since gotten one more round from them. At the time of writing, we are waiting on another round of ducks from a supplier in Oregon that conserves and

Raising baby ducks with the help of azolla, a powerful nitrogen-fixing fern containing 24 percent protein that can double its weight every seven days

breeds Ancona ducks, which are renowned for their foraging skills and general vigor. We hope to crossbreed the ducks we have so far (Indian Runner, Welsh Harlequin, Gold Star Hybrid, and Khaki Campbell) with the Anconas to begin eventually to arrive at a duck most suited to this piece of land. Without knowing the species mix this will be, we can specify the traits of this ideal desired duck:

▶ Active forager
▶ Cold hardy and healthy; low maintenance
▶ Good layer but decent meat producer
▶ Good mothering instinct
  (so we can perpetuate the flock)
▶ Good predator awareness
▶ Flightlessness or general disinterest in flying

The ideal bird is, of course, not a reality but a goal. Much like the quest for the perfect disease-resistant, vigorous, fast-growing, huge, luscious tomato, so, too, is the world of animal breeding. In all breeding and animal selection, we must prioritize and at times compromise. For us the compromise is in the meat-producing aspect of the breed, as that is a secondary and even tertiary goal to eggs, slug reduction, and fertility production/cycling. Foraging is a crucial trait, and all breeds we've had seem very adept at the fine art of billing around through grass, leaves, and water digging out slugs, snails, and probably many things we can never figure out because we are unable to observe them closely enough.

When researching ducks you'll find that some are rated as fantastic foragers, while others are poor in this job. I think this rating system is somewhat misleading, as I've never seen a hungry duck that doesn't walk around looking for and finding food. (Same with chickens—more on that in a moment.) Much of how an animal behaves—especially its tendency for forage—is highly based on how much it is fed, the land it has access to, and how it is raised. All of our ducks have always foraged without prompting. This brings me to an important point that bears mention: animal breeding tends to be overemphasized when it comes to this or that "special" breed. While breed characteristics

Cocoa Chanel and her babies, McQuacken II, and Willa enjoying the middle pond in the rain.

are real, it is often the differences in management that are the most effective way of integrating animals into a farm. A farm that manages a "poor" breed of animal well is better off than one that manages a special heritage breed poorly.

I also never feed any of our animals as much as "the literature" typically recommends—with one notable exception: pregnant sheep. With this approach even our meat birds have never been underweight and are much healthier and more active at harvest than is typical of their breed (last year the crop was Cornish Giants).

Cold and general hardiness of any breed of animal (and plant for that matter) is of prime concern for us in this intensive and complex permaculture system. It is far too easy and common to be devoting proportionally unbalanced amounts of resources to specific components of the system, and animals, in particular, can be resource sinks (the opposite of what any component in a functioning ecosystem should be). For us, sheep fit this category to some extent—more on that below. All the ducks we've used seem extremely cold hardy—we've never had a cold-related problem in temperatures as low as –20°F in most winters, using the protection of a barn open on one side to the weather in one year and a small insulated hay-bale bunker in the other two years.

Of the couple of dozen ducks we've had on-site, we've only had one real health-related problem—a foot injury from the Welsh Harlequin named Willa Cather stepping on a sharp object (likely due to dogs chasing her). All in all, the ducks have proven to be

as maintenance-free as I can imagine an animal—no medicine, no special coddling, just a nice nest at night and predator protection, and 90 percent of the work is going out to collect the eggs in the morning.

While we do not focus on egg-laying specifically, the breeds we've used seem to be good layers, providing on average about an egg a day per female for two-thirds to three-quarters of the year. It has been impractical to tell with our current and past setups who is actually laying, so I cannot comment on the laying performance of one breed versus another.

That brings up a good point—the differences between breeds actually seem slight as a whole. All forage well, all are flightless, all go in at night without prodding, all are relatively predator aware as well. The male Indian Runner seems somewhat more predator aware than others, however, and is often seen scanning the sky for birds of prey while the others are foraging around him. Runners are also the fastest bird so are likely to get out of a predator's way faster than others, making them safer only if in a flock of slower birds, their speed still being far too slow to outrun a fox, a weasel, or a coyote. Any predator protection a duck has—which is slight to begin with—is not truly found in running, however; more so in flying. My ducks do fly if charged by a dog or scared up quickly, but the larger breeds of Harlequin and Campbell seem slower on the draw than the more nimble Runner.

All our ducks have shown a strong and surprising disinterest in flying, even though they could all fly

Willa, Amore, Luna, Pax, Bandit, and Happy

to New York State if they liked. The few times I've seen them actually take flight with more than a large bound I have been impressed with their ability and the strange lack of using it. This is helpful because we like to manage what ponds they have access to at different times of year. If they flew, no fence—especially the small 18″ high metal 2″–4″ varieties we use—would keep them out.

One notable difference between breeds does seem to be in mothering instinct, with two different Indian Runners showing greater tendency to go broody (sit on eggs) reliably. Currently, we have a chocolate runner, Cocoa Chanel, which has been tending a clutch of about a dozen eggs for nearly three weeks, with ducklings just now emerging. She is a strong mother, staying on eggs all day, only sprinting out each evening when the other ducks come into the barn to furiously forage some food for about ten minutes. She covers the clutch with hay and runs around quacking while she searches for food as if to ward off potential egg eaters from the nest she just left temporarily.

The challenges we've run into with ducks thus far are threefold: seed starting and vegetable integration, nutrification of water systems (but a good thing when in the right location such as crucial fertility for the rice production), and predator protection. Ducks love to poop in water and, indeed, seem to hit the eject button whenever they get floating. They also love to tear up aquatic plants and seem to favor plants only when in water, compared to when they are on land. For these reasons they really make a mess of water systems, which in general we like to have low in nutrients for raising fish and for swimming. This requires a calm pond with little edge disturbance where we promote wetland plants and no extra nutrients to be added to the ponds. Ducks also love to eat seed and will disturb small seedlings inadvertently with their feet.

Until this last year I always told people how great ducks were in terms of not disturbing established veggies. Well, scratch that one from the list of positives! In general this has been true, but this year we have experienced a few weeks of the opposite conditions. We had the driest, hottest summer in a long while here in Vermont, and by late July the ducks started eating full-grown cabbage leaves and messing with some other established veggies. The damage was not nearly what a chicken can do, but it made me realize, again, that simply because an animal has never done something before does not mean it won't happen in the future. I surmise that the slug population withered away this summer, and by midseason there simply was a major food shortage for the ducks. Like any other animal, they had to seek alternatives to their usual approach, which meant switching from slugs, snails, and other small creatures to veggies, in part.

We have found a need to keep them out of newly seeded veggie gardens (a good reason for using transplants), cover cropped beds, and ponds. At the same time we promote their use of small pools, which the ponds feed occasionally, for fertigation uses. (See the "Keyline Agriculture and Fertigation" section later in this chapter, on page 147.) Because of their flightless tendencies, we've had success keeping ducks out of ponds and vegetable gardens with one-foot- to eighteen-inch-high chicken wire fencing lightly staked with fiber posts, wood stakes, or small saplings sections. This fence goes in quickly, lasts a long time, and can be pulled up easily in the winter. Ducks could barrel over the fence if they really wanted to, but it simply redirects them. They also have enough land—ranging over three or so acres of swales, pools, sometimes ponds and perennial planting areas—to be content and find enough food. If we blocked many of the routes they take while free ranging throughout their day, they would certainly get used to simply trampling or hopping over the one-foot-high fence—so it's more of an encourager/discourager than anything else.

That said, such fencing takes time, is an input, and is not failproof. Ideally, an animal system would require no fencing—that seems to be a concept rather than a reality, however; I've never seen a truly free-range animal system with veggies and small perennials being produced in the same area. The ducks' range does seem to vary greatly across the course of the year and enlarges with the number of individuals present and over time—the longer the ducks occupy the site, the more comfortable they get pushing their range ever farther. Adding dogs to the site also seems to expand their comfort zone and range accordingly.

Ducks, in general, are about as perfect an animal for this farm and homestead as we could hope for, and as such are the reference point for us when thinking of how well other animals fit into the socio-ecosystem here. The only other animals we've kept for any length of time pale in comparison to the sensibility of keeping ducks in this landscape.

We have found that raising ducks outdoors in fresh air and a more natural environment from as young an age as possible is very helpful. They can be put outdoors in very warm wind-protected microclimates for brief periods of time within the first week of being born.

Cocoa Chanel, a particularly feisty Indian Runner, and her new babies that were born on the farm. We helped her hatch them by sneaking away new eggs each day after a clutch of twelve was formed.

CHICKENS

We have not kept chickens for long periods of time, only to raise ten- to fourteen-week meat birds for two years and once, a few years ago, for eggs. However, in that small period we've learned some interesting things about chickens that I've never found in books on the subject, and some of which are counter to the common wisdom. These areas of surprise for me (having read up on them a bit before diving in) have included pest management, fattening the animals for slaughter, understory management, using mobile shelters with the birds, and using animals on pasture and in the vegetable garden.

Pest Management

We've had little to no success with chickens eating "June drops" under our plum trees to reduce curculio infestation. This was a big reason we got chickens initially. Also, birds under fruit trees—great idea in theory and sometimes in practice, but some chickens can make their way into roosting in low branches and will break them if the trees are young.

Fattening

The prevailing numbers for amount of grain needed to raise a given meat bird are not accurate (big surprise). We raised Cornish Giants one year in twelve weeks on less than half a bag of grain per bird, with birds weighing in at seven pounds or more finish weight. These birds also were also decent foragers during their whole lives and mobile and walking well at slaughter time, counter to what is standard for the breed.

A major surprise with this "industrial" poor-foraging bird also became apparent. Because these birds get so big and thus are unable to range far distances, they actually have an advantage in a complex polyculture: You can keep them in certain zones without intense fencing needs. This is a huge advantage. This year we raised Kosher Kings—lithe, serious foragers. This means they need serious fencing since they cover large distances (moving from a nice pasture zone into our zone 1 veggies with ease) and they fly, easily leaping two- to three-foot fences. The Cornish Giants could be fed in areas we wanted to disturb and fertilize and would only migrate away from those zones lazily, generally staying where they were needed. The Kosher Kings found their way into the worst locations constantly and proved to be many times more work for the result.

I will probably not go back to one of the highly recommended "free rangers" with our current setup and can't imagine how they would be effective in an intense polyculture without major fencing infrastructure and time related to fence management and moving the flock. What an incredible surprise this discovery has been for

us—that a nonheirloom breed of bird known for poor foraging abilities was actually much more suited to our farm than those known to be great in free-ranging situations. Again, free ranging is an idea, not a reality, in a polyculture that's not loaded with fencing. I must note that the Cornish Giants were fine foragers—they found lots of bugs and had a great scratching ability; they just applied it to more localized areas—a great benefit for the permaculture homestead. Also, after eating a couple dozen Cornish Giants and a dozen Kosher Kings, I cannot say that one breed tasted better than the other.

Understory Management

Unlike larger grazing animals, chickens have been perfect for grazing zone 1 intensive gardens, including the common vegetable bed cleanup and under raspberries, perennial flower patches, currants, gooseberries, and so on. Tossing some grain into these areas encourages a rampant scratching, feeding, weeding, and fertilizing of these hard-to-tend zones, with the chickens doing a combination of great work for you while feeding themselves (aside from the small grain input). I would consider a free-ranging laying flock as a pretty optimal solution to the seemingly endless challenge of how to deal with grass encroachment into the basal areas of perennial plants. Chickens are the perfect size to do this job, and unlike ducks, their scratching nature and abilities, along with their insect-eating capacity, match this need perfectly.

Pasture Renovation

Here, chickens are a scratching, fertilizing, seedbed-making machine. A primary reason we are doing meat birds for our second year in a row was the sward of grass in about a one-eighth-acre area in zone 1 where last year's meat birds spent their last week of life. Despite being an area filled during house construction, composed of formerly patchy horrible soil, this area greened up into the thickest sward of clover, vetch, and grasses I've ever seen on the property. I realized come May of this year that this was the work of chickens. I decided that we couldn't *not* run meat birds again—the land-restoration service they provided was simply bar none. I think this is largely due to the seedbed-making

The "Dawn Treader"—our mobile chicken nighttime housing and weather protection. It probably would work great on lower angled, smoother land—the small percentage of land that is "agriculture land" in this region.

service they yield through the scratching effect—in combination with their manuring (with very rich droppings), the way in which they transform land to produce a super dense sward of forage is astounding.

We are running forty meat birds this year in rotational grazing via electro-netting with a movable coop for nighttime protection and shade to bring this incredibly restorative influence to the larger property acreage. According to one of our clients who has run many chickens for a number of years on formerly poor pasture, chickens also up the pH of soil quite rapidly, thus reducing or eliminating liming needs.*

In or Near Vegetable Gardens

Simply put, don't do it, unless you have fencing. We've found that chickens and vegetables are simply not

---

* Increase in pH from chicken grazing is one of the few items in this book that we have not actually tested, but we have noticed evidence of this. Additionally, a client's nearby farm reports this to be the case where they free range hundreds of birds each year in a CSA.

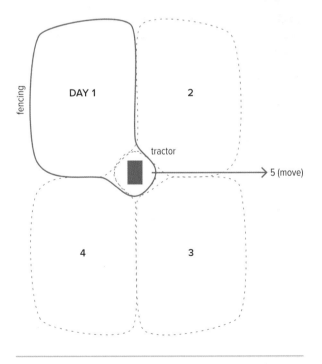

The general movement pattern for grazing chickens in our fields using the mobile coop and electro-net.

Akira and Jangles hanging out with Levon, the newly arrived lamb

compatible without a lot of work—like goats and young trees. They oppose one another and are not mutualistic. Like goats, which are the antithesis of a tree, chickens are the antithesis of vegetables. Vegetables are tender and require newly disturbed or worked soil; chickens love to scratch such areas, thus uprooting veggie starts. Chickens eat vegetables, ducks do not. So the challenge for us was simple: Either fence the chickens or fence the vegetable beds. We took the latter path in our last meat-bird cycle. This season we plan to rotate chickens for pasture renovation in electro-netting, then free range them in zone 1 for final fattening in the last week.

Dogs: Guardians and Companions

Starting four years ago with Akira, we began keeping dogs at the homestead. A primary motivation, beyond companionship, was to keep deer away from our perennial plantings in zones 2, 3, and 4. While having Akira here has proven enormously helpful in that regard, it has not been without effort, and some deer pressure remains on trees in the further reaches of the farm site. Deer are smarter than we often give them credit for, and at the end of the winter, when they are very hungry,

they venture into the farm, even as close as zone 2 areas at times, to snack on our tree crop buds. If Akira is not outside or is sleeping or has not been traversing these areas much, they will become emboldened over several weeks and do some damage.

Having dogs is one thing; distributing their presence across the landscape is another. Some dogs will do this on their own, while Akira, being a female home-oriented dog, does not. Despite being raised with many walks around the property edge, she has never become a dog to wander all but zone 1 and 2 on her own, with forays into zone 4 and 5 near the house to defecate. Due primarily to the continued deer pressure and the need to distribute a canine presence farther into the landscape, we acquired a second dog named Jangles in the spring of 2012. We're planning on his male nature to help expand the dog range on-site and hope, also, that his propensity for wandering will draw Akira farther out into the site. After almost a year, this instinctual drive is showing itself and he already ranges much further out than Akira, but still stays generally on the property. He, we hope, will also be a good mouser to keep the rodent populations down and be as alert a bird-watcher as Akira.

Sometimes Akira will chase the ducks but just briefly and for fun.

Our dogs have lived up to their breed description as being incredibly attentive, trainable, and loyal, as well as gentle and easily trained to protect and tend to livestock on the farm.* They are not, clearly, a livestock guardian dog, the values of which could be used on this farm from time to time. Their greatest single functional value beyond being blessed, loving companions is their ability to keep birds of prey away from the ducks and chickens; they are also good at general farm watching to reduce crime (they bark whenever anyone is near), and simply their presence keeps deer out of zones 2, 3, and sometimes 4. They have also proven effective at reducing groundhog damage to gardens, as their primary resting area is in the zone 1 gardens, thus providing a serious deterrent to groundhogs coming into this area. While they would be hard-pressed to catch and kill a groundhog, they give chase at any sight of them.

The most important lesson I've learned with dogs on the farm is the need for their presence and attention.

Merely having a dog guarantees nothing; the dog must be outside in the right place at the right time with the right instinct and the right training to offer the service you'd like him or her to provide. I will be working harder to raise our new male dog as more of an outdoors dog than our first female. Gratefully, her instincts are to guard, so getting some of this service has come easily. She tends to pick high spots to rest where she can see and hear much of the inner area of the site. She is attentive to the sounds ducks and chickens make and knows an alarm call when she hears one. One shrill "quack" that is beyond the volume or tone of the ducks' normal conversation will send Akira sprinting and bounding in the air toward the ducks, looking up in the sky and across the landscape to see

> **Merely having a dog guarantees nothing; the dog must be outside in the right place at the right time with the right instinct and the right training to offer the service you'd like him or her to provide.**

---

* The breed of our dogs has been deliberately left out in an effort to help preserve the breed's value, which results in large part from an absence of overbreeding, resulting from a lack of popularity of the breed.

what the fuss is about. More than once this has sent a bird of prey flapping to another location.

SHEEP

We've raised sheep for almost three years now in an effort to reduce the heaviest brush in our abused and abandoned fields while building soil and producing better forages, which will offer more yields of meat, milk, fiber, and soil in the long run. Sheep are challenging to fence (not as bad as goats but much more difficult than cows) and require electro-net or at least four strands of wire or line. Their influences have been significant but not without a lot of work in the process. From moving their electro-netting, water, battery, and charger every day or two to checking in on them and challenges associated with lambing, perennial plant damage, predator control, and a score of common diseases, sheep are a problem-prone farm animal relative to many others.

Fly-strike, parasites, foot rot, clostridium, tetanus, bloat, mastitis, prolapse, and various mineral deficiencies are just a select few of the most common health issues sheep experience, unlike cows, ducks, and chickens, which are much lower maintenance. Look up "sheep diseases," and you'll literally find multiple-pages-long A to Z lists. This makes sheep a much more difficult proposition and less functional as a whole than other animals, unless you're a sheep farmer.

This does not mean sheep aren't a viable short-term tool for land restoration and transition—here they have done good work, and our pastures, only recently laden with brush, fern, and moss, are starting to look somewhat grassy. Could this transformation be accomplished more readily and with less work? Probably, but I am not sure, as we have not tried any other ruminants. I have a feeling that cows and chickens might be a better combination and require less stringent time-intensive fencing, and we plan to experiment with them next year.

Grazing and Perennial Food Crop Integration

Developing a synergy between growing soil, animals, and plants has become perhaps the primary land optimization challenge for the WSRF. The challenge is simple but was unforeseen in the early years of this homestead farm. This challenge occurs because of a few basic processes and goals.

1. This part of the world wants to be forest; trees will grow on almost all pieces of land if left unmanaged.
2. Land wants to grow grass on the way to being forest or if it is grazed by animals.
3. We need annual and perennial plant crops in the system for food, medicine, and fuel.
4. We need animals in the system for fertility enhancement and cycling along with food, fiber, and medicine. Plants will suffer in the short and especially long term because of fertility shortages if we don't grow animals.

One of the most nutrient dense of all foods to be produced here at the homestead: sheep's milk

So therein lie some basic design problems. Having open land requires animals because simply mowing does not cycle fertility well and produces few synergies (most a one-way flow of resources). Growing perennial plant crops requires grass management in the understory because in this part of the world the areas beneath perennial plants will grow grasses and other herbaceous perennials at least until they are shaded out. So, even if you don't want to grow animals, you need to either mulch or mow the understory—grasses will slow drastically the rate of perennial plant growth.

Thus, in the early years we mulched heavily and did some scything. However, we soon realized that it was unrealistic for both labor and material sourcing reasons to keep up with such a level of mulching as the planting areas of the farm kept growing. It was impossible to get that much land mulched enough to keep the grasses back for long enough to establish healthy perennial crops. Mowing with a scythe or motorized equipment was fine for harvesting hay, but we needed to cycle fertility back into the soil—only animals do that well.

So after a few years we realized the need for animals to graze the perennial areas (on top of their easy-to-recognize need to restore degraded field). The design solution emerging from this is simple in goal: cycle understory through animals; challenging in process: keep said animals from damaging perennials. Unfortunately, this is not a simple task; most grazing animals love perennial plants as well or will at least do some damage to them while grazing an understory. Our attempted solutions to this challenge have involved sheep, goats, chickens, and ducks, along with people armed with weed whackers, scythes, and backpack sprayers. I can claim no total success at this challenge, having arrived at multiple ways to achieve the desired result but all requiring more labor than is realistic to apply over the long haul at the scale of this farm. I will now explain our attempted solutions, with their results and next steps.

Graze Away from Perennials, and Mow Understory with a Scythe

This is reliable and predictable, but the biomass must be removed and composted or used for mulch. That's fine in an area with decent soil but is not good for building soil—it's extractive of fertility. This has the disadvantage of providing singular yields via direct and heavy but healthy labor. We have done this for a few years and are reducing the number of times that we take this approach lately, but we still perform this approach in zones 1 and 2.

Graze Away from Perennials, and Mow with a String Trimmer

This is also reliable and predictable and has the advantage of being able to cut areas that are hard to access by scythe (inner corners and the like where a scythe swing or arc is impossible or impractical), along with the positive of chopping up the biomass such that it can be left in place to decompose. This has the obvious disadvantage of requiring a breakable machine running off-site inputs and negative health effects from running it. This also has the disadvantage of providing singular yields via direct and not ideal labor. We have done this for a few years and are reducing the number of times we take this approach lately, but we still perform this approach in zones 1 and 2.

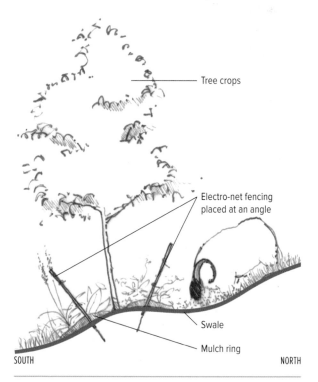

GRAZING PERENNIALS
within a swale-mound system

Tree crops

Electro-net fencing
placed at an angle

Swale

Mulch ring

SOUTH NORTH

The usual way we fence movable electro-net around our swale-planted tree and berry crops when grazing the area

LEVEL OF DISTURBANCE NECESSARY TO TRANSITION LAND FROM ONE STATE TO ANOTHER

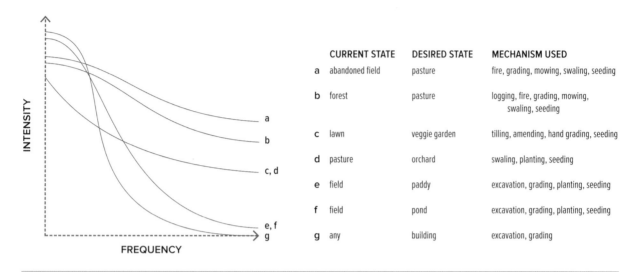

	CURRENT STATE	DESIRED STATE	MECHANISM USED
a	abandoned field	pasture	fire, grading, mowing, swaling, seeding
b	forest	pasture	logging, fire, grading, mowing, swaling, seeding
c	lawn	veggie garden	tilling, amending, hand grading, seeding
d	pasture	orchard	swaling, planting, seeding
e	field	paddy	excavation, grading, planting, seeding
f	field	pond	excavation, grading, planting, seeding
g	any	building	excavation, grading

Usually, the more degraded a piece of land in this climate, the more powerful, careful, and oftentimes frequent the disturbance necessary to heal it—very counterintuitive. Note that seeding is a form of disturbance.

Grazing Paddocks That Include Perennial Plantings

But we spray each planting with deer repellent to keep animals from browsing too hard. This works great! And it's terribly labor intensive. This could be very practical at the right scale with the right labor available but not for the WSRF as of yet.

Grazing Paddocks That Include Limited Perennial Crops

This list includes only hazelnut, seaberry, other species, and tree crops that are unpalatable enough or tall enough to avoid serious browse harm. This works well but is tricky—relying upon timing of animals in paddock, the individual habits of those in the paddock (not just species and breed), what the animals had eaten in their last paddock, and time of year, among other factors. This is a good solution if you have crops of the right size and species but requires constant vigilance, as their habits and other factors are always in flux.

For instance, our sheep didn't touch hazelnut bark or leaves last year when they were fenced in with a hazelnut hedge in August. This year we fenced them in to the same area in May; they went right for the succulent young hazelnut leaves. So do they like young hazelnut leaves or were they simply in need of that type of nourishment? Unknown as of yet, but that shows the kind of dynamic factors at play with animals—they are not predictable. And predictability becomes of new value when you return to a paddock a day later to see that they've reduced the entire year's growth of one of your crops (or worse when they girdle trees) that they weren't "supposed" to eat and never did before.

We plan to keep using this system when possible but have learned to watch them for a few minutes to see if they feel like eating the crop they aren't "supposed" to eat normally. We have found with seaberry and some other crops that they will eat the perennial but only after they've exhausted other forages and browse, so it's all about timing for movement, as is often the case but hard to do precisely. We are leaning ever more to larger plants such as seaberry and elderberry because of the need to graze nearly the entire landscape—size is a reliable way to manage browse levels: If the plant is more than five or six feet tall, animal browse (except goats, for the most part) can only get so bad if the animals are small and moved often. Some browse stress on the plants can actually be beneficial as well.

MAXIMIZING SITE AWARENESS

Keeping animals on the farm has made me acknowledge the importance of this principle. I've had sheep stuck in electro-net in the front yard that I've managed to free within minutes simply because they happened to be in areas that I look out onto from my studio at least hourly and areas that I can hear via open windows. Coyotes attacking the flock in the lowest area of my lower fields might not be heard at all and only discovered the next morning when I'm doing chores.

This principle is manifested through both the *layout of elements on a site*—for example, grazing and animal nighttime quarters in relationship to the human's sleeping quarters—as well as through the *behavior on site*; for instance, working and sleeping inside with the windows open, not closed, looking outside often, listening carefully at all times, and paying attention to what you hear. It is surprising just how many "alarm" calls you can pick up on.

As I write this I am listening to our three-week-old meat chickens and one-week-old ducklings free-ranging the zone 1 areas around the buildings here. With each day I seem to pick up on particular sounds they make (or stop making) when a hawk flies close overhead, when my new puppy gets too close and needs to be called off, or when one falls in a post hole left open in the ground. I've already likely saved more than a few lives by running outside to find both my own puppy and a neighbor's dog in the early stages of closing in on a chicken, falcons close to the flock, and one duckling eighteen inches down in a narrow post hole. Being in a constant state of listening to the site as a whole becomes instinct when keeping animals, especially baby animals. Examples of maximizing site awareness include the following:

VIA SITE DESIGN

▶ Placement of access ways, buildings, and outdoor living spaces (zone 1 areas) such that these spaces look out upon the widest area of the site possible. This often means placing zone 1 areas on slight high points and planting them in such a way as to allow view corridors.

▶ Positioning of animals and other sensitive elements (seed-starting areas) in view and listening range of zone 1.

▶ Orienting indoor spaces toward the outdoors: Kitchen sink and primary interior work space such as a desk should always look onto and, ideally, listen in on zone 1. A window in front of the kitchen sink that is low enough to see through and into the landscape beyond is a baseline advantageous design strategy on most sites—it's hard to imagine an unsuitable application of this pattern. Sleeping spaces are particularly appropriate for this design pattern, given the number of nocturnal predators.

VIA BEHAVIOR ON-SITE

▶ Avoiding excess noise in these areas so the rest of the site can be heard.

▶ Sleep with the windows open as much as you can.

▶ Look (and see!) and listen. Pay attention to what you are hearing and seeing. It is surprising how many people go through their day not actually paying attention to what their ears are picking up.

▶ Walk outside often; don't hole up indoors.

Finally, it is important to note patterns in animal behavior—where they tend to rest and walk, sizes of groups, and so on. One day while touring a group here, I saw that the sheep had gotten out of their paddock and were wandering, but only three of the four of them. I knew instantly that something was wrong but did not say so aloud. Noting that one was missing, someone in the group said, "Maybe she's just finding food somewhere else." I knew immediately that this person never spent time with sheep—they don't walk far off on their own; at least, my sheep never do. That pattern was never in evidence. After a few minutes of looking and calling for the sheep, we found it twenty feet from the others, floating upside down, dead in the pond. The sheep always stay together—that's the pattern, and if that pattern changes, it should be noted immediately.

Grazing Paddocks with Perennial Crops Fenced Out

This works great, is reliable, and avoids browse damage but is labor intensive and sometimes impossible to do, given existing perennial plant layout. Grazeability was not a design parameter for the farm when we started planting, but now it is. We made a lot of spaces that are hard to fence neatly as a result. Think hedges, not patches.

TRANSFORMING A WATER-LOSING LANDSCAPE TO A WATER-HOLDING LANDSCAPE

EXISTING HYDROLOGICAL PATTERN
Inherited from previous landowners; land had been continually cleared for grazing and views; heavy loss of topsoil with little to no infiltration and large amounts of site runoff

DESIRED HYDROLOGICAL PATTERN
Harvest and retain water on-site as much as possible; water stored uphill for potential energy and gravity-fed systems; slow, spread, and sink water into ground to build soil; establish site resiliency to retain large storm events and spread them across long periods of time

WSRF HYDROLOGICAL FEATURES
Includes ponds, keyline swales and ditches, paddies, and passive graywater systems

~150 ft

800 ft

N.T.S.

NORTH

SOUTH

The overall approach to water management at the farm always involves lengthening the distance water must travel before it leaves the site.

Grazing Paddocks Include Perennial Crops, but They Are Fenced Permanently

This works great! But it is expensive and labor intensive to set up. It's also hard to prune, harvest, mulch, check basal areas, weed, and otherwise maintain the plant. We've done this primarily with metal 2″–4″ tree cages with two stakes on either end woven through the mesh. We use four-foot-tall fences of twelve to fifteen feet in length to make circles around each tree or shrub. These reliably keep deer away as well.

The primary impediment to this is cost—such a system is often much more expensive than the plant at about $2 per stake and $12 to $15 for the metal, plus labor. This summer we will be trying to permanently fence existing plantings within paddocks by using electric polywire of either one or two strands strung between fiber posts, the way many fence cows around here. This will likely work very well but is only workable when fencing a line or hedge of plants, not individuals.

Graze Animals That Only Forage, Not Browse

The simplest solution of all. We plan to try geese for this, and it will likely work but not in brushy areas, as geese only like relatively fine, succulent forages. I am still looking for the mini cow that stands only two feet tall, can't reach up high, and hates all perennial leaves but can eat a wide variety of forage and is low maintenance and weather hardy.

Keyline Agriculture and Fertigation

Keyline activities, such as plowing and ditching, convey water from the valleys, where it collects, toward the ridges, which are the driest areas of the landscape. As discussed earlier, keyline agriculture was conceived of in the drylands of eastern Australia, largely by P. A. Yeomans, and it is especially powerful for regenerating arid lands. However, it probably has strong applications in cold climates, where the combination of soil compaction, wet-dry rhythm of weather, and steep sloping land creates water-limiting conditions where precipitation moves across, not *into*, the soil. If drought conditions become more consistent, keyline approaches will also be crucial for lower-angle landscapes.

The biological climax and dieback action described in the tall-grass grazing section above is probably the most potent soil-building tool for application across very large areas of the planet, but its effectiveness can be limited by soil compaction and especially by water availability. Soil compaction, fortunately, can be addressed by subsoil plowing via a Yeomans-style keyline plow. Other keyline methods of agriculture are aimed at water capture so that the root-dieback action can occur in lands where significant slopes cause water to run off so quickly that only minimal amounts actually enter the soil.

This water-managing aspect of keyline agriculture is foundational and addresses the inconsistencies in water availability across a field. To understand the importance of pulsing water into a landscape when the weather would not naturally do so, you should know that many landscapes are periodically in a period of drought—even places that receive forty to sixty or more inches of rain per year. If and when that rain is cyclical, as it is in most continental climates (not as it is in some maritime climates), even high-rainfall areas that never see true drought experience periodic drought lasting a couple of weeks at a time during the growing season. This short-term droughting greatly limits plant growth and root penetration, thus greatly limiting the productivity of pasture and the amount of soil building action that can occur. Pulsing water into a field at times when plants are just about to enter a limiting phase of growth because of drought (drought stress) reduces this slowing of the system.

At its fullest extent keylining involves storing water high in the landscape, usually via a pond; subsoil plowing (which loosens the soil rather than turning it, as does a moldboard plow) in a slightly downward-trending pattern from the valleys out toward the ridges of a field, bringing water from the wettest areas to the driest, and flooding the landscape after a grazing rotation, distributing the manure/bioinoculants across and into the landscape. Think of a rain or flooding event washing nourishment across and *into* the landscape.

Since roots only want to penetrate relatively loose soil where oxygen and water are present, keyline agriculture leads with the water, mechanically allowing water to enter areas of the soil not previously available. Water leads, roots follow, soil organic matter is deposited, and carbon is banked in the soil. Flood prevention, climate stabilization, farm fertility, drought resistance, crop nutrient density, and myriad other benefits result. Deep, healthy soils support resilient ecologies and culture—most other functions can only be built atop its solid foundation.

At the WSRF we employ keyline approaches by utilizing several keyline ditches (I consider a ditch a channel conveying water, while a swale holds it on true contour) that are angled at a 1 to 3 percent grade from the primary valleys toward the single primary ridge on the ten acres. These ditches catch both surface water from snowmelt and rain along with, occasionally, overflows from the ponds. These pond overflows are occasional because we do not want to soak the fields below the keyline ditches all the time, just intermittently to achieve the more moist/less moist cycle, not a saturated/dry–saturated/dry cycle typical and suboptimal. (See "water management" drawing, on page 146, for specifics on this.)

To achieve the intermittent function, we open and close the pond overflows with lumps of clayey soil placed into one side or the other of the water channels. Damming up the keyline ditch causes water to flow more directly downhill from the ponds, while damming up the "natural" downhill flow causes water to more slowly flow along a slight downhill path out toward the ridges. This simple "valve" is free and unbreakable. When the water flows toward the ridges, it is, in all but the most severe rain events, infiltrated before it actually hits the ridge. In this capacity you can think of the landscape as a water-absorbing net that is activated or can be "turned on" to full capacity by putting a shovel load of clayey soil in a water channel. When in the "on" position, the landscape is equipped to absorb all storms up to four to six inches without letting most of the water flowing into the site on the surface and onto it from the sky go off-site.

The keyline ditches at the WSRF are between three and four years old and have been built over a staggered period of time. Interestingly, they function in similar yet different ways in the two main locations they occupy—the central part of the site and the northern central edge of the site. The former location is the driest part of the site, which is underlain by ledge covered only in zero to twelve inches of silty subsoil. The latter location is perennially damp, and for eight to ten months a year, the water table is within a few inches of the surface or at the surface. When this bottom wet keyline ditch is in such a wet condition, it is simply a nutrient-distribution channel bringing excess nutrients from areas upslope—mainly the barnyard zone—across a wide area planted in perennials and growing pasture in the lower field.

In the droughty area the keyline ditch percolates water very quickly, providing what I like to call "curtain fertigation," in which water and nutrients are swept down the ditch, quickly migrating into the slope as the ditch travels the length of the ridge. This operates like a curtain drain in reverse and fertilizes the formerly dry slope as it waters it.* The productivity of that slope since the keyline ditch has gone in is astounding, having gone from an area that did not produce enough biomass to warrant a mowing, a scything, or grazing for over ten years to an area that can now be grazed at least three times in a year. By managing water in a keyline manner and through on contour swales in this same area, we've managed to increase the productivity in this dry, infertile, ledgy acre by at least thirty times its former condition.

Leach Field Cropping: Making the Most of a Faulty System

Cycling fertility optimally on-site entails that we scavenge all possible sources of organic matter and fertility with an eye particularly focused on the lowest hanging fruit. Where are the easiest, most practical sources of fertility? Cycling existing nutrients that are now "lost" or underutilized is the most effective starting point. Luckily, such nutrients are literally in the front yard. Since "plugging the leaks" is usually the most effective starting point for optimizing a system, ensuring that all fertility loops within a site are closed (cycled and not lost) is a good place to start. Landfilling food scraps, cardboard, and other organic matter; tossing animal

---

* Rate of percolation varies greatly with soil type. Ideally, you can infiltrate a gallon or two per minute for every hundred or two hundred feet of ditch if you are working with a handful of acres or so. The more land you are attempting to fertigate and drought proof, the slower infiltration you want. Our relatively tight clayey soils generally allow us to convey water a long way along a ditch. Sandy soils make it difficult to convey water and easily end up draining too much water in one location in the keyline system described above. In that case, puddling the swale in over time and tightening up some of the soils with animal action can help.

Leach fields can be some of the most productive spaces on a property—they don't lack for nutrients!

manures over the bank; and other ways by which people discard potentially valuable nutrients represent leaks in the would-be soil-building system. Plugging such leaks requires cycling nutrients on-site and turning waste into food and is the foundation for a viable and effective soil-building system.

Obvious nutrient sources are well known and utilized by many, from food scraps to garden residues (plant parts, immature fruits, and so on), lawn clippings, leaves, and other yard debris. All of these should be turned back into soil, of course, as quickly and thoroughly as possible via composting or by feeding animals. There are, however, other sources of fertility often overlooked on the homestead site, the most potent of which are the nutrients we ourselves emit on a daily basis. Human effluent can be captured by a composting toilet or humanure system, in a urine watering can, or, most commonly, by an in-ground septic system.

Aside from solid waste and heating, a typical septic system represents the greatest loss of energy and nutrients from a typical homestead. Modern septic systems are the product of well-meaning regulation gone awry. A high proportion of septic systems don't function as they are intended to, because of clogged main lines, and operate for decades in a seemingly functional state while varying amounts of untreated wastewater leak into the local watershed. If you can smell your leach field at any time across the year, it's leaking untreated black water, most likely into the nearest river. *Leach* is what these

> Aside from solid waste and heating, a typical septic system represents the greatest loss of energy and nutrients from a typical homestead.

systems were designed for, and leach they do, silently seeping valuable nutrients and water away from the site and into the watershed, where they do only harm.

Even when septic systems function as intended, they are grossly unsustainable, consuming priceless nutrients and water while producing nothing valuable on the other end. They prop up bureaucracy and industries from petroleum to plastics while simultaneously leaching value from the home economy as the People pay Industry to take our fertility from us. Indeed, the modern septic system and leach field perpetuates an anemic citizenry and empire.

Recapturing much of the concentrated nutrients and water from the home septic system simply involves growing plants in the leach field, which can be harvested, composted, and returned to the soil as fertilizer. Such "fertility farming" should be applied where nutrients are excessive in the landscape—for example, sewage treatment areas and fertilizer runoff zones. Fertility farming and bioremediation go hand in hand to counteract the industrial economy, which tends to mix nutrients and toxins together. Gardening or farming your leach field can take many forms, and we have yet to figure out the optimal ways of using these increasingly archaic systems (composting toilets and humanure piles are the most appropriate ways to harness human nutrients).

Depending on what is most needed in the system, and on the history of the leach field, one can grow either food or fertility crops on the leach field. For biomass production fertility crops such as grasses or comfrey are allowed to grow tall, then harvested with a scythe and used as a compost amendment or a mulch around vegetable beds and fruit or nut trees or as animal fodder. Food crops should be plants that do not produce on-ground or in-ground fruits such as squash or root vegetables. Unless you know the history of the field's inputs and can be sure that it contains no heavy metals or other bioaccumulated toxins, it is safest to grow only fertility crops.

If you do grow food crops, bear this in mind: Plants cannot move bacteria and other organic pathogens through their tissues, so you can't get *E. coli* from sunflower seeds or tomatoes perched above your leach field. Plants *can* bioaccumulate heavy metals (usually in their tissues and not seeds, though research is inconclusive) and other inorganic compounds. If in doubt, consider testing the plants grown in your leach field.

At the WSRF we have experimented with a variety of plants on the leach field, including squash, corn, amaranth, and sunflower. We were surprised during the first year of production in our leach field that despite amazing growth (and later realized that *because* of amazing growth) the plants never matured seed. Sunflower and corn just grew taller each week but never produced seed. We realized at the end of the season that it was likely too much nitrogen in constant supply keeping the plants in a vegetative phase constantly—great for beauty but not for eating. The squash, however, seemed to do fine and even stored pretty well, so the following year we planted only squash. This year we'll just be scything the field for biomass to add to our compost piles. Next year we'll likely do squash or amaranth again, as both of those proved viable in such constant nitrogen environments. Before producing value on your leach field, the following points should be kept in mind:

▶ Use "heavy feeders" (plants that require lots of nitrogen): Corn, squash, sunflowers, and grasses are all good selections.
▶ Don't use trees or other deep-rooting perennials, as they can, reportedly, clog up the distribution pipes and can topple over, exposing the field's inner components and causing damage.
▶ Consider forage crops such as sunflowers or corn if you keep animals.
▶ Don't cultivate or dig deeply in the leach field.
▶ Plant the field early, as high nitrogen loading in the field can significantly delay flowering. You may need to grow multiple, successive years of heavy-feeding crops before nitrogen levels are low enough (nice problem to have!) for flowering to come on time.
▶ Keeping urine out of the leach field and saving it for direct fertigation use during the growing season is one easy way to avoid the nutrient loss of the septic system. Human urine contains a near-perfect spectrum of plant nutrients (not surprising, given

the coevolution of humans and plants) that, when watered down at a ratio of 1:10 to 1:40, is ideal plant food for the vegetative stage of growth.

► Think of the leach field as a transitional and salvage resource. A composting toilet is far more regenerative and affordable over the long haul, requires little to no energy to operate, is totally maintainable, requires no heavy equipment or dump-truck loads of material to construct, and allows the use of 100 percent of its inputs. If you're building new, consider putting the $10,000 to $25,000 required by a leach field to better use, such as more insulation for your building, tools, a masonry oven, reskilling courses, and countless other useful post-oil resources.

Taken as a whole the strategies above can be combined in unlimited ways—along with others that may be more suitable to your site—to capture and cycle as much fertile, soil-making energy and materials as possible on your site. Soil fertility enhancement, along with and built upon the even more foundational basis of a healthy water system, is a precursor to resiliency over the long haul. It is also the foundation upon which a regenerative and healthy human-land system is built. These strategies need to be experimented with on each site and will emerge to synergize in unpredictable ways—always at least slightly differently from one site to another.

The good news is that the results come surely and sometimes quickly—give your efforts three growing seasons. If you don't see results within that time frame (or sometimes a lot less!), then change your approach radically. Some systems will respond more quickly than others—for example, developing a lush pasture should visibly be happening within three years, whereas soil enhancement and woody plant growth from chop-and-drop nitrogen-fixing plant guilding will likely take longer. Remember to stack the approaches you take as well: Combine them; never rely on one approach to increasing fertility if possible. Hard work, proper positioning of elements, vigilant management and some patience will no doubt yield visible (and tasteable) differences in the revitalization and productivity of your landscape.

Witnessing these and health increases also happens to be one of the most rewarding endeavors of all—like watching your child surmount some challenge that previously stymied her; seeing the system's fertility and vigor increase has often given me the acute feeling that the system is increasingly able to be more productive, resilient, and beautiful than I ever thought it would. The land's ability to perpetuate this cycle of health—truly the essence of reproduction and fertility—becomes clear. It can take you by stunning and joyful surprise.

Chapter Five
Food Crops

In the hills north of Delhi, outside Dehradun, India, there is a mix of grains and pulses that occupy an ecological and ritual niche in the landscape as they produce food and sustain the soil and ultimately the culture. In the mountains of the Andes, the Quechua people rely on guilds of the hardiest tubers (oca, ulloco, mashua, achira, bitter potato, maca, and, of course, our ubiquitous potato), which grow at altitudes of thirteen thousand feet and with proper processing can be stored for many years. From these arose the genetic diversity that allows us to grow innovative and disease-resistant potatoes worldwide.

Between the Tropic of Cancer and the Tropic of Capricorn, the Polynesians voyaged in enormous canoes navigating the expanses of the Pacific and locating small spits of land, where they settled and prospered. With

The author amid a guild of comfrey, wine cap mushrooms, clover, and dock under a canopy of plum at the homestead Photograph by Brian Mohr/EmberPhoto

them they carried twenty-four plants, known as the "canoe plants of the Polynesians."* These species were the basis of their food, medicines, building materials, dye plants, fiber plants, and plants for ceremony. They carried the potential to settle anywhere in the tropical/subtropical Pacific in their canoes, with enough plants to adapt to drought, different soils, tropical monsoons, famine, and war. The Polynesians voyaged throughout the tropics, selecting plants that were resilient, nutritious, and adaptive. They collected from no less than Africa; subcontinental India; South America; Melanesia, including New Guinea and Vanuatu; Indo-Malaysia; and Polynesia. Navigating the frontiers of climate change and peak oil, current cultures have the opportunity to learn from traditional peoples intelligent and innovative history. The foundation for responsive, biodiverse, and resilient agroecosystems that can respond to climate change and the disintegration of centralized food and energy systems can be built on the deliberate development of new totemic species.

As cultures have evolved, so have the plants they depend on for food, medicine, and fiber. Yet since the Industrial Age humans have lost countless useful plant varieties. As this diversity is lost, so are options for an attractive living future. Often called "guilds," specific groups of plants that work in unison to provide the needs of their cultural stewards were the source for much of the food and materials people needed to sustain their cultures. People tended and bred plants as if there were no line between the forest and deliberately planted areas, or forest gardens. This has been true from North America to the equatorial tropics and across the globe as well.

Forest gardening has provided a complex web of foods that provided unique and varied foods, craft materials, fibers, psychotropics for ceremony, dyes, and building materials. Human needs were provided for in part by forest gardens; therefore, less land needed to be cleared for annual grain crops. The tending of these forest gardens defined the culture and in some cases maintained the living matrix of ecosystems and agroecosystems that supported life. Within each culture, totemic species were used that were honored and respected for their role as the staff of life that ensured survival in an unknown and capricious world. Certain perennial plants have proved so successful at our research farm that it is worth covering their habits, yields, and interactions in the system in particular detail.

Perennial Plants and Resiliency

Perennial plants are growing to become the base load engines of our regenerative land system at the Whole Systems Research Farm. These permanent producers only need to be established rarely—once every couple of decades to every century or three, depending on species—yet they can produce annually while building soil health and requiring little or no fertility inputs. Because perennials are established only once per decade or century compared to annuals, which must be established once every year, they are able to put more energy into larger seed yields relative to annuals, which must spend a much higher proportion of their lifetime simply becoming established.[†]

In addition, the roots of perennial plants inhabit deeper layers of the subsoil horizon with each passing year. A landscape covered in a mantle of perennial plants has the capacity to transform ever more subsoil (mineral soil) into topsoil (organic-matter-rich material) with each passing year. A landscape of annual plants functions only to a very shallow layer—typically, the top six to eighteen inches of the earth's surface, depending on soil quality, aridity, species, and other factors. Many perennial plants penetrate two, three, six, twelve feet or even farther into the earth. This rooting capacity brings organic matter, water, and biological activity—the basis for organic soil formation into the subsoil. When we harness this mechanism, perennial

---

* I am especially grateful to Chris Shanks of Project Bona Fide in Nicaragua for highlighting the global significance of Polynesian plant use and for teaching me an enormous amount relating to the vast and nuanced ethnobotany of first peoples in general.

† While this is an important general pattern, it is crucial to point out that some annuals, such as many salad greens, whose entire plant body is edible do have tremendously high ratios of edible/harvestable energy relative to time required for establishment.

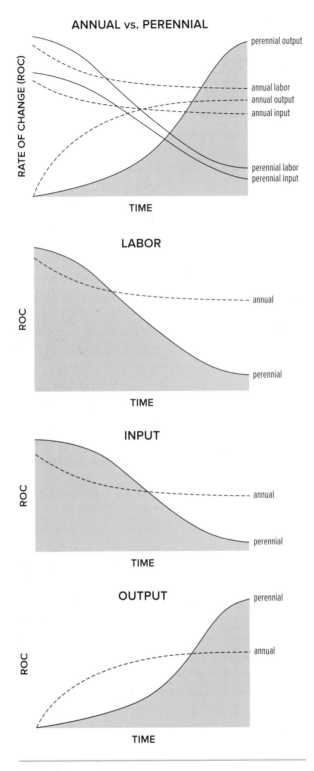

ANNUAL vs. PERENNIAL

RATE OF CHANGE (ROC)

perennial output

annual labor
annual output
annual input

perennial labor
perennial input

TIME

LABOR

ROC

annual

perennial

TIME

INPUT

ROC

annual

perennial

TIME

OUTPUT

perennial

ROC

annual

TIME

Annuals and perennials often entail inverse labor input and output relationships while perennial systems continually improve performance over time.

plants allow us to farm more deeply the earth beneath our feet, thus doubling, tripling, or more the amount of mineral and other resources we are able to draw on in our job of growing value from the intersection of sun, soil, water, and living organisms. Not surprisingly, perennial-based systems such as a woodland/savannah system with three dimensions of crops, from grasses to shrubs to trees, and grazing in the understory typically captures between three and seven times the amount of solar energy as a field of annual crops.*

Their ability to grow deeply into the soil horizon allows perennials another advantage, which is often their most crucial advantage as the climate becomes more variable: drought resistance. Deeper roots mean a much higher ability to mine deeper water tables and moisture that evaporates from the surface downward. Many arid areas of the world have been made more brittle and even created deserts because of a lack of adequate and appropriate perennial plants and disturbance mechanisms such as farming, which has often stocked the ecosystem with annuals. In addition to drought resistance, perennials offer a high degree of flood resilience: They can often withstand seasonal inundation (if not exposed to high-flow velocities), whereas many annuals die or are rendered unusable (due to contamination) if they go underwater.

Equally important is perennials' ability to grow tall above the ground, allowing us to farm additional vertical space into the atmosphere. This vertical tendency offers a complementary value to the drought resistance achieved by their deep-rooting ability: Perennials actually harvest and increase the moisture available in a site and hold moisture in via shading. The ability of perennials, most notably trees, to pull moisture out of humidity and to actually promote cloud formation over a landscape through evapotranspiration and structural texturing is why a forested area always receives more rainfall than the same landscape in the same region without forest. This is why humanity has made many deserts through deforestation.

---

* Diverse perennial systems capturing three to seven times the solar energy of annual cropping systems has been found through various studies and is well documented especially by Mark Shepard at his New Forest Farm in Wisconsin.

In much of the world, it can be clearly said that "losing our trees means losing our water." Losing trees, of course, also means losing the buffering effect on massive rainfalls and flooding as tree leaves reduce the erosive, percussive force of raindrops as they slow, spread, and sink surface water rather than promoting sheet flow off the landscape and, consequently, disastrous flooding. Haiti is one of the best examples the world has to offer for deforestation begetting a wide range of ecological and social systems failure: With the disappearance of tree cover comes drought, flood, and massive soil loss. With those catastrophes come social system dysfunction—the history of humans abusing land and ending up in a stricken society seems to bear this out repeatedly.

Along with moisture-harvesting abilities are other microclimate-buffering capacities of a plant—or wall of plants—including the ability to reduce drying and stressful winds; slow and deposit snowfall, which is beneficial to tree, grass, and soil health; and serve as sun traps, increasing the radiant heat available in a landscape to promote ripening of a crop and offer season-extended outdoor use areas for people. Related to the vertical ability of perennials is the sheer size and biomass potential of certain perennials. You can't turn any annual crop into fuel or structural material without lots of processing to combine thousands of smaller plants into something offering structural or energy yields. Trees, on the other hand, offer such yield in their raw form—a woodland is a living lumberyard that needs minimal processing to be useful. A forest is also a direct source of fuel that can also be of value with little processing and energy expenditure. You simply can't get such yields with an annual plant.

Finally, there is an increasingly important advantage to perennials that is just becoming better understood: Plants tend to accumulate toxins most acutely in their vegetative tissue, not in their seeds. Most perennials

> **Perennials actually harvest and increase the moisture available in a site and hold moisture in via shading.**

Reliance—a hardy peach at the Whole Systems Research Farm, three years after planting

offer us an edible yield of seeds, nuts, or fruit, which accumulate less toxic buildup of metals and inorganic chemicals. This tendency also complements perennials' tendency to more densely accumulate nutrients due in large part to their inhabiting a wider spectrum of the soil horizon and accessing a greater range of nutrients as a result, thus being able to make those nutrients available to people in the form of food.

To summarize, the reasons perennial crop plants are a crucial and foundational part of a cold-climate landscape are diverse and include the following:

▶ **High return on investment (ROI)** in energy, time, and materials, as they are only planted a few times per century. This stems from a generally high

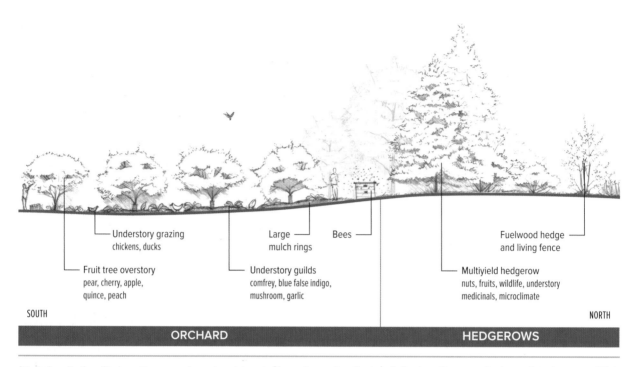

A typical application of tree- and berry-cropping systems integrated for maximum microclimate and other benefits across a landscape Illustration courtesy of Whole Systems Design, LLC

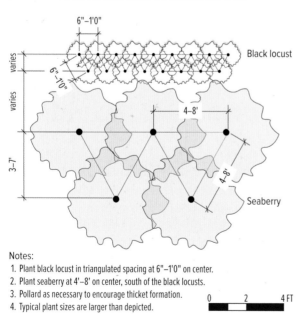

LIVING SECURITY FENCE DETAIL

Notes:
1. Plant black locust in triangulated spacing at 6"–1'0" on center.
2. Plant seaberry at 4'–8' on center, south of the black locusts.
3. Pollard as necessary to encourage thicket formation.
4. Typical plant sizes are larger than depicted.

Black locust affords some of the fiercest living fence one can grow in the cold climate region of North America—probably rivaled only by the slower-growing hawthorne. Illustration courtesy of Whole Systems Design, LLC

growth-to-establishment ratio. This allows more energy to go into yields (reproduction, seeds, fruit) and less, proportionally compared to an annual, into the organism's establishing itself.

► **Deep soil penetration:** Allowing us to build more soil and access more nutrients and water.
► **Climate resilience—drought and flood:** A greater ability to both avoid and bounce back from climate stress, including heat, lack of water, and inundation.
► **Microclimate enhancement:** Moisture harvesting and holding, windbreak, snow fencing, sun traps.
► **Structural yields:** Timber, fencing, fuel.
► **Human health enhancement:** Toxicity avoidance and nutrient density—seeds and fruit tend to accumulate toxins less acutely than vegetative tissue, and perennials access a wider range of soil nutrients than annuals do.

PERENNIAL CROP DISADVANTAGES

► **Climate:** Many yields from a perennial crop depend upon flower survival, which is becoming an ever greater challenge as global weirding produces early-season heat waves, often causing perennials to flower ahead of their normal period, damaging or destroying those flowers—and the fruit or nut crop they would yield (not to mention harming already stressed pollinators).

▶ **Breeding cycles/genetic agility:** Some annuals allow multiple seed-production cycles per season, which allows faster breeding of more adaptive strains of plants that can increase their fitness as the climate, pests, and atmospheric toxicity and other conditions shift.

▶ **Slow yield:** Annuals can give us a large yield from seed within months, not years or even decades of seeding. Perennials by nature require a longer lead time to get established and offer yields. The one-two perennial-annual punch of growing a lot of annuals while simultaneously planting perennials is key during site establishment.

▶ **Space:** Simply requiring more room in which to grow can be a downside for people living in urban or dense suburban spaces.

At the Whole Systems Research Farm, many of the perennial cropping systems are beginning to reach maturity, including species such as apples, pears, mulberry, plum, peach, hazelnut, elderberry, seaberry, blueberry, honeyberry, aronia berry, grape, *rubus* species, and many other berries in particular. These species are fairing very well, and when planted on high points and given enough care (mostly consisting of mulching and deer protection), they perform as one would expect. Many of our other tree crops are very slow to establish—these include almost all tree nuts except bur oak. These systems take significant vigilance to keep the deer off for the many years the individuals remain below browse line—up to five years for many of these trees in poor soil and/or zone 3 or 4.

The biggest lessons we've learned on the perennial woody cropping front—aside from the need to graze beneath them to suppress grass and fertilize (as expanded upon greatly in chapter four)—have to do with (1) simplifying layouts and access, (2) ensuring protection from deer, and (3) ensuring fertility. The last two of these considerations can be summarized fairly

simply: Without a lot of labor on hand, caring for plants in zone 3 or 4 is a ton of work, and we don't often keep up with it to the extent that would be optimal. "Optimal" here means mulching to ensure growth and keeping deer away. We have learned the hard way—after losing dozens of trees over the years—that it's simply not worth planting a tree in zone 3 or 4 without heavy mulching with manure bedding and woodchips for at least the first two years (better if done for three) and a proper deer fence.

The first of these considerations—layouts and access—refers to the need to continually have decent access to the plants in the ground for care to happen. It's very easy to "plant yourself out" and over time make it harder and harder to access earlier years'

Fifty black locust seedlings—five to fifteen cords of fuelwood within about twenty years' time. This tree offers by far the fastest return on investment in this climate when it comes to transforming sunshine into usable fuel.

TREE PLANTING DETAIL
for bare root trees in early spring

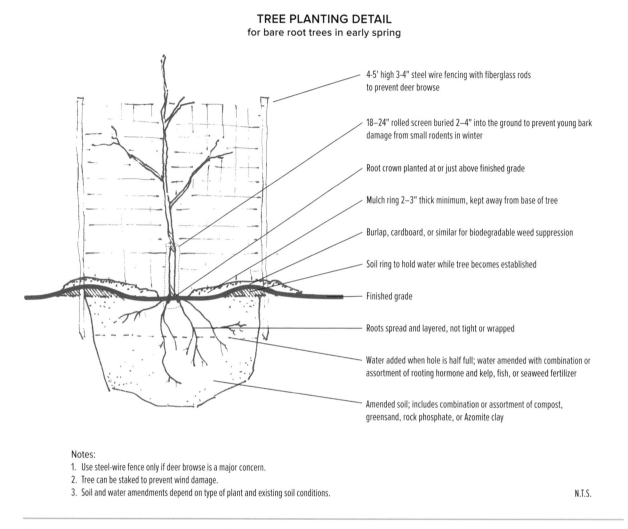

4-5' high 3-4" steel wire fencing with fiberglass rods to prevent deer browse

18–24" rolled screen buried 2–4" into the ground to prevent young bark damage from small rodents in winter

Root crown planted at or just above finished grade

Mulch ring 2–3" thick minimum, kept away from base of tree

Burlap, cardboard, or similar for biodegradable weed suppression

Soil ring to hold water while tree becomes established

Finished grade

Roots spread and layered, not tight or wrapped

Water added when hole is half full; water amended with combination or assortment of rooting hormone and kelp, fish, or seaweed fertilizer

Amended soil; includes combination or assortment of compost, greensand, rock phosphate, or Azomite clay

Notes:
1. Use steel-wire fence only if deer browse is a major concern.
2. Tree can be staked to prevent wind damage.
3. Soil and water amendments depend on type of plant and existing soil conditions.

N.T.S.

Our typical method, which has evolved as the most effective, simplest, and most cost effective approach we've used for planting when significant deer protection is needed. When a plant allows for a five-foot tree tube, such as a nut tree, we use only the tube.

plantings with a cart or tractor, or sometimes even by foot. The need to graze near most trees and shrubs also plays in the layout considerations hugely. If I could redo the farm, I would vastly simplify the planting patterns and try wherever possible to lay out everything on contour in hedges. Even larger trees like apples and mulberries could work well in a hedge, I think.

Hedging is not only beneficial for all the obvious reasons of creating corridors and microclimate effects, but it is especially attractive from a "fence-ability" perspective. I can weave the electro-net (and probably future polywire for cows) around hedges with relative ease. The patch pattern of a tree here, a shrub there, and "Oh, wait, there's another one here!" is a downright pain in the butt when it comes to many forms of management, especially fencing. In permaculture we seek the complex route—yet the most complex systems become almost unmanageable.

On some fronts, and theoretically, unmanageability might be the desired state: the archetypal food forest, maintaining itself—all we need to do once it's established is walk through and forage. I fell in love with this idea, and it got me into permaculture fifteen years ago as a college student. But, the fact is, I have yet to see a system even remotely close to this idea on a larger scale than the small backyard of maybe a quarter acre. I don't mean to say that this scale is unimportant, but

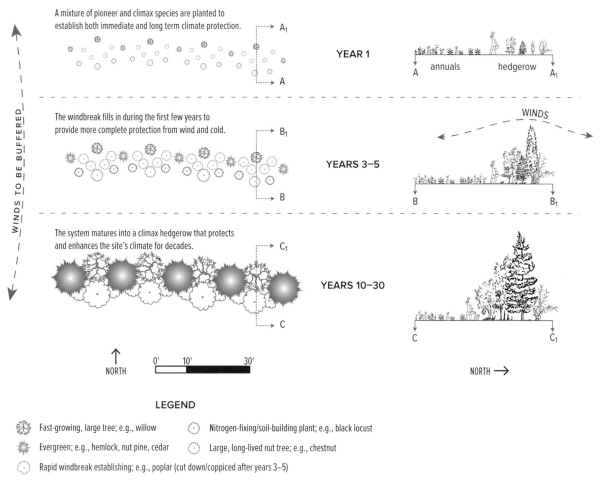

WINDBREAK PLANTING LAYOUT

A mixture of pioneer and climax species are planted to establish both immediate and long term climate protection.

YEAR 1

annuals   hedgerow

The windbreak fills in during the first few years to provide more complete protection from wind and cold.

YEARS 3–5

WINDS

The system matures into a climax hedgerow that protects and enhances the site's climate for decades.

YEARS 10–30

WINDS TO BE BUFFERED

NORTH   0' 10' 30'

NORTH →

LEGEND

Fast-growing, large tree; e.g., willow

Nitrogen-fixing/soil-building plant; e.g., black locust

Evergreen; e.g., hemlock, nut pine, cedar

Large, long-lived nut tree; e.g., chestnut

Rapid windbreak establishing; e.g., poplar (cut down/coppiced after years 3–5)

An example of a microclimate-enhancing configuration that is useful on nearly any site Illustration courtesy of Whole Systems Design, LLC

I happen to live on ten acres and need to figure out how to manage those ten acres with a high degree of restoration, productivity, and resiliency.

Engaging in that has made me realize that at many points the system needs to be simplified to function optimally. I started my first fruit and nut trees ten years ago—black walnuts, plums, and pears in fairly sophisticated guilds, including comfrey, baptisia, mushrooms, clover, and other plants. Over the years such guilds, without massive hand-tending, are simplified—the baptisia gets scythed by accident because the grasses were overtaking the tree and needed to be mowed. The mushrooms died back because the wood chips were

not kept up with. Sophisticated guilds are a great idea if you have five or twenty trees to take care of. But if you have hundreds, there needs to be management simplification so grass does not overtake your trees.

For us the answer to that has become plainly obvious: grazing. Without grazing the answer would be lots of human labor (and you'd still be short nitrogen). This may be very climate specific, but it's the hard, dark truth of "food forest" development that I have bumped into in this region. Time and time again I've seen far more food per area come out of a well-managed thousand-square-foot garden than a poorly run fifty-thousand-square-foot farm. My experience has shown me that manageability

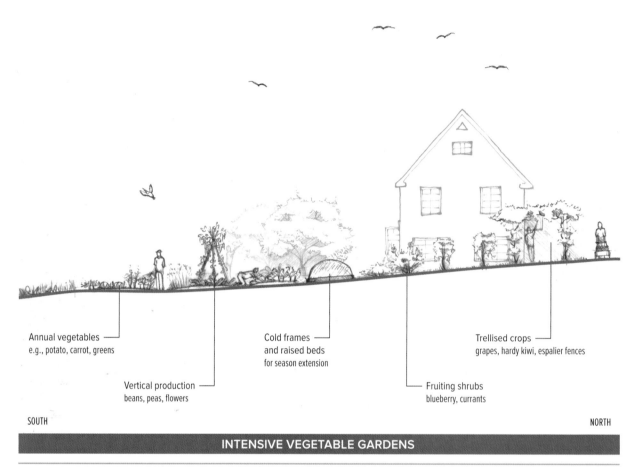

Annual vegetables
e.g., potato, carrot, greens

Vertical production
beans, peas, flowers

Cold frames
and raised beds
for season extension

Fruiting shrubs
blueberry, currants

Trellised crops
grapes, hardy kiwi, espalier fences

SOUTH

NORTH

INTENSIVE VEGETABLE GARDENS

A microclimate-enhancing layout of garden spaces increasing in height to the north/right of the drawing. This pattern should be consistent across all cold-climate regions of the world.

is key, and labor needs are the primary limiting factor to that need's being met. Therefore, the layout consideration is of prime importance when positioning tree crops in the landscape.

Staple Crops: Paddy Rice, Meat, Eggs, Fruits, and Nuts

I write this chapter in late February, a fitting time to think about the crucial role of those special crops that you can depend upon from harvest to harvest. Those of us inhabiting very cold climates (USDA hardiness zones 5, 4, 3, 2) are attuned to just how short a production season it is up here. We need to grow twelve months of food, fuel, and medicine in just three to four months. Sustaining 100 percent of the year on what is produced in roughly 30 percent of that year represents a primary challenge to inhabiting a cold climate, and much of the entire homestead and farm design hinges on this basic imperative. This challenge demands that we couple high productivity during the short growing season with reliable storage. If we do that well—which takes a number of years to establish as a pattern—we can achieve a highly self-reliant resource relationship.

> Sustaining 100 percent of the year on what is produced in roughly 30 percent of that year represents a primary challenge to inhabiting a cold climate.

The list of reliable staples is not long relative to the myriad foods we can cultivate. Staples form the foundation, roof, and walls of our food supply; without them everything inside the building falls apart. Of course, other nutrients are crucial, but a solid bedrock of storable calorie crops comes before all else. Without it everything else is icing on a nonexistent cake. To be on this short list of robust calorie crops, a food must meet the following criteria:

▸ **Storable for at least half a year,** at least until late spring the following season, when the cycle of production begins again.

▸ **Productive and processable at scale,** to grow enough to meaningfully feed those raising the crops; for example, a potato versus a tomato. You can and store enough potatoes to easily get you into spring if they are grown and stored well, but to do so with tomatoes requires an enormous input of either time or energy in canning. The crop must represent an efficient return on time and material investment into the crop; this input-output ratio should be one of the primary determinants as to whether you should grow the crop.

▸ **Nutritive and edible in quantity,** to keep one healthy *and* sane for months on end. You can eat potatoes for two meals a day or even three if you get creative—they cook up in so many diverse ways, and they are relatively nourishing. Try eating daikon radish for two or three meals a day. Although daikon can keep as well as potato, they don't cut the mustard in this respect.

▸ **Light or restorative on the soil and/or low in fertility needs;** for example, dry beans. A legume that fixes nitrogen into the soil, can store well, and can be grown in large quantity, dry beans in their culinary aspects lack only the same level of use as potatoes—though they can be and are eaten for multiple meals a day across large areas of the world. Rice is also a perfect example of a crop that can be produced perpetually—even more so than beans, in all likelihood.

▸ **Tradeable/desirable by many:** If you and your family like it, but no one else does, the crop has limited potential when you end up with too much to

Potatoes cropped in terraces amid the rice paddies at the Whole Systems Research Farm

A hearty meal in March made mostly with staples harvested and stored from the previous fall

consume in a year. When you are planting a staple crop to depend on, you need to plant more than you might need because of possible pest and disease pressure. So if you can't trade or sell the surplus you will inevitably end up with in a good year, it's a sub-optimal situation. Turnips might fit this category in some areas along with rutabagas or less usual crops that neighbors don't know enough about to want. As resiliency-minded gardener-farmers, we must always keep in mind what we're going to do with excess crop, as we must always aim for production of a surplus, given the vagaries of climate, pests, and innumerable variables outside our control.

► **Genetically diverse:** The proliferation of genetically modified organisms in recent years and the continual monoculturing of the global food supply make this an increasingly important criterion. A certain crop, such as a specific potato variety, may fit all the above criteria, but say it's being grown at large scale by tens of thousands of growers in your area of the country. What if a genetically engineered variety of it is being sown across the country? As this becomes increasingly likely as the years go by, we need to consider what species and varieties are being affected by this and seek always to cultivate those that fall outside the realm of genetic commodification. The Irish Potato famine is a perfect example of the lack of resiliency resulting from losing the diversity of what was once a reliable staple crop.

1816, THE YEAR WITHOUT A SUMMER

Resiliency planning, development, and operations require planning for worst-case scenarios, but we humans tend to have short memories indeed. We look back at the past with rose-colored glasses, perhaps a necessary psychological mechanism, but physically, it's certainly a maladaptive one.

"Those who forget the past are bound to repeat its mistakes." The Year Without a Summer is a particularly informative planning example, as it highlights the effects of a wholly natural chain of events: volcanic eruptions. The year 1816 saw a killing frost every month of the year. People died in a snowstorm in July in northern Vermont. And this was all before human beings started really tampering with the climate in earnest. This was merely the result of a series of volcanic eruptions occurring the year before in 1815, many of them on the opposite side of the planet. A year without a summer is not "likely" to happen again—it's guaranteed to. It's merely a question of when, not if.

As with any decision in life, we can plan for inevitable events, or we can ignore them and pretend that they won't happen again. While the latter is certainly the modus operandi of the Modern Age, it's not a terribly resilient way of engaging the future. So how to plan for the inevitable vagaries of Earth's dynamic, plate-tectonic-driven influences (not to mention any other change agents such as global trade resource supply disruption)? The following strategies serve as a general overview. Write them off as tinfoil-hat approaches at your own peril:

► Store months of food, preferably a year's supply or more. It's easy and not that expensive, and in the end it saves you money (food costs are always increasing).

► Diversify, diversify, diversify. Note the crops that failed in 1816 in New England: vegetables and grains, annuals, fruits and nuts. Pastures, on the other hand, probably had a decent year, given that moisture levels remained high (evaporation stayed low), and the animals grazing on such pastures can handle frost easily enough. Grazing systems may have actually benefited from such a catastrophic year. That's the power of diversity. Sure, you won't live on meat and milk alone; that's where the stored food comes in. Grains and beans from the year—or five years before—added to the animal-based diet would do wonders to round out the survivability of a year like 1816. Add some greens to the mix—kale, chard, arugula, and a host of other cold-hardy greens don't care about a little frost. Now you have not only a survivable but a thriveable way to get through a particularly dynamic year like 1816. Whether it happens again in 2014 or 2114, it really doesn't matter. You're prepared. Your surplus can be sold or traded for value. You are a resource to your community if a food disruption only lasts a week or the entire growing season.

Rice maturing in the homestead paddies in early September

Paddy Rice

"Rice is the most important grain with regard to human nutrition and caloric intake, providing more than one fifth of the calories consumed worldwide by the human species."

—**BRUCE D. SMITH**,
*The Emergence of Agriculture**

Rice may rise to the top of the list of staple crops in this region for a wide variety of reasons—it's a top performer in all of the above categories, and this should be no surprise: Rice feeds much of the world. Rice cultivation represents the largest single crop in calories harvested across the globe but ranks third in amount of land area devoted to a crop. Why? For one, a rice paddy

is a fertility trap. Paddy rice is the only grain that humans have managed to grow successfully in the same location century upon century without destroying the land's (and water's) ability to produce the crop. This is made possible by a water- and gravity-based nutrient distribution system rather than a mechanical system in a terrestrial crop situation. Rice is nutritious enough to live well on with the addition of some protein sources and other fresh vegetable sources for the nutrients it lacks. It is also adaptable enough culinary-wise to be able to represent a large part of any meal of the day. Rice as a grain outcompetes all annually cropped foods except dry beans and other grains in its ability to be stored for many years under the right conditions. Rice is so storable, in fact, that viable individual grains have been found centuries after they were harvested.

At the WSRF we are now in our fifth year of rice production, the first year of which was performed in

---

* Bruce D. Smith, *The Emergence of Agriculture* (New York: Scientific American Library, A Division of HPHLP, 1998).

five-gallon buckets growing our seed crop for the following two years of paddy production. We are growing rice on what comprises most of Vermont and indeed much of the planet's surface—sloping land with very poor soils. Our challenge is no different from what inhabitants of hill and mountain country have faced for millennia: to grow a climate-durable, reliable staple crop from year to year, century to century, on the same plot of land without diminishing that land's ability to keep producing. For the most part this experiment has failed, and societies have been forced to move on to new lands from generation to generation. Where it has succeeded it's done so by employing several principles:

▶ Slow and infiltrate surface water (usually achieved with swales, terraces, and paddies).
▶ Grow on contour, never shunt water downhill.
▶ Grow the most reliable vigorous genetics possible.
▶ Grow intensively, and always use biological labor instead of technical inputs.
▶ Capture as much nutrient as possible, and return all nutrients back into the system.

Of all the examples of proven successful approaches to hillside staple farming—from the potato culture of the high Andes to the chestnut-swine dehesa system of the Iberian peninsula and the terraced paddy rice systems of northern Asia, it is the rice-producing paddy approach that we have decided offers us the most immediate yields and application possibilities on Vermont's hillsides. I must note that we have nutteries planted of chestnut, oak, walnut, hazelnut, and other staples, but the yields of those systems will always lag far behind (in time frame of bearing) those of an annual, grain-based crop. I see these tree crops as an essential backdrop and foundation of a highly productive, more intensive cropping system.

Our rice production system is fairly simple and makes use of the above principles at many intersections. The system consists of five paddies, stacked directly above one another, with water fed to the rice via gravity from a holding pond located about ten vertical feet above the top paddy. This pond collects surface flow overland and is harvested via gutters from the house and farm buildings. The pond then harvests sun, which serves to warm the water, aiding in rice growth, and serves as a storage mechanism between rain events; this pond alone could water the paddies for the entire summer if needed even if no rains came, assuming the winter snowmelt had filled the pond completely. Given our very wet climate, our pond in the past five years has been 100 percent full or nearly full almost every week of the season, so it's a good bet as a water source for the rice.

The water leaving this pond then flows to a small pool—the fertigation input pool—to which ducks are allowed access and other nutrients, such as chicken house bedding and human urine, are introduced. From this manure tea pool, we have a source of warmed, nutrient-rich water located above the paddies that is then fed via three-quarter-inch poly tubing as needed to the rice paddies below. There are two other ponds-pool pairings that also feed into this system in series farther up the slope as well.

Taken as a whole this form of combining irrigation with fertilization (fertigation), combined with growing in a detention basin (paddy), is for us the crucially important aspect to rice production and why it can be maintained perpetually from year to year. By combining fertigation with detention-based growing, we have a system that can capture easily most if not all of the nutrients flowing across the homestead and utilize them in a cropping system where the nutrients are totally captured.

Our rice paddies have only overflowed on rare occasions (in tropical storms), so we don't lose nutrients—they all go into the rice plants and paddy soil. On the rare events when the paddies do overflow, the water is shunted into a series of back-and-forth swales, the tops of which are cropped in elderberry, pear, apple, mulberry, oak, hickory, walnut, chestnut, and many other tree crops. As the water flows through this lower field of swales, it is infiltrated and captured by soil or roots. During the growing season water never flows off-site via the surface of our landscape except in four- to five-inch rain events or larger.

At this stage in our rice-growing experiment, it is clear that rice can be produced intensively and successfully in this very cold (zone 4) climate with no off-site

THE FERTIGATED LANDSCAPE

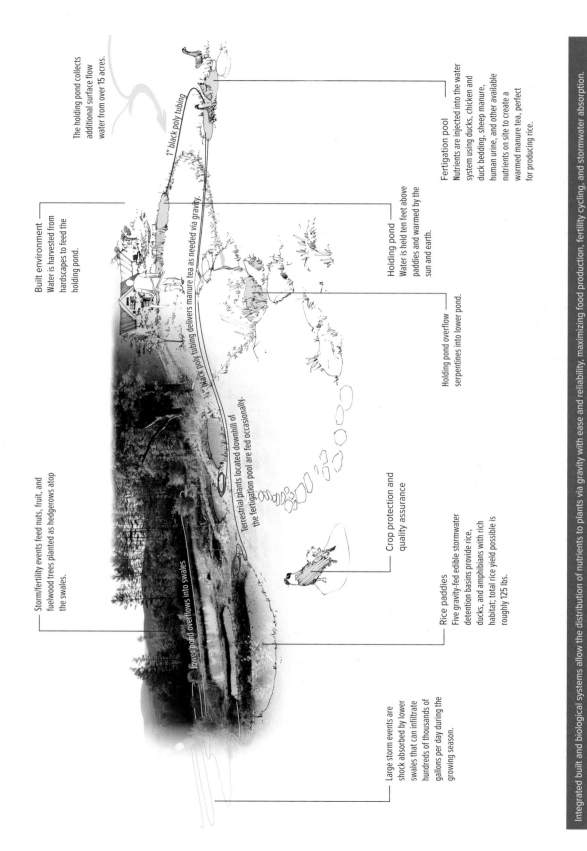

The holding pond collects additional surface flow water from over 15 acres.

1" black poly tubing

Built environment
Water is harvested from hardscapes to feed the holding pond.

3/4" black poly tubing delivers manure tea as needed via gravity.

Terrestrial plants located downhill of the fertigation pool are fed occasionally.

Holding pond
Water is held ten feet above paddies and warmed by the sun and earth.

Fertigation pool
Nutrients are injected into the water system using ducks, chicken and duck bedding, sheep manure, human urine, and other available nutrients on site to create a warmed manure tea, perfect for producing rice.

Holding pond overflow serpentines into lower pond.

Storm/fertility events feed nuts, fruit, and fuelwood trees planted as hedgerows atop the swales.

lower pond overflows into swales

Crop protection and quality assurance

Rice paddies
Five gravity-fed edible stormwater detention basins provide rice, ducks, and amphibians with rich habitat; total rice yield possible is roughly 125 lbs.

Large storm events are shock absorbed by lower swales that can infiltrate hundreds of thousands of gallons per day during the growing season.

Integrated built and biological systems allow the distribution of nutrients to plants via gravity with ease and reliability, maximizing food production, fertility cycling, and stormwater absorption.

Rice plants growing quickly but yet to head out (put on seed) in early summer.

inputs. It is a well-suited fertility-cycling crop that can handle extremes of both drought and flood with ease, given the nature of its culture. The challenges to its production that we are facing include weed control—aquatic weeds are moving into the system and reducing yields—and we have also experienced severe wild bird and domestic (our own beloved ducks) damage. We plan to introduce our ducks earlier into the paddies next year to reduce the weed pressure; ducks find most aquatic plants palatable but not rice with its high silica content. We also will be netting the crop or otherwise deterring wild birds much earlier in the season next year.

Locating a Paddy

"I have a really wet area—can I grow rice there?!" I often am asked this excitedly after presentations and in workshops. The answer is—you guessed it—it *depends*. You don't need a wet area to grow rice—you just need to be able to get water to the paddy—ideally, via gravity. A paddy can be made in an existing wet area, but it's challenging—moving muck is no fun, and it's hard to make a durable land shape in mud. If an area is wet, it will need to be drained well enough to dig and sculpt before a paddy can be constructed. But wetness is only one factor in growing rice—the other primary criteria for siting a paddy is sunshine: You want as much as

possible and need a full day, similar to what you'd want for hot-loving plants such as tomatoes. If you have an area of land that is very sunny, you can get water to it, and the land will allow you to shape or sculpt it (it's not ledgy and bouldery), it is time to consider other factors including steepness, soil type, access to zone, fertility sources, and pest protection.

You wouldn't want to make your first paddy on land that is steeper than a 30 to 40 percent grade, as it gets tricky to shape land at that angle or greater. If, however, you're gung ho about it and up for a challenge, paddies have been made on an 80 to 90 percent grade or more in Asia. Just remember that they knew things we don't. I'd recommend a nice mellow sloping area of a 10 to 20 percent grade for your first paddy project—some slope is nice to have, though not essential.

The soil type of a possible paddy location must be considered because a truly sandy soil will make paddy management difficult—but probably not impossible. You want to be able to keep paddies full at times to suppress terrestrial weeds. If the soil drains rapidly, this can be a challenge and will require a lot of water over time. However, it is possible to seal paddies over time just like ponds—even free-draining silts tend to "puddle in" over time, especially if they are disturbed while inundated often. We have made a paddy for a client on very well drained land and used a few bags of bentonite clay to help seal it up. The paddy still drains more quickly than ideal but seems to be silting in and sealing up over time.

Pigs would probably be effective at helping this process—just like gleying a pond. They would want to, of course, be introduced to the paddy right after construction and before planting. Pigs as well as other animals could perform key roles in the paddies after the harvest as well, but we have not experimented with this—except for the ducks, which don't disturb the paddy too deeply, although they do help to level it out and silt it up over time. The sheep show no interest in even touching the paddy but happily graze to the edge. A nice heavy soil like a silty clay makes an ideal paddy.

Ensuring that your paddy is fairly accessible is a good idea; consider rice as you do your vegetable garden: It may need nearly as much care at times of the year, and it's simply really cool to be around it—amazing, in fact (it's a water garden)—so keeping it close at hand is a good idea. This will pay off in spades when it comes time to fending off birds or other seedeaters. Remember, too, that rice is a grass and is happy with about as much fertility as you can reasonably put to it, so try to locate the paddy in an area downhill from fertility. This is especially important if the operation takes on significant scale and a lot of material needs to be moved.

Pest protection for rice involves—at least for us— keeping birds away. The best way to do this seems to be being able to be near the paddy as much as possible. While this isn't practical all the time, locating a paddy in an area where a dog can be leashed, kids can play, or other human presence is can be a real asset. I have heard from a friend who visited a rice-growing region in Indonesia where the families kept birds away from the rice by pulling string that was rigged long distances out over multiple paddies, connecting tin cans that would rattle with each tug on the string. I might try that here next year to help ward off our incessant bird population.

Constructing Rice Paddies

Although overall labor requirements of rice might in the end be low per yield, the crop (to grow "wet" rice) does require the development of significant infrastructure compared with most vegetable gardens. While this can be daunting to many at first, it is not difficult to do on a small scale, and any able-bodied energetic gardener can make a small paddy by hand with shovel and rake in a day if he puts his mind and body to it. Our paddies are a little larger than hand-digging would be practical for—especially when a compact excavator lies at the ready—so we built them with a thirty-five-horsepower machine with a blade on tracks.

Building rice paddies is no different from making a terrace, except that the terrace is contained on the downhill side by a small berm. We aim to be able to raise the water level of the paddy up to eight inches, although in practice it is almost always much less than that—normally one to to four inches deep. To get eight inches of water retained in the paddy, you need a berm that ends up lying at least twelve inches higher than the bottom of the paddy. When you make a berm for a pond or paddy, much settling occurs over time, so I build the berms fourteen to sixteen inches higher than the bottom of the paddy is initially during the time of construction. It's important to remember as well that the berm only settles and lowers over time, while the bottom of the paddy actually rises and fills in slowly (or quickly, depending on water management) because it is a depression acting as a silt trap (the main reason it's able to perpetuate its fertility).

The construction process begins the same way a terrace is built—cutting (digging into) soil begins on

The author making the homestead's first rice paddies with a 9,500-pound compact excavator. This photo shows the puddling-in process for sealing paddies in loose soil.

the higher area of land (if on a slope), with the material filled (moved) onto the lower side. This cut-fill process levels the area. As in pond building be sure to scrape and save aside any topsoil from the area; you'll want it later for putting back into the paddy or using elsewhere on-site. If the paddy is on level ground, then a simple depression is made with land higher on all sides. A paddy in such an area would be challenging to drain for harvesting (seemingly nonessential but makes it easier), but this should not discourage you if that is the only location you have.

Be sure as you cut and fill that the area filled is left a little higher than the area cut, since that side will always settle more than the undisturbed fill below the cut area, leaving the downslope side too low. This happens with many terracing projects. Make sure to

GROWING RICE IN BUCKETS

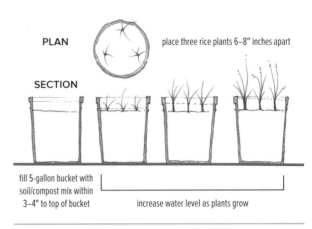

PLAN — place three rice plants 6–8" inches apart

SECTION

fill 5-gallon bucket with soil/compost mix within 3–4" to top of bucket

increase water level as plants grow

Growing rice in typical five-gallon buckets or the like is an easy way to make more seed to eventually plant a paddy. One 5-gallon bucket can produce enough seed to grow 50+ pounds of rice.

Rice test plots in our USDA-sponsored research

save enough material to make the berm; it takes more than you might anticipate, though it doesn't need to be wide—just big enough to walk along is fine. Remember also that nowhere in the construction site should the grade exceed a slope that can be stabilized—usually about 100 percent or 45 degrees, depending on your soil type. Those new to earthworks will usually try to sculpt the berm too steeply, like a kid making sand castles—you can't get away with near-vertical walls of soil and not have them slump over time, even though it's tempting. You could edge a berm by retaining it with stone and save soil or fill by doing so—that would look really nice as well. Be sure to put an overflow spillway, just as you would with a pond, if your paddy is of any consequential size from a water-capturing standpoint.

As with all earthworks, the day the soil is done being moved is the day seed should be spread around the site. We spread the same mix we use with all our pasture renovation work (see chapter four for details on the seed mix). Depending on the time of year, we'll seed immediately and rake it in lightly with a soft rake, then usually shake hay lightly across the surface to keep moisture in and reduce erosion while the seed takes hold. If you're on a steep slope, be extra careful in this site-stabilization step, and consider watering the site if necessary to get a jump on stabilizing the soil. If paddy construction happens late in the year, use annual rye if it is appropriate for your climate—there's no faster way to hold soil in this climate than annual rye, and it's an important tool for this reason alone. We don't cover crop the inside of the paddy, just the outside, although we have experimented with white clover as mentioned, with no success.

We leave the inside of the paddy bare and ready for planting—you want it as weed-free as possible, of course. Some opt to manure their paddies, but we do not add any fertility. We do, however, shake some topsoil, if any is around from the scraping process, back into the paddy when the site is leveled.

Starting the Rice Crop: Seedling Rearing

In the first year of growing rice, we were only able to procure enough seed to warrant growing about twenty plants—not enough for a paddy per se. Fortunately, I had been to the Akaogi Farm in Putney, Vermont, where they had been experimenting with rice for a few years already. They showed me that you can easily grow rice in five-gallon buckets. We started the first rice crop that year in a similar manner and in all following years, since it has worked very well, though it is somewhat labor intensive, and as we've scaled up the planting area—now to about fifteen hundred plants—we are looking for faster rice-rearing methods.

The season for us begins in April when the rice is germinated. We've begun this process between the tenth and twenty-fifth of April each year, and the crop has always been ready before first frost—and that's the ticket: Rice is a long-season crop, and we need to expand the frost-free season by a solid month or more to get the seeds to mature before the first fall frost. Rice cannot likely endure even a mild frost, which often sets here around late September. Even with the shortest-season variety we can get our hands on—Hayayuki*—a seed from the northernmost large Japanese island of Hokkaido, our frost-free season is not long enough to sow seeds outdoors. So we start all seeds indoors, then set them out under cold frames until sometime in May, when the last frost seems to have passed.

We begin this process by soaking the rice—usually about three cups' worth for fifteen hundred seeds (in five paddies). The seed doesn't need to be hulled for this process to work, which is handy because hulling is the most challenging feature of rice processing bar none (more on that below). The seed is rinsed at least once per day for about a week, until a radical (tiny rootlet) forms, which indicates that the rice is viable and is ready to be seeded out. The next step in the process is no different from what most gardeners are familiar with: seeding out in cell trays—fifty to one hundred count are the sizes we've tried so far, with 50 seeming too soil greedy and bigger than needed and 100 seeming too small.

We will likely move to 72s, if we even use cell trays in the future at all; we might opt for community sowing in open flats, as wheat grass or similar garden crops are often grown. The rice is laid out in the trays, two to three seeds per cell, which will be thinned to one or two plants per cell if they all prove viable. We have found that

---

* Means "early snow" in Japanese

Rose Robataille, Whole Systems Design rice researcher, seeding fifty-count cell trays.

Jackie Pitts seeding out rice seed in our first year of paddy production.

planting two rice plants as one seems to be okay—though literature often says this lowers overall yield. Once seeded and pressed in with the help of a chopstick, the cell trays are watered heavily and set out in the cold frames. In the first year of our rice research, we made small containers of 1″–4″ wood to hold rubber membranes filled with water. In this way we made mini paddies for the baby rice to be reared in. This was somewhat labor intensive, and the cells often floated too high or sunk too low, threatening to drown the small seedlings.

In the last few years, we've abandoned that approach and raised them just like normal veggies except watering them more heavily than you would most vegetables. This method seems to work just as well as the mini paddies and is a lot less work—though if you go away the flats can dry out much more easily. These seedlings are kept in the cold frames for about one month, managed in such a way as to get them maximum light and heat. Rice is a lot like growing a watermelon—it wants a long, hot season and plenty of water. Achieving this involves a dance with the cold frames, venting them enough not to fry them but making sure to close them at night. At this stage rice has proven so far to be absent of the damping-off and disease pressure that commonly affects a lot of vegetable seedlings.

After about two weeks the rice plants begin to get quite robust, and fertilizer in the form of liquid fish emulsion, water/urine blend, or manure tea from the barn is added to the daily watering. Sometime in early to late May, when the weather forecast looks good and warm and the danger of frost seems to have passed, we move the trays of seedlings (about twenty last year) into the paddies.

Planting Out the Paddies

Before planting the rice paddies, we make sure to prepare them—though, as with most of our early efforts at any new research project, we attempt to do as little as possible to find the easiest, quickest, lowest-labor way of achieving the end result. For rice this level of paddy preparation is probably not enough! We have taken the approach of doing some quick hand-weeding and long-handled-tool weeding along with some paddy leveling (through simple walking disturbance) prior to planting. We spend about ten minutes or less in each paddy knocking back the thickest weeds that have developed over the past years—mainly cattail—and try to knock down high areas of the paddy into lower-lying ones. Each year the paddy levels out better, and I imagine there won't be much leveling to do next year. It's key to

Rose Robataille planting one of the paddies with five-week-old transplants. The paddy soil condition is slightly drier than optimal in this image.

get the paddies as level as possible during construction, certainly the most difficult part of the entire process.

We plant the paddies on a nice, cool, cloudy day, as you would want for any transplanting. We work backward from one edge, working in a line, carefully trying to avoid stepping on neighboring people and seedlings. It gets congested, so plan it well. We don't string out lines to guide us but just estimate the spacings—using from eight to fourteen inches in the past. We're starting to feel that about nine inches seems right for us—though that could change over time. We have found that a small dibble aids the process and that the moisture level of the paddy is important to get right: too dry and it's hardpan, too wet and the seedling wants to flop over.

This is by far the most labor-intensive part of the whole rice-growing process and represents the limiting factor to producing rice, as far as I can tell (without mechanizing the process, which I have no intention of doing at this scale). It usually takes us about a day with three people working hard to plant fifteen hundred seedlings, though we often do at least part of the planting with large groups during workshops. As with veggies, I like to root dip the seedlings with some nourishment before planting and have found that just

Planting rice in the paddies seems best done when soil is dry enough to hold the plants upright but not so dry that the ground is hard—though I've met Bhutanese immigrants in Vermont that swear by planting rice in very wet ground with some standing water.

Rice seed emerging in late July. The seed has still another six to eight weeks from this point before it's harvest ready.

dipping each flat into a manure tea basin works well to give the rice some help during the transplant shock.

Rice Crop Management

The full swing of the growing season might be the best part—as good as the harvest or eating it—it's simply magical to watch the golden green grasses rapidly filling the water basins and the wetland ecosystem of the paddy emerge. Dragonflies, frogs, salamanders, and dozens of aquatic creatures and flying insects come upon the scene over the summer to feast on the production of the new ecosystem that was created. The main tasks during the growing season involve water-level management and fertility injections via the influx of water. I like to give the rice a lot of nutrients until about July 4—putting manure tea into the water about once per week or two depending on weather and how the crop looks—the hotter it is the better the growth potential, so the more fertility I'll get into the paddy. I will water once to twice per week to keep the paddies wet—but not necessarily deeply flooded.

A nice stand of rice in our first year of paddy production with plants almost at harvest stage

In our third year of growing rice, the birds discovered this valuable new food source. Since that time they have become a major problem and represent the only pest we've encountered so far. We have attempted to meet this challenge with ribbons waving in the breeze, visiting the paddies more often, and with fishing line strung across the paddies. None of these approaches has managed to deter them enough to avoid massive crop loss, however. Though I'd like to avoid it, we may net the crop next year if we cannot achieve some level of crop protection via fake snakes, scarecrows, more dog presence, or other less material and time intensive means.

We are right now in the process of tabulating results from the USDA-granted research study we have been conducting on the influence of water levels on the rice crop. The project involved testing the often-discussed variable of water level in the paddy. Although we'd been testing both no-watering and high-watering approaches for three years before this study, we were able with this grant to test the influence of water in a controlled bucket-based setting using soil mediums that were the exact same from bucket to bucket. We tested four strains of rice across three gradations of water level: (1) moist (like vegetable garden bed), (2) soaking wet but not standing water in the bucket, and (3) standing water about one inch up the plant stem through the entire growing season.

The results, interestingly, seem to indicate that, all things being equal, rice seems to show no preference for very moist or inundated conditions, while normal vegetable bed (drier) conditions seem to reduce yields. Note that we studied yield of seed, not overall plant size or vigor. This result is interesting in that inundation appears to be more of a weed-suppression need than a management strategy for the rice in of itself. This is not an unknown finding in the world of rice, but it adds data to the approach used by Fukuoka in the past and empirical results we've seen in the past four years.

Harvesting and Processing

Between September 6 and September 20, our rice crop has matured each year. This process seems to last a long

At four to five thousand pounds per acre and perpetual fertility possibilities, it is easy to see why rice feeds more people than any other crop on the planet and long has.

An early summer garden filled with food and no bare soil

longer—and that seems to be the case. One of so many examples of how the highest quality foods are truly those we produce ourselves.

Annual Vegetables

Though as a permaculturist (and historical realist) I aim for my focus to be "permanent producers," vegetable gardening still comprises the bulk of my time on the homestead during the summer, aside from moving and protecting plants from animals. This is a factor that is constantly changing as I work toward less of a focus on annual agriculture every year. If this system keeps improving, we will spend a small fraction of our time gardening annual vegetables in another five to ten years as the perennial systems mature and we get better at working with them. While ours is certainly not a perfectly optimized farm, the lessons we're learning continue to prove valuable to us and others pursuing an optimized situation. Annual vegetables simply provide high and often reliable yields when compared with

fruits and nuts in this climate. Only animals compare with reliability in this regard. This is not to say that overall reliability in perennial systems cannot be high—but that situation requires decades to develop. We're getting there, but are not there yet.*

Thus far we have had some great fruit years and some bad—there's no predicting it. So we must stock our eggs in as many baskets of self-reliance as possible. For us that means vegetables, animals, fruits, nuts, and fungi. And when it comes to veggies, we aim for high diversity because some years will be hot and dry, good for corn and squash, while some years are cool and wet, better for the cruciferous family of crucial veggies. Again, there's no predicting the growing season's weather, so we must stock the gardens with an array of food, assuming that if we're lucky (avoiding hailstorms,

---

* Readers interested in seeing a perennial system that is closer to being "there" should visit Mark Shepard's New Forest Farm and the Bullock Brothers Farm—both some of the oldest and most mature perennial food ecosystems on this continent.

freak frosts, and pests) and work smart and hard (and the groundhogs don't mount an all-out midnight assault) we might get 75 to 90 percent of what we plant. We don't aim to depend on that high of a yield, though.

Increasingly, with the way the climate is shifting, I am learning to plan on getting half of what's put in the ground. This conservative approach helps ensure (not guarantee) that we'll get enough food for the entire year if the growing season is poor. If the growing season proves disastrous—for instance, an event like the 1816 Year Without a Summer—then we'll be hungry and would have to revert to hunting but should not starve, even assuming systems failure from a global food supply perspective (which I *am* assuming, because it's better to plan on that breaking than that it working perfectly forever). Nothing works perfectly forever.

Although I grow a relatively high diversity of veggies, I have actually moved away from trying an array of small-quantity foods to growing high volumes of very reliable foods. In this way I still grow a high diversity of foods but put little effort into most of them—just growing in small numbers, the same way the perennial crop systems here include such "nursery" and "seed bank" plants. The basis of doing this is simple: potato, winter squash, cabbage, garlic, and carrot-radish-turnip more reliably produce more calories per area with lower input (including labor) than other crops I have tried. I supplement these staffs of life with greens, hot peppers, peas, corn, beans, tomato, melon, eggplant, beets, sunflower, amaranth, and others, but I don't grow those in any large quantity or devote much time to their culture. Distilled into a three-pronged strategy, my vegetable system has emerged to look like this:

1. **Staple crops** (reliable, high calorie, storable): Potato, winter squash, garlic, carrot-radish-turnip (kimchi crops) supplemented with my other staples (rice, chicken, ducks, sheep, and hunted meats)
2. **Nutrient crops** (nutrient dense, storable): Greens such as arugula, kale, cilantro, chard; garlic (note it appears twice); superfruits such as seaberry, elder, aronia, and many others; duck eggs, meat, sometimes sheep milk
3. **Fun crops/taste supplements**: Hot pepper, dill, bush beans, melons, tomato

A typical midsummer salad—the kind of homegrown meals that are achievable within a handful of months of moving to a piece of land

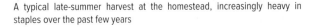
A typical late-summer harvest at the homestead, increasingly heavy in staples over the past few years

SURMOUNTING THE HUNGER GAP WITH KRAUT-CHI CROPS AND OVERWINTERING

Green cabbage—one of the staples we are working with in our focus on the longest storing varieties. Red cabbage has proven more successful on this front for us but takes much longer to ripen.

In the last few years of growing an ever larger percentage of my calorie needs across the entire year—now to roughly 75 percent—the need to preserve food easily and quickly while adding nutrient density if possible and putting as little energy into it as possible (*unlike* canning), the value of lacto fermenting has been mindblowing. There's nothing else comparable when it comes to rapidly preserving massive quantities of food and doing so in a way that actually enhances nutrient density, all while preserving it for very long periods of time.

Turnip, radish, carrot, and cabbage—some of the best but certainly not only—easy to grow, abundant, and reliable staple crops have risen to the top as "must grow" veggies. Add garlic to this mix for small-quantity health additions into the kraut-chis*. These crops are all extremely hardy and reliable in this climate, are light on the soil, and can be stored on their own until it is time to make them into kraut/chi for long periods in a cool/ ground situation. These crops are also somewhat nutri-

ent dense and otherwise very health promoting in their own right; for instance, daikon radish, which is particularly balancing for the body. We grow these crops on a rotation, like everything else on the homestead, and focus especially on the summer–planted, late fall, and even early-winter harvest.

This very late crop is aimed at the hardest food self-reliance window of time during the year—the "hunger gap"—usually from March into mid-June. This period of time is the part of the year most lacking in food produced on-site. The lateness of this window—extending all the way into June—has been made clear in recent years and is due to the lag between plantable conditions and harvest time—usually many weeks. So planting in May (the biggest planting month here) doesn't give you a real crop of food until late June, as even the shortest crops take forty to sixty-five days. Some of the hardiest crops can be outdoors in early to mid-April, yielding some fresh food coming in during early June, but they are mainly only greens.

We get a jump on this partial hunger gap solution from season extension under cold frames (and will even more with a future greenhouse), but that won't shift real caloric production more than several weeks or so. Plants need to be outdoors in the ground for real growth; you can only get them so big in a resilient (read: passive) greenhouse. Thus, the strategy of overwintering has emerged as a crucial way to address the primary food security challenge in this northern cold climate—the late-winter to early-summer hunger gap.

Overwintering annual vegetables has become a reality here in just the past couple of years, so I am no expert on it, and much exists on this topic in the literature and from homesteaders sharing their success online and in groups. Seek them out because you will want to become an overwintering expert to be sure. What I can offer are a few simple approaches that have been working for us. Carrots, turnip, daikon, arugula, and kale have worked the best for us so far. A midsummer sowing of these crops (timing can vary by a few weeks, it seems) has yielded a maturing crop come deep snowfall of early winter.

At this point we do one last walk about the property before the first large winter storm, harvesting turnip and radish especially. Though I've seen these plants easily survive temps of 20°F, then keep growing when it warms

---

* Sandor Katz, author of numerous fermentation books, most notably *Wild Fermentation*, first used the term "kraut-chi," to my knowledge, and his books are the best source I have found for all things fermentable. I consider a kraut-chi to be a fermented mix of cabbage and other vegetables that can't be truly called a kimchee or a straight sauerkraut.

up, they get buried in snow and are impossible to find. I also broadcast them all over the site (see chapter four for more on this), so I do not have specific beds in which to look for them under the snow—that's my approach for arugula, kale, and carrots.

This super-late-season harvest then is brought in, and some of it is made into kraut-chi right away. That kraut-chi is stored in the fridge, although we are looking into fermented storage in cold winter root cellar conditions this coming winter (will be reported on in next edition).

The bulk of this harvest then goes into cold-cellar storage, to be eaten fresh, used in chicken soups, and for the most important latest season-extending food, made into kraut-chis in April and May (culling from the best that has survived to that point, as some goes rotten before then). We also harvest carrots to add into this mix in the early spring before it warms up too much (the last harvest of the season aside from garlic). This leapfrogging of the most storage crops from super-late harvesting to storage to fermenting to storage allows a calorie and nutrient extension from midsummer sowing to early summer eating the following year. This represents one of the most important food security pulse spreadings we have been working out in the project so far.

We will continue to tweak this approach, add more species, and especially figure out the raw-material and kraut-chi storage challenges of this system in the future. We're far from having it mastered but know that it works well enough to be something we can rely on and use heavily for the long haul. These kraut-chis are eaten in May and June, from food sown nearly a year before, when a whole new round of food is almost beginning to come in—it completes a circle, which yields a satisfaction that's hard to describe. The rewards of this full-circle sowing, storing, processing, and feasting are one of the gifts I have come upon so far.

Carrots, daikon radish, cabbage, jalapeno pepper, garlic, and salt form the large quantities of lacto-fermented krauts and kimchis we have come to rely on as a major staple food. All the ingredients except salt are easy to produce reliably in large quantity.

POTATO

The potato is of particular importance because it's such a crucial long-term store of calories and, at least until recently in this climate, has been utterly reliable. All of that has shifted in the last few years as late blight has become a yearly reality even in dry growing seasons. This is the disease that was behind a million Irish starving during the potato famine; coupled with the hyperdependence on single varieties and likely major lack of crop rotation, they saw massive crop failure. My focus at the WSRF in the past two years has been potato blight resilience via seed saving and selecting against blight, rather than simply maximizing spud production. The official recommendation regarding blight is to buy certified potato tissue from sources that can guarantee no blight on their spuds. Such sources are not in every state or even every region in this country, and providers often sell out of this material as blight has proliferated recently. Rather than buy into yet another dependency here, I've done two things related to potato blight: (1) I save my own seed potatoes and manage carefully for blight during and between the seasons, and (2) I grow yacon, a potential backup potato.

Selecting Out the Blight

If you want potato self-reliance, just as with any other crop, you need to save seed (in this case a root) or be able to get it readily and locally.* Given that your food self reliance is sometimes only as dependable as your seed production self reliance, your ability to grow your own seed should not be underestimated. Buying certified virus-free and blight-free spuds is not going to cut it if you find yourself in situations where sourcing potato seed from afar is not possible. And doing so is also not going to eliminate blight in your area as is commonly cautioned—because there's simply no way that every blighted potato in each town, never mind in each county or state, is going to be destroyed and not be allowed to overwinter. Potatoes come up as volunteers all over the place on many properties, so to think that by destroying your plants and not planting your blighted spud seed somehow you'll be reducing blight is silly—like recycling is going to fix the resource-wasting problem. We have spuds coming up in beds that hadn't been planted in potatoes since two years earlier—they appear to almost perennialize in this way. One real solution to blight might be much more active, hands-on, and have to do with selecting out disease, breeding, and/or working with the plant in more nuanced ways.

So somewhat counterintuitively, I am experimenting with the opposite approach: Save the harvested crop, which ends up including some blighted spuds, and cull them often throughout the storage season, constantly pulling out the blighted ones and chucking them in the woods (though composting them might actually not contribute to more blight, in practical terms). By spring I have culled all imperfect spuds, mainly by eating them before they get really bad or culling. Come planting time in late May or early June, I have had for two years about one-sixth of the spuds I put up in the autumn before. Last year I lost about half the original crop planted to blight, such that actual rotten potatoes were found in the patches—I got to them too late.

This year I have already started culling imperfect plants in July, using Carol Deppe's roguing approach (see appendices for her book) and will continue to do so until late summer and fall—the "normal" potato harvesting time. I am leaving most of the spuds in the ground where they can cure—the goal being simply to not let blight into the spuds themselves. By the end of the summer and early fall, I'll have harvested all the spuds in the same way as last year but am hoping I will have none rotting in the ground, as I did last year. I will then continue to cull as I did last year, saving and planting only the best ones each year. I started this process using four different varieties and so far, the red-skinned spuds that are ideal for baking appear to be least blight prone, whereas a purple variety appears most blight prone. I would recommend starting your potato gardening with as many varieties as possible, however. Will this approach eliminate blight by actually selecting out the blight-prone varieties? Maybe. I wouldn't bet on it, but it's still useful if that does not happen.

---

* The official recommendation of agriculture Extension and similar agencies is to destroy the blight pathogen by burning all blighted material or landfilling it in plastic bags, since the disease can overwinter in my location in the protected warmth of a compost pile. It cannot according to official recommendations, however, overwinter in the open in this climate.

Here's why: I am learning how to live with the blight—how to plant, manage, harvest, cull, and reseed accordingly. By doing this year after year, I may find that it's possible to practically still get a reliable *and* storable (that's my main concern) crop of potatoes. The important thing to remember here is that late blight is just that, it's *late in the season*. As long as it continues to arrive in late July or August, which has been the case and is expected with this disease and how it functions, this system could work—tubers are of decent size and abundance even in just late July. The potato being the amazingly productive plant that it is, the amount of calorie-dense food this plant can make in even somewhat marginal soil and water conditions inside of two to three months is truly mind-blowing.

So far we are six months into storing the second year crop of blighted and heavily rogued potatoes and are finding the odd spud in the root cellar with blight.* These, however represent a small fraction of the overall crop, which seems to be holding up perfectly. We're cooking and feeding the blighted potatoes to our ducks instead of the recommended landfilling of these foods. What if the late blight turns into midsummer blight or we see the rise in cases of early blight? While early blight now exists in the region, it's not nearly as threatening—yet, anyway—as late blight. In the event that early blight becomes a major pressure, this approach may have to change or may not work at all, and this is why I am growing an alternative to the potato—the yacon.

Yacon

Yacon and other lesser-known roots and tubers play a crucial role as backup crops to the more common, and thus disease- and GMO-prone, crops. Yacon is one of the many "lost crops of the Incas" and like the others represents spectacular advances in food system evolution by those living in the Andean Highlands for thousands of years. The Incas bred these tubers from their wild ancestors and facilitated a wild diversity of them—from

Yacon—a potential potato replacement and tasty staple food addition to cold-climate self-reliance options

which, in the industrial food system, we draw on a tiny fraction, giving us our white and russet potatoes.

Yacon is like the potato in that it is a calorie-dense tuber, is adaptable (probably) to our climate and very storable and can put on a high amount of edible biomass in short order. Yacon is arguably tastier than a potato—almost like a cross between Jerusalem artichoke, potato, and water chestnut. It's delicious in a stir-fry and livens up a soup much more than spuds do. Yacon grows similar to a potato in that it's a vigorous large plant but is actually twice the size of potato plants or larger—ours reached easily five to six feet tall. The plant's habit is almost like a Jerusalem artichoke but with a broad wide-leafed form. They seem to want to be planted with plenty of room, at about twelve- to sixteen-inch spacings or wider.

We are now about two months into storing our first yacon crop, and it seems to be storing like a potato—still hard and looks perfect after a fall and early winter in the root cellar. I've stored it in damp sawdust, unlike the way I store potatoes, because it seems much more inclined to

---

* I rogue potatoes—pick through them very carefully and discard those with imperfections or any sign of disease. Carol Deppe's book *The Resilient Gardener* outlines this process in detail.

dry out—more like a carrot or radish in this regard than a potato. I'll know more come spring as to the results of storing it for the winter, which is the intended goal.

Growing Food as a Response to Toxicity

The greatest service which can be rendered any country is to add a useful plant to its [agri]culture.

—THOMAS JEFFERSON

The plants we grow on the homestead represent a small sampling of the options available for adapting to the increasing rate of change and adversity brought about by both natural cycles and the terminal phase of industrial empire. Plants such as honeyberry, aronia berry, hardy kiwi, and a couple of dozen other fruits; shiitake, wine caps, and other mushrooms; styrian pumpkin (for seed); chestnut, nut pine, and a score of other nuts and seeds; along with a selection of powerful vegetables and animal foods will need to be harnessed as a new era of land-based toxic resistance is mounted.

Increasing our prospects for survival and thrival depends upon expanding our sustenance possibilities—the options that each food-fuel species, variety, and production system represents. Every food and fuel plant, animal, and fungus species (and variety) represents options for enhancing the fitness between humans and their environment. The development of this cornucopia allows us to expand the length of our growing season, the range and density of our nutrition, and the variety of our fuel sources. Taken as a whole, this expanding diversity allows us to increase the resilience of the human ecosystem and its ability to cope with change. All resilience strategies are valuable to the degree to which they allow us to cope with change. The following section is an overview of a selection of living organisms that are particularly helpful in allying ourselves with the improvement and maintenance of healthy body-mind systems.

LEAFY GREENS

Ranging from spinach to arugula to the oft-cited kale, these vegetables (and some are perennial) can have some of the densest concentrations of nutrients per

Elderberry, wine cap, seaberry, bush cherry, currant, blackberry, raspberry— a small sampling of some of the more medicinal yields of the more than two dozen or so fruits and fungi we grow at the homestead

Dark-green vegetables are becoming an increasingly large part of my paleo-leaning dietary and land-use approaches.

THE ENDLESS ARUGULA BED

Silvetta arugula growing vigorously in March after overwintering in dormancy for three months

Silvetta arugula is almost impossible to beat for an early- and late-season crop in this very cold climate.

The power of overwintering plants so they can begin growing again in the spring was fully realized at the WSRF in the late winter and early spring of 2012, beginning with a late September 2011 sowing of arugula from seed in a 3′–8′ raised bed. This bed was established in September 2011 with a local college course group that was visiting the farm for the day. The students and I filled a slew of newly built raised beds on the south side of the WSD workshop and seeded them immediately, most with cover crops, one with garlic, and one with arugula. The remaining days in September and much of early October were mild and rainy—perfect to start the arugula, which had been broadcast and spread by hand.

By the time the real cold of winter set in around mid-December (late compared to normal), the arugula had reached about three inches in height and filled the surface area of the bed. At this point the plants stopped growing. They went through the winter in this condition covered by one layer of greenhouse film draped over metal "quick hoops" (one-eighth inch or flexible metal rods bent into a half hoop over beds). During the coldest time of the year, the arugula died back to near the surface of the soil. Although I thought they might have died, by February the tiny plants had greened up.

A heat wave in March (seemingly more common each year now) caused the plants to begin rapid regrowth, and by mid-March we were eating the sweetest, most flavorful arugula I'd ever tasted. We then proceeded to harvest this bed in cut-and-come-again style about eight times, taking a week or less between harvests. Finally, at the end of May, the arugula began to become slightly bitter and very spicy—perfect for pesto. We then harvested the whole bed at once and ran it through the food processor with olive oil, sunflower seeds, pepper, and some green garlic pulled from the burgeoning garlic beds.

Looking back, I figure that this bed produced the equivalent of $250 to $350 worth of leafy greens if bought from the local co-op at the going late-winter/early-spring rate of $8 a pound. Beyond those savings is an equal or greater value: These greens tasted far better than arugula from the market—and of course they did! Every time we ate these greens, they were mere minutes old. Fresher, nearly free, and from the remineralized soils we made, such greens are guaranteed to be nutrient dense. The value of taking five steps out the kitchen door to harvest a salad while the rest of the meal is on the stove is also immeasurable.

Pesto—heavy in garlic—one of our favorite ways of using arugula

calorie and should form a large part of the basis of a diet that aids one in adapting to the adversities of the present day. Leafy greens are quick to establish and are fast growing, allowing you to get this foodmedicine source into your diet even in the first growing season at a new location. Since many aren't perennials or long lived, and since some establish in as fast as four to six weeks, you can easily produce them in areas you may not occupy for a long period of time.

Additionally, most leafy greens are very cold hardy and can overwinter easily, thus producing earlier than most if not all other crops in the spring, and many can yield well into the deep winter, even in this very cold climate. Kale, for instance, in Vermont can usually be harvested through December and often into January, and it's improved by frost. Surprisingly, it's the midsummer heat that challenges greens production most, so choosing a shade-protected spot for late June through August production is key. A bed of leafy greens should be part of all of our diets, and for all of the reasons above, on top of the fact that they are about the easiest crop to grow, there's little excuse not to.

It's important also to note the use of perennial and nontraditional leafy greens—some of these are actually more nutrient dense than typical garden crops, having some of the highest concentration of nutrients per calorie of any food. Some of these include lamb's-quarter, Turkish rocket, nettle, skirret, and dandelion. Some of these plants are even easier to produce than the common garden crops listed above and don't even need to be cultivated—just foraged or, at most, established, then promoted and foraged for.

APPLE (*MALUS* GENUS)

The apple makes this short list of most important foodmedicines to grow, process, and store on-site due not primarily to its nutritional characteristics but to its incredible versatility, hardiness, and reliability. The apple has been and should continue increasingly to be a totem tree of climate places, much like the coconut is in the pantropical world. Likely, the main reason that the apple has not made a mark quite as widespread as the coconut is due to lack of utilization and dependence (currently) upon it.

The apple hails from Kazakhstan in the Caucasus Mountains of Eurasia and has been distributed across North America for less than a few hundred years. Despite such a recent arrival, the apple has already become embedded ecologically and culturally across the cold temperate region of North America, where it thrives most. No fruit tree is as hardy to varied conditions or as reliable in this region as the apple. In addition to this ability to live and produce, it is a highly broad-spectrum nutrient source.

We grow a wide variety of apples, but they only represent a tiny fraction of the more than five thousand named varieties that have been grown in the United States alone. Varieties at the farm here include Bethel, Stone, Lobo, Honeycrisp, Winesap, Yellow Transparent, Ashmead's Kernel, Sweet Sixteen, Tolman Sweet, Roxbury Russet, Liberty, Prairie Spy, Northern Spy, Rhode Island Greening, Mutzu, Gravenstein, Wolf River, Honeygold, Bergundy, Freedom, and many grafts of locally found trees, among many other named varieties and selected wild cultivars we've propagated from local trees. It is too early—being only seven years in at the longest—to say which varieties are best for this site.

SEABERRY (*HIPPOPHAE RHAMNOIDES*)

Thought to originate in Eastern Europe and Siberia, the seaberry, also known as sea buckthorn, is now found in large expanses across Eurasia. Seaberry's Latin name means "shining horse"; legend has it that Genghis Khan fed his army's horses seaberry before they entered into battle. It is a large shrub, growing to about ten to twelve feet wide and twelve to eighteen feet high (variety dependent) if left unpruned. The plant produces a bright orange berry that ripens in late summer. Seaberry is exceptional in that the plant fixes nitrogen in the soil, thereby increasing a key soil nutrient that almost all other fruiting plants actually deplete.

Parallel with this soil-restorative function, seaberry also aids in cellular restoration, a function thought to stem from its large spectrum of essential fatty acids (unusual for a fruit) and micronutrients. The restorative quality of seaberry has been known for decades, if not centuries or longer, in Russia, where the plant pharmacopeia is highly evolved. Russian doctors have

Seaberry, an antioxidant- and bioflavonoid-rich fruit that fixes nitrogen and builds soil fertility while thriving in poor soil

Seaberry, elderberry, and aronia berry: potent health tonics and abundant even when grown on poor soil Photograph courtesy of Costa Boutsikaris

Cornelius Murphy and Jackie Pitts, grafting seaberry in March—not an easy task

The juice we make from seaberry is one of the most potent medicinal foods I have had the pleasure of tasting.

administered seaberry to people facing environmental stresses, including cosmonauts, Olympic athletes, and those suffering from radiation poisoning.

Seaberry contains about fifteen times the vitamin C of the same quantity of oranges and is extremely high in essential saturated and polyunsaturated fats, carotenoids, amino acids, and many micronutrients. Of particular interest are seaberry's likely anticancer benefits, which are currently being researched. Unfortunately, seaberry's legendary healing properties have led to overharvesting in many areas of Europe. Today, the plant is considered endangered in Hungary, though China has more than six hundred thousand hectares of it planted in dry regions of the country.

Seaberry is uniquely valuable in cold-temperate climates for several reasons: It fixes nitrogen; it is highly drought tolerant and tolerant of poor dry soils; it is extremely hardy and able to tolerate high winds, salt, and cold to about zone 3 (around –40°F). Seaberry will not tolerate wet soils and does poorly in the shade (needing sunlight for at least three-quarters of the day). Most sources list seaberry as being intolerant of clay, but our research has shown that it may be grown on compacted clay soils if planting depth, amending, and mulching strategies are done specific to these conditions.

Whole Systems Design is testing seaberry in hedgerows as windbreaks and living fences, and in conjunction with sheep and goats to determine browse resistance, palatability, and use as a fodder crop for livestock medicine. So far, the results are promising and show this plant to be of low palatability to much grazing, hardy as a hedge fence, and of strong medicinal value for people, and likely animals as well. We've found best success with this plant on the tops of swale-mounds and on convex rises of land in sunny locations regardless of soil type.

We have been grafting male tips to grafted female plants to ensure pollination. Grafting of seaberries is a challenge unto itself but seems like a practical way of propagating specific genetics. We also raise dozens of seedlings, which we are testing for taste, nutrient content, ease of harvest, vigor, and more.

We are also testing various harvesting methods for seaberry because it is such a difficult plant to glean from, given its plentiful thorns. So far results are inconclusive, but we are finding that a combination of pruning the branches, then stripping the berries off with gloved hands; stripping berries in this way without pruning and harvesting individual berries; and harvesting berry clusters in small handfuls (again with gloved hands) are all practical and apply specifically to each variety we grow. The nuances of this process and which varieties lend themselves to each method are covered in some of our workshops but are too lengthy to cover in this broad work. Fortunately, seaberry offers a very long harvest window—up to a month or longer in which the berries ripen for the picking—giving us ample time to harvest a winter's worth of medicine. We will be producing seaberry-specific writing within the next few years when results have stood the test of time a bit longer.

To extend the value of seaberry and elderberry (below) across the year from the point of harvest, we are experimenting with various approaches, including drying, making tinctures and oxymels, juicing and freezing, freezing whole berries and making kombucha. It is too early to determine which methods are most practical, but it looks as though each will have a role and be useful to some extent. Seaberry oxymel—a blend of apple cider vinegar and honey with the juice—is emerging to be our favorite method and seems to be highly potent and preserving of seaberry's innate medicinal qualities.

Seaberry will likely be of great import in a post–peak oil cold-climate homestead and economy for its medicinal, food, soil-restoration, and animal-fodder/medicinal values. There exist few better ways to extend abundant nutrients produced in the growing season into the dormant season than by drying and juicing nutrient-dense produce (berries in particular). Seaberry may be superior to most other crops (currently or potentially in use) in this regard.

ELDERBERRY (*SAMBUCUS CANADENSIS*)

People across the cold-climate world have been cultivating and wild-foraging elderberry for millennia. Elderberry grows as a vigorous clumping shrub, six to twelve feet in width by six to sixteen feet in height if left unpruned. You can think of elderberry as a shade-tolerant seaberry that can also tolerate moist (though not inundated) soils, even with large amounts of clay and shade present. We grow elderberry most successfully (as is true with most plants) on mounds in wet areas. These high areas are situated above the water table, yet surrounded by areas of high water table, and this seems to offer the ideal growing environment for elder—the fruit has access to consistent moisture but is not swimming in it. Its small, dark berries are harvested from clumps that hold the flower heads in midsummer.

Elderberries, like other strongly pigmented berries, are rich in bioflavonoids, phytochemicals, and other antioxidant-containing compounds, as well as vitamins and minerals. Many parts of the plant are

Elderberry harvest in late summer—proving to be one of the most reliable staple medicines on the homestead Photograph by Connor Soderquist

useful medicinally, including the berries, flowers, and bark. Interestingly, the flowers contain compounds used as compost accelerators commercially, so the odd flower or two in your compost heap could be of significant value, like comfrey—famous for its compost-enhancing qualities.

Elderberry—like seaberry and many lesser-bred varieties of minor fruits—is nearly disease-free and can thrive with little care if planted correctly and weed suppressed in the first year or two. Pruning needs are minimal, and elderberry is easy to harvest.

CURRANTS AND GOOSEBERRIES (*RIBES* GENUS)

The *Ribes* genus represents one of the only vigorous, reliably easy-to-grow superfruits possible in the cold-temperate climate that, like elderberry (but as substantially), can tolerate shade. They are happier in full sun in northern New England but can do well in half-day sun, the likes of which will lead to very poor harvests of blueberry, raspberry, blackberry, seaberry, and almost any fruit tree. These plants are hardy in a wide range of soils, though they cannot withstand droughty, sandy environments like many nitrogen-fixing plants have a tendency to be adapted for. *Ribes*

Currants and gooseberries comprise an increasingly large portion of our harvest but the cultivation demands of these crops discourage us from planting huge numbers of them.

species have been held in high regard in most Northern European countries for generations, as they can tolerate significant cold (hardy to zone 4, easily), cloudiness, and lack of heat for ripening, as well as diseases that predominate in cool, moist climates.

Ribes species are compact shrubs easily kept at three to four feet wide by three to five feet tall and can be integrated into small spaces easily. They do well as small hedge borders and are used as such throughout many Scandinavian cities in small urban lots. *Ribes* fruit is generally tart, although gooseberry is less so than currant, with black currant being the most tart compared to red and white. While generally new to most North American growers, *Ribes* species have adapted well in New England, depending on soil and sun situation. Whole Systems Design's Research Farm has planted *Ribes* in various configurations and finds that these species generally favor mounded sites with consistent moisture. This can be achieved in on-contour swales.

We have found the black currant more reliable than reds and whites because of both bird predation and sawfly. In the last two years, birds have taken a liking to the green, unripe fruit, and I have discovered, to much dismay, abundant fruit sets of red and white currant disappear overnight, leaving bare fruit stems. Usually, this has happened within two to three weeks of ripening from late May to mid-June. If you grow currants, be on the lookout for this challenge.

Gooseberries have so far proven immune to bird predation, but I wouldn't bet on this staying true—the birds seem to learn what's food over time as they adapt to the emerging landscape around them. Gooseberries do seem just as susceptible to sawfly as currants, however. Sawfly has defoliated the leaves of most if not all of our currant varieties later in the summer, usually in July. We have had complete success in avoiding the sawfly by applying a British (Brits love currants) acquaintance's advice, which was to put a thin layer of wood ash around the base of the shrub. Apparently, the sawfly cannot cross this fine ash layer or doesn't like to. This trick has worked two years in a row now.

Like all other colorful fruits described here, the *Ribes* species are high in vitamins, minerals, and antioxidant, cancer cell–resisting factors. Its fruit is enjoyed by

Golden oyster, blueberry, autumn berry, black currant, Korean bush cherry, raspberry, and gooseberry—an array of medicinal-quality foods (except the raspberry—that's just for taste)

those fond of tart flavors, although some varieties are sweeter and not so tart. The tartest varieties, like certain black currants, are particularly useful for jams and jellies and are generally considered to be the most medicinal, as evidenced by their dark pigmentation.

Like most other fruiting plants, currants and gooseberries seem to do best at the WSRF in mostly sunny to full-sun locations with good fertility. They do not seem to like moderately wet or very wet areas or much shade here. This runs counter to much information out there, which may be suited to warmer locations, which says that *Ribes* species are fine in half shade. Only elderberry, in terms of fruiting plants, can take much shade in this climate as far as we've experienced thus far.

SHIITAKE

Of all the fungi we can ally ourselves with, the shiitake mushroom is the highest on my list in terms of overall health benefits relative to practicality of production in a resilient homestead in my location. Shiitake is roughly 30 percent protein with a large array of micronutrients. The medicinal nature and, especially, the health-bolstering addition to a diet provided by an organism such as shiitake should not be surprising when you consider that unlike most foods in our diet it grows entirely on wood as a substrate. Energetically, mushrooms are unlike plants at deep levels—this alone should prompt us to include such diversity-opportunity in our diet.

NATIVE TO WHEN: "INVASIVE" SPECIES, SPECIES-ISM, AND OPTIMIZATION

In human-inhabited systems high levels of biodiversity and structural diversity, along with productive yields, are necessary ingredients in a healthy ecosystem. (Permaculture and regenerative land use is not a framework for managing "wild" lands, but it does call for a wild zone—zone 5—in any site large enough to accommodate such.) Permaculture, as I understand and practice it, *is* about enhancing, not just sustaining, the health of the ecosystems from which we derive our sustenance on Earth. This is not possible by placing the needs or wishes of one species above the health of the whole system. To that end, permaculture regards humans as participants in their ecosystems—not as tyrants, beneficent kings, or evildoers, but as "natural" a participant as an ant, a bear, or a beaver, and capable of both beneficial and destructive ecosystem membership.

Permaculture as System Designer

Permaculture focuses on providing for basic human needs in healthy and regenerative ways that don't depend on distant destruction of ecosystems to provision ourselves. Permaculture is not the lay-environmentalist approach of sitting back enjoying the view of green hills while forests across the globe are razed to provide for the resource demands of our lifestyle. Permaculture is not armchair environmentalism at all, but gardening that embraces the entirety of a complex, biodiverse, and ever-changing living world.

And a world with humans in it. It does not, in general, see plants or other organisms that have been in a place for ten or a hundred or three hundred years as fundamentally more "natural" or proper in a place than plants that are recent arrivals. It asks first, "What does a plant do, and how does it relate to other plants, the soil, and human needs?" Not, "How long has it been here?" It never views a plant, an animal, or another human culture as evil or alien. It works from a perspective of inclusion, rather than exclusion, and recognizes that all members of a living system are connected. It sees synergy, not conflict, everywhere.

Permaculture design never seeks to *eradicate* and simplify any part of a system—just the opposite. It works with ecosystems for what they are—constantly evolving assemblages of species and shifting relationships between all pieces of the ecosystem. It asks us to find ways, the most synergistic ways, to fit into this changing web of relationships. Permaculture also sees a need to adapt to and respond to emerging challenges, such as a more rapidly shifting climate, increasing biospheric toxicity, mass extinctions, social system and human health declines, and other current challenges—challenges that require us to respond to, not retreat from or ignore, changes that are underway.

In response to these challenges, permaculture design promotes an exceedingly high level of biodiversity in systems. For instance, permaculture work involving the exchange of seed and plants advances *ex situ* conservation goals so that species threatened by climate and other changes in their historic locations can survive in new, more suitable locations. This work also results in continual increases in food, human health, and ecosystem health possibilities found in the increasing diversity of ecosystems and synergies present there. Permaculture sees human and ecosystem health as mutually dependent. Permaculture, in contrast to conventional "natives first" gardening, embraces the legacy of food systems diversity we all benefit from during each meal (unless you live on groundnut, hazelnut, venison, bison, or certain berries), and it actively expands that diversity to enhance human and other living system health.

Permaculture sites become wildlife restoration zones as a matter of course. Much of the food we promote ends up feeding "wild" life because of the sheer diversity of foods present and the "wildness" of the site itself. This is a far less managed approach than most any other codified gardening and farming systems. Additionally, a permaculture incorporates earthworks, such as swales, ponds, and terraces, and mixes tree crops with annual crops. This structural diversity actually creates far more opportunities for species of concern, such as songbirds and amphibians, than all other forms of more simplistic gardening and all forms of annual-only organic agriculture. However, this should not be a surprise. Permaculture emerged from direct and participatory observation and engagement with diverse ecosystems across the globe. It is not modeled simply on what an ecosystem happened to look like in a particular year, say, 1492.

Humans in the Equation

Many people today, particularly "environmentalists" and "conservationists," have a particular bias for the

Table 5.1: Plant Origins

Species by Latin Name	Common Name	Origin
Amelanchier alnifolia	Downy serviceberry	North America
Amelanchier canadensis	Shadblow serviceberry	North America
Aronia melanocarpa	Black chokeberry	North America
Asparagus officinalis	Asparagus	Eurasia, cultivated
Beta vulgaris	Beet	Mediterranean
Brassica juncea	Mustard	Eurasia, cultivated
Brassica oleracea	Broccoli	Cultivated origin, Eurasia
Caragana arborescens	Siberian pea shrub	Siberia, cultivated for centuries
Caragana microphylla	Small leafed pea shrub	Siberia, cultivated for centuries
Castanaea dentata	American chestnut	North America, primarily the East Coast
Capsicum annuum	Peppers	Central America
Citrullus lunatus	Watermelon	Africa
Corylus americana	American hazelnut	North America
Cucurbita pepo	Summer squash	Central and South America
Daucus carota	Carrot	Northern Europe
Fagus americana	American beech	Midwestern to eastern North America
Fragaria spp.	Strawberry	Cultivated species are from North and South America
Fraxinus americana	American white ash	North America
Glycine max	Soybeans	Central Asia
Gymnocladus dioicus	Kentucky coffee tree	North America, primarily the East Coast
Helianthus tuberosus	Perennial sunflower	North America
Hippophae rhamnoides	Siberian seaberry	Eurasia, cultivated for centuries, many varieties
Ipomoea batatas	Sweet potato	Central and South America
Juniperus virginiana	Eastern red cedar	North America, primarily the East Coast
Lactuca sativa	Common lettuce	Central Asia
Lycopersicon esculentum	Tomato	South America
Malus domestica	Apple	Central Asia, specifically Alma Ata, Kazakhstan
Morus rubra	Red mulberry	Midwestern to eastern North America
Myrica pensylvanica	Northern bayberry	North America, primarily the East Coast
Ocimum basilicum	Basil	Tropical Asia
Phaseolus vulgaris	Common dry bean	Central America
Pinus koraiensis	Korean nut pine	China and Korea, wild harvested for millennia

Species by Latin Name	Common Name	Origin
Pisum sativum	Snap pea	Mediterranean
Populus × canadensis	"Prairie sky" poplar	North America
Prunus domestica	European plum	Eurasia, probably a hybrid of 2–3 *Prunus* species
Prunus persica	Peach	Central Asia, domesticated 8,000 years ago
Prunus virginiana	Pin cherry	North America
Pyrus communis	European pear	Eurasian origin, in use for thousands of years
Ribes uva-crispa	Gooseberry	Eurasia, cultivated for centuries
Ribes × spp.	Jostaberry	Cultivated, generated in the horticultural trade
Robinia pseudoacacia	Black locust	North America
Rubus spp.	Raspberry	Cultivated species are pan-temperate hybrids
Salix 'Flame' Red	Flame willow	Cultivated, generated in the horticultural trade
Salix 'Flame' Yellow	Yellow flame willow	Cultivated, generated in the horticultural trade
Salix purpurea 'Eugenii'	Dwarf purple osier	Northern subarctic species, circumpolar
Salix purpurea 'Nana'	Dwarf purple osier	Northern subarctic species, circumpolar
Sambucus canadensis	North American elderberry	North America
Solanum tuberosum	Potato	South America, Andean altiplano
Spinacia oleracea	True spinach	Eurasia, cultivated
Taxodium distichum	Bald cypress	North America
Tsuga canadensis	Eastern hemlock	North America
Vaccinium spp.	Blueberry	North America, cultivated blueberries are hybrids
Viburnum opulus	Highbush cranberry	Central Europe, used for centuries
Viburnum trilobum	American highbush cranberry	Northeast United States
Vitis vitifolium	Grape	Cultivated species are hybrids from Europe, North America

North American ecosystem as it was assembled just before European contact. Indeed, seeing "natives" as fundamentally more beneficial than newcomers to an ecosystem depends on this notion. At what point is a given ecosystem ideal in their minds? Was it before "native" societies cultivated the Three Sisters from Mesoamerica? Was it only after the ice sheet retreated from New England and the last version of hardwood forest blanketed this region? Was it before "native" peoples promoted vast forests of chestnut and oak and managed landscapes extensively with fire, or before those "artificial" disturbances? Do nativists' visions of an ideal ecosystem include seven billion human beings or other emerging conditions? And if so, where and how should they derive their sustenance?

Permaculturists are answering this challenge to Earth's ecosystems by cultivating systems that produce as much food, energy, materials, medicine, energy, wildlife habitat, water purification, carbon sequestration, pollination, and other ecosystem services as possible in the smallest amount of space possible for the longest amount of time possible. Doing this requires that we engage the continuous forces of change and partner with other species and whole ecosystems to promote resilience. Permaculture both acknowledges and works with the process of change, whereas, surprisingly, many forms of "conservation" seem to be focused primarily on maintaining specific species and ecosystem arrangements as they were at one idealized time in the past. These attempts to maintain (with great frustration) an unchanging romantic notion of species assemblage are currently retarding real progress toward enhancing the health of living systems on Planet Earth. It is time we looked at these systems for what they are and are not.

Defining "Native"

In contrast to many native-plant fundamentalist statements made over the years, consider the following facts about ecology and ecosystem dynamics, along with some of the ways in which reality simply does not mesh with the many assumptions made by the Nativistic War-on-Alien-Invader ideology at large:

▶ Humans are now an active influence in most, if not all, ecosystems on the planet and have been so for many thousands of years. During this time humans have been moving plants and animals both for daily survival and for trade. Most of the diversity of our current "local" food system is a direct result of this: the potato from South America; corn, beans, and squash from Mesoamerica; the honeybee and the earthworm from Europe; and the apple from southwestern Asia, to name a few. When using the term "native," what year do we use to determine whether a plant is "from here" or "an alien"? If we choose European contact as a starting date, we ignore a multiple-thousand-year history of anthropogenic plant dispersal that was highly active before Europeans began to settle the "New" (actually very old) World.

▶ No plant community is permanent: not knotweed, not barberry, not white pine or goldenrod or any other dispersive plant. Plant succession and ecosystem change are wholly "natural." Why when it involves human activities is it automatically "unnatural"? The goal of a truly sustaining and regenerative working land use is to promote a high biodiversity ecosystem that offers large yields of biomass while cycling fertility on-site; while slowing, spreading, and sinking water; and while performing other key ecosystem services, such as soil building. To do this we need to look at what functions the plants provide, not only if they have been located in a place for a hundred or five hundred years. That's an important factor, but only one of many criteria as to whether a plant should be promoted or discouraged in an ecosystem.

▶ Any plant, whether it has been in a region for ten years or ten thousand years, has the capacity to influence a site to the point that other species are reduced in abundance: Witness "native" white pine and goldenrod, both of which force out numerous species across New England because of their ability to compete in abused sites, their generalist nature, and their fast growth. Is that not a destructive pattern? Is this destructiveness negated simply because these plants have been here for five or ten thousand years?

▶ Dispersion, growth, and decline of individuals and species are basic phenomena of all ecologies in all places. The idea that it is "unnatural" when a plant or animal moves from one region to another because of humans is rooted in an ideology that sees humans as separate from the rest of the living world. Why is it natural if a bird or an ocean current moves a plant, but not if a human does? Are humans fundamentally bad or destructive? In a constantly changing world of land-use shifts and climate changes, how will species survive if they are supposed to "stay where they are from"? Movement of organisms is crucial to keep pace with global changes, if biodiversity is an aim.

This does *not* mean that we fling seeds of various plants wildly across the globe without analysis of what would be helpful where. Instead, it means that we evaluate how to feed seven billion humans while honoring and also feeding the thousand trillion other lives that exist in the land community. Feeding one's self from a monoculture in Iowa or Mexico while devoting time to spraying Japanese knotweed with toxic chemicals will not get us where we need to be. The true origin of most of the foods we take for granted each day is often surprising to many. See table 5.1, to see how we have long been and are daily the beneficiaries of global food species and variety exchange.

Human beings have been on Planet Earth in current form for roughly fifty to a hundred thousand years. Our ability to remain here well into the future depends greatly upon our ability to participate within the living world of which we are part and parcel. Being a nonparticipating observer attempting to maintain the world around us in a static condition is simply not an option. What is the nativist approach for engaging the world in a sustaining and regenerative manner such

that we can provide for ourselves and those that might come after us while allowing the full flourishing of the rest of the living world? Are nativists suggesting that we live on and from an economy based on ecological communities as they were for a period of time in the mid part of the second millennium AD? If so, how do they see a hunter-gatherer culture reemerging that operates a functional food system without honeybees, earthworms, apples, potatoes, pears, cherries, kale, cabbage, carrots, onions, garlic, sheep, cows, beans, wheat, and other annual grains? In their vision of a "native" world, would they like to see all but the first (indigenous) peoples removed from the ecosystem in which they have artificially been introduced—including us Europeans in North America?

VAVILOV CENTERS
WORLD CENTERS OF ORIGIN OF CULTIVATED PLANTS

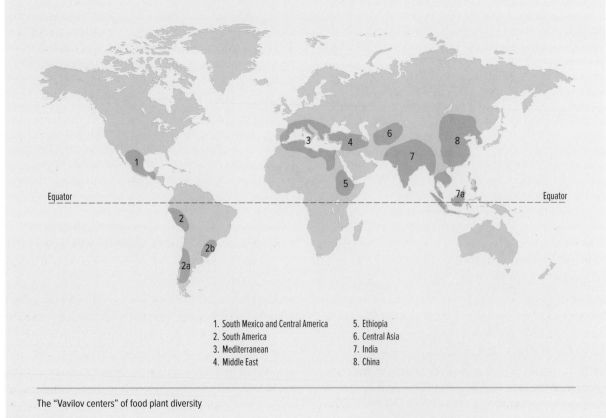

1. South Mexico and Central America
2. South America
3. Mediterranean
4. Middle East
5. Ethiopia
6. Central Asia
7. India
8. China

The "Vavilov centers" of food plant diversity

Perhaps people as a whole are not part of the vision, since obviously there are few "native" people on this planet today. Ultimately, people are not part of the native fundamentalism view; at its core that view is as antipeople as it is antialien. If nativists have a solid plan for this vision in action, I'd enjoy seeing it and have yet to—it's not as simple as "re-wilding." Here at my own farm and homestead, we are witnessing the rapid increase in both biodiversity and biomass on this formerly abused and abandoned Vermont hill farm, which, without human-assisted healing, would continue to be far lower in diversity, soil health, and wildlife value than it would if left fallow in continued abandonment. Human presence can be regenerative, not just less bad. That's the good news and powerful leverage that permaculture harnesses.

Defining Systems That Unite

At its basis, the native-plant ideology is predicated on more than the simple misconception that biological communities are static or that they have been in "ideal" states at some point in the past, only recently "disturbed" by human beings. It is also built upon a fear of nature ("taking over, invading"); the desire to control its evolution; and nostalgic, deeply emotional beliefs that stem from a paradigm that sees humans as fundamentally separate from the rest of the living world.

Such a paradigm is counterproductive in a time of urgent ecological and social issues that require unified and integrated solutions. After all, Earth is a whole and interconnected system—it must be regarded as such if we are to find a synergistic way to fit within the patterns of the system we call home. Divided and fragmented approaches, including wars on specific plants (and cultures), have rarely, if ever, worked. It's time to focus squarely on integrated strategies and lay aside the emotional baggage and the unscientific, unhelpful mental habits of the "nativist" approach.

Fortunately, the results of ecosystem regeneration in such participatory fields as permaculture and agroforestry are emerging and stunning. In contrast, "species eradication" and other such fear-based, hypercontrolling, and divisive efforts are failing as reliably as they line the pockets of Monsanto executives, pollute soils and groundwater, and further alienate us from the living world of which we are a part. It is time to replace eradication with transformation. Killing one part of the system without addressing the entire structure of the system is a doomed approach from the start—a failure of design much like today's "health" care system.

It is surprising what's possible when we work from an angle of inclusion and partnership in human-ecosystem relationships, rather than domination. When we treat all life forms with respect—waging war on none—we begin to gain deep understanding that only comes through reverence and partnership. Only then will there be prospects for dwelling in beneficial relationship with the rest of nature. Humanity's prospects for developing a positive presence on Earth depend on inclusion, rather than exclusion, synergy rather than simplification. There are no evil plants, just dysfunctional human designs.

We have found shiitake to be the most reliable long-term producer in the fungi kingdom with which to partner for health. Sugar maple is regarded as a good substrate, and we've had the best success with this species, though we have tried hop hornbeam (and have had stump inoculation work occasionally) and red maple (second best). Red oak does not grow on this site but is the preferred substrate of shiitake growers continent-wide. We have worked with red maple for shiitake production for about five years and have had success with solid flushes of dense mushrooms. However, the bark tends to fall off the logs quickly—within two years or less, the log becomes light and rotty, and the lifespan of the logs appears very low.

I cannot say that we've exhausted red maple logs with 100 percent assurance, but it seems clear their life span will be far less than sugar maple. This may be a reconcilable disadvantage, however, given that red maple can be produced two to three times faster than sugar maple (we coppice and pollard it for rapid regrowth within pastures), thereby allowing much higher feedstock capacity of the substrate itself. Add

Shiitake mushroom has emerged to be one of the more reliable, abundant, practical, and important sources of health from our homestead.

Drying is an ever more crucial method at the homestead, and shiitake dries quickly in the sun, actually improving in nutritive quality, with vitamin D content becoming very high after a day in the sun—especially if dried with gills facing up.

that to the fact that sugar maple is far better firewood and building material, and you can see why a red maple pollard/coppice perpetual mushroom substrate system might be a good approach.

HAZELNUT (*CORYLUS CORNUTA*)

Of all the powerful health-promoting plant species available to us, only a small number produce foodmedicines dense in proteins and fats. Hazelnut is one of these, and perhaps the only one that is productive in a short time frame, a result of its shrub formation; all other nuts in this climate are tree form and longer to bearing.

Hazelnut oil is arguably the most valuable foodmedicine produced by this plant, although the nut is of high value to humans and wildlife alike. The oil is dense in essential fatty acids, while the nutmeat is high in protein and fats. An abundance of minerals and micronutrients is also made available from the soil, rain, and sunshine by hazelnut.

Although this plant has been native to the New England region for centuries and possibly longer, its cultivation commercially or even by the modern small-scale subsistence grower has been almost nonexistent. We are experimenting with growing hazelnuts in hedgerow patterns to figure out its optimal site, soils, harvesting, and interactions with other species. Hedgerows offer microclimate benefits, enhanced yield density, and snow-fencing functions. This medium-to-large shrub seems to favor a multiple-stem habit for maximum yield; well-drained loams are ideal, but a large array of soil types is possible, given appropriate amendments, earthworks, and water availability.

Hazelnuts almost ready for harvest

A Siberian nut pine planted at one of Whole Systems Design's project sites

PINE NUT
(*PINUS CEMBRA* **AND** *PINUS KORAIENSIS*)

Over the long haul it is hard to beat a foodmedicine that concentrates protein, fat, and micronutrients as intensely as pine nut trees do. Though they do not bear significantly for a decade or three from planting, the pine nut can withstand long drought, difficult soil conditions, and extremes of temperature with relative ease. Like many of the most potent foodmedicines, the pine nut can thrive in adverse conditions—aiding the medicine-making ability (potentization) of this species. In Russia it has been common for Olympic athletes to be fed a diet rich in pine nut oil because of the nuts' phenomenal spectrum of fatty acids and micronutrients.

We have had small crops of both these species at the WSRF, and where we have planted more of them, at Teal Farm in Huntington, Vermont, crops are emerging on trees ten to twenty feet tall and between ten and twenty years young. Though we have not tried putting *Pinus cembra* in wetland locations, it can purportedly take such challenging terrain, making it highly valuable because of the abundance of inundated, hard-to-use land in the cold-climate regions of the world.

THE CHESTNUT GROUP

Chestnuts have been recognized by many as the finest nuts in North America for eating. The chestnut is a starchy, high-carbohydrate nut that has been and is used as a staple by many past and present cultures around the world from Europe to Asia. The American chestnut's recent demise has created a black hole in the productivity of the North American forest that could be

Chestnut, wine cap mushroom, and pine nuts—all possible to grow in this climate, all potent supporters of health in an increasingly toxic biosphere. Pine nuts grown in central Vermont (not on the WSRF site), courtesy of Nicko Rubin at East Hill Tree Farm. Ours should be ripe within another decade or so.

THE ROLE OF HIGHLY POTENT NONLOCAL FOODMEDICINES FOR RESILIENCY

There are specific, superpotent foods and medicines that aid health to such a large extent that they are worth acquiring now, while it is easy and relatively affordable to do so, regardless of where they are from across the world. Many of these are only sourced from relatively few areas across the world and are worth obtaining now, while they are more accessible than they have ever been in human history. Only those that store for relatively long periods of time are worthy of this list. Such foodmedicines include the following:

▶ Clean, mineralized salt, especially from the Andes and Himalaya. These salts were formed from ancient oceans before they were polluted with high levels of industrial pharmaceuticals, endocrine-disrupting polymers, radioactive isotopes, heavy metals, and other effluents of modern society.

▶ Seaweeds, including alaria, digitata kelp, kombu, nori, dulse, and others. These sea vegetables are rich in a broad spectrum of minerals, trace elements, and micronutrients that aid in body functions. Many of these, digitata kelp in particular, are loaded with the element iodine—a key way to help maintain thyroid health in the face of an increasingly radioactive environment. However, please note that since Fukushima I have personally avoided all seaweed and seafood from the Pacific and would recommend these are sourced from the Atlantic only.

▶ Blue-green algae from verified metal-free sources. Those in sunnier locations might have success with growing blue-green algaes as well.

filled by the development of new crossbred trees that have American, European, and Asian genetics. Many efforts have been made in this endeavor, giving hope for a multitude of agroforestry systems in which the American chestnut is reintroduced and again becomes a staple of diet and economy.

We have planted chestnuts for six years with some degree of success. They are slow on this site's generally poorly drained soils (purportedly they like gravelly sites most) and that's the main challenge. The classic nut tree description of "sleep, creep, leap" applies well to all nut trees and certainly our chestnuts. We plant them only in the "highest and driest" locations we have, yet even in these favorable spots if we do not mulch them each year, new growth is barely existent, limited to two inches, maybe six inches in wet years. Neighbors in our valley have, however, experienced relatively rapid success, with chestnuts bearing heavy crops and actually naturalizing in their forest inside of ten to fifteen years. Readers interested in cultivating this supremely important (and once unbelievably prolific in the American landscape) food crop should tap the resources of Oikos Tree Crops, New Forest Farm, The American Chestnut Society, the Northern Nut Growers Association, Saint Lawrence Nurseries, and Badgersett Research Group among others.

Varieties of chestnuts that can be grown in zone 4, Northeastern United States are as follows:

Castanea dentata
(American Chestnut)

Castanea dentata × mollissima
(American/central Asian cross)

Castanea mollissima
(Chinese chestnut)*

Castanea seguinii × mollissima
(dwarf hybrid of two Asian species)

Castanea crenata
(Korean chestnut)

Castanea pumila hybrida
(single-trunked selection of the chinquapin)

Castanea pumila
(multiple-stemmed chinkapin)

Castanea sativa × mollissima
(central Asian/Chinese cross)

---

* We started planting many Chinese chestnuts early on but after seeing their tendency toward an open spreading form and seeing the oldest known planting of them in this area, about twenty miles from here, we have stopped planting this variety. Major structural damage from ice and snow loading was visible in the older trees at this regional site and confirms the need for a strong central leader tree in these very snowy and icy climates.

Food Processing and Storage: Spreading Abundance across the Entire Year

I like to call my home region a "storage climate." Here in the northeastern United States, we can easily grow far more food than we can eat for two to three months a year. However, given the long, cold winters, that short window of production needs to sustain us across the remaining three-quarters of the year. Though short-growing-season climates are an extreme example of the need for storage, the same pattern—and design challenge—exists in all climates where abundance occurs in pulses, and distribution of that abundance across time has always been a lynchpin in the long-term

sustainability of a people in a place. The actual level of production is often less a limiting factor than the effectiveness of storing that production. All things being equal, it's often better to produce less but store it more effectively than to produce more but ineffectively store the harvest over the long year. In this way the same limiting factors apply in the cycling and optimizing of food systems as they do in the utilization of energy: Storage is more of a challenge than production.

Our harvest season is short and intense, as are periods such as logging, planting, pruning, grafting, weeding, mulching. There is only a limited window of time during which many activities on the homestead make sense—those activities represent the scheduling design challenge around which any activity that can

A typical early-summer harvest of shiitake mushroom and garlic scapes—both qualify as foods and medicines

happen across a wide range of periods, such as bucking firewood, should occur.

At the WSRF we have tried various strategies over the years following a theme of migration away from canning and movement toward the most passive approaches such as dehydration, lacto fermenting, and root cellaring. We are no experts at food processing and storage, and there is much information specific to this subject, including Mike and Nancy Bubel's and Sandor Katz's books on the topic. The information here, however, is from our direct experience of what works well in our particular cold climate.

For instance, we see canning as a good way to put up special treats, such as some pickles and hot sauces—things that are added to food but not food calories in and of themselves. This runs counter to many traditional, somewhat self-reliant Vermonter lifestyles of past generations here, where it was not uncommon to put up hundreds of cans for the long winter. I have spent time with a local man in his eighties who recalls his mother's putting up a thousand quart-size Ball jars for the family. That's three jars a day across most of the year. I don't see the sanity in getting calories from such a method. The laboriousness of boiling that much food and water alone during the hottest and busiest time of year does not make much sense and seems to be only a last-resort option when other strategies of putting up massive quantities of food are not available—but they are!

During the first few years of homesteading here, I was always bothered by the major canning operation,

Storing our elderberry as syrup (with honey mixed into the juice) in sterile canned jars for the long term

TIME ALLOCATION ACROSS ONE YEAR AT WSRF

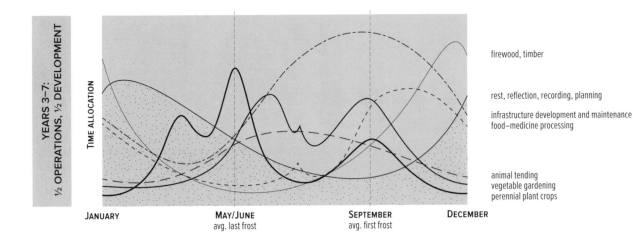

YEARS 1–3:
¼ OPERATIONS, ¾ DEVELOPMENT

TIME ALLOCATION

firewood, timber

rest, reflection, recording, planning

infrastructure development and maintenance

food–medicine processing

vegetable gardening
perennial plant crops

JANUARY   MAY/JUNE   SEPTEMBER   DECEMBER
                 avg. last frost   avg. first frost

YEARS 3–7:
½ OPERATIONS, ½ DEVELOPMENT

TIME ALLOCATION

firewood, timber

rest, reflection, recording, planning
infrastructure development and maintenance
food–medicine processing

animal tending
vegetable gardening
perennial plant crops

JANUARY   MAY/JUNE   SEPTEMBER   DECEMBER
                 avg. last frost   avg. first frost

YEARS 7–10:
⅔ OPERATIONS, ⅓ DEVELOPMENT

TIME ALLOCATION

firewood, timber
food–medicine processing

rest, reflection, recording, planning

infrastructure development and maintenance

animal tending
vegetable gardening
perennial plant crops

JANUARY   MAY/JUNE   SEPTEMBER   DECEMBER
                 avg. last frost   avg. first frost

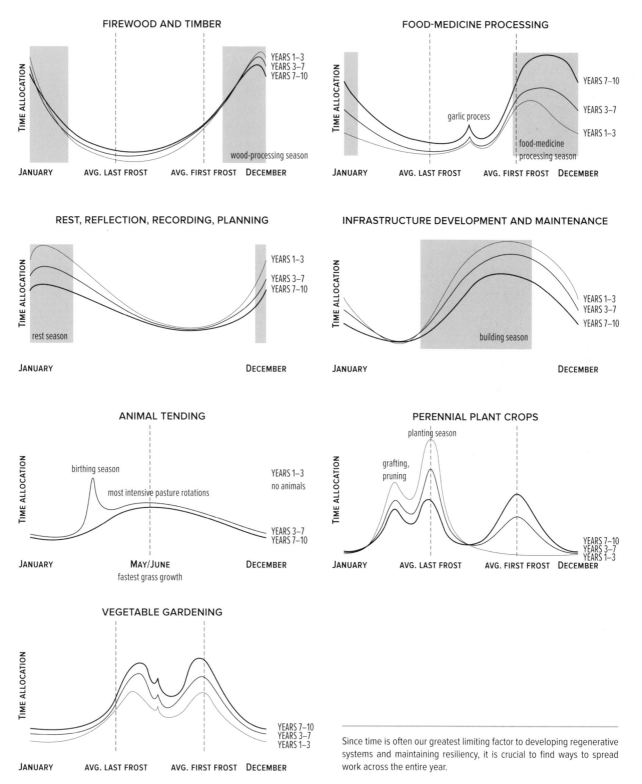

TIME ALLOCATION ACROSS ONE YEAR AT WSRF

FIREWOOD AND TIMBER

TIME ALLOCATION

YEARS 1–3
YEARS 3–7
YEARS 7–10

wood-processing season

JANUARY | AVG. LAST FROST | AVG. FIRST FROST | DECEMBER

FOOD-MEDICINE PROCESSING

TIME ALLOCATION

garlic process

YEARS 7–10
YEARS 3–7
YEARS 1–3

food-medicine processing season

JANUARY | AVG. LAST FROST | AVG. FIRST FROST | DECEMBER

REST, REFLECTION, RECORDING, PLANNING

TIME ALLOCATION

YEARS 1–3
YEARS 3–7
YEARS 7–10

rest season

JANUARY | DECEMBER

INFRASTRUCTURE DEVELOPMENT AND MAINTENANCE

TIME ALLOCATION

YEARS 1–3
YEARS 3–7
YEARS 7–10

building season

JANUARY | DECEMBER

ANIMAL TENDING

TIME ALLOCATION

birthing season

most intensive pasture rotations

YEARS 1–3
no animals

YEARS 3–7
YEARS 7–10

JANUARY | MAY/JUNE | DECEMBER
fastest grass growth

PERENNIAL PLANT CROPS

TIME ALLOCATION

planting season

grafting, pruning

YEARS 7–10
YEARS 3–7
YEARS 1–3

JANUARY | AVG. LAST FROST | AVG. FIRST FROST | DECEMBER

VEGETABLE GARDENING

TIME ALLOCATION

YEARS 7–10
YEARS 3–7
YEARS 1–3

JANUARY | AVG. LAST FROST | AVG. FIRST FROST | DECEMBER

Since time is often our greatest limiting factor to developing regenerative systems and maintaining resiliency, it is crucial to find ways to spread work across the entire year.

not clear at the time why. In retrospect it was likely several things. Harvest time is one of the most beautiful and busy times of year. August, September, October—these are stunningly beautiful days with a crispening air, the bulk of the harvest coming in, frost soon to arrive—gratitude, urgency, and abundance all rolled up into a couple of intense months. Spending long hours in the kitchen boiling water and putting up relatively small amounts of food for each massive pot of boiling water (and energy input) seems even crazier to me now than it did then. The harvest time is a time to be outdoors. It's still swim season, the beginning of some hunting seasons, foliage season. Not a time to be slaving away over a stove. Besides, think of the energy input in physical terms alone: Boiling three or more gallons of water to put up maybe three to five quarts of food. The energy exchange is a poor one for food but seems acceptable for diet supplements like sauces.

PROCESSING STRATEGIES

As we learn to produce abundance reliably (see chapter four and this chapter for food-production techniques), we must also learn to extend that abundance across the year. The most resilient techniques for processing and storing foods are those that are:

▶ **Most healthful:** retain and even enhance the life-sustaining qualities of the food. An example would be lacto-fermented versus canned, the former being living and as such more life-enhancing than dead canned foods.

▶ **Most passive:** the approaches that can be performed with minimal time, energy, infrastructure, and material inputs, both initially and in the long term. A root cellar is clearly superior to a refrigerator in this aspect.

▶ **Most multifunctional:** All things being equal, the best method of processing is one that harnesses an energy stream or activity already taking place; for example, dehydrating on a woodstove while the stove is used to heat water and warm interior space. If you're already using the woodstove for those latter needs, other values can be harvested virtually for free. Drying apples in the rafters of a wood-heated kitchen is a great example of this strategy.

A friend, Richard Czaplinski, putting up some of his many apples to dry in the rafters of his kitchen ceiling

▶ **Most time efficient:** All things being equal, the faster a method can put up abundant produce the better because, simply put, time is your most often and greatest limiting factor, especially during the busy harvest season. Equally as important as speed of processing is when the processing can occur; for example, red cabbages, which can keep until late winter, then be made into kimchi and keep another four to six months. Any processing method that you can apply well after the harvest season has in and of itself a high value in this regard. These strategies apply across all of the primary ways we store caloric and nutrient value, including drying, canning, pickling, smoking, lacto-fermenting, in oil or in vinegar, living (vegetable in the ground, animal on pasture or in the barn), and tincturing.

VERY LONG-TERM FOOD STORAGE: AN INSURANCE POLICY

LONG-TERM FOOD SUPPLY
staple food/crop storage

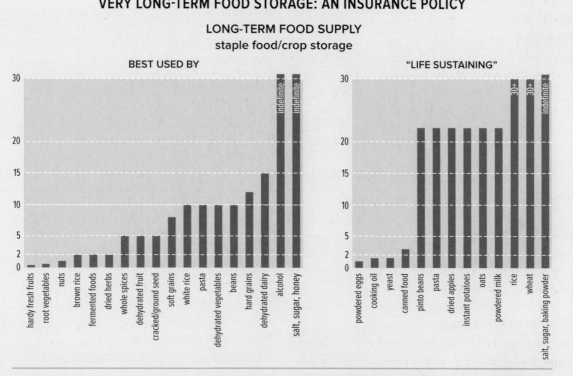

An overview of some of the most important long-term food stores sources: https://www.usaemergencysupply.com/information_center/storage_life_of_foods.htm; Nutritional Adequacy and Shelf Life of Food Storage by Dean Eliason and Michelle Lloyd copyright © 2005 Brigham Young University

Three modern technologies that have emerged in recent years allow us to put up a baseline stock of food for insurance purposes—for particular use in an emergency event that lasts awhile. These three tools are Mylar bags, oxygen absorbers, and plastic five-gallon buckets. Easily available from many emergency preparation suppliers, the bags and absorbers can be used with certain stable foods—dry beans, grains, salt, and sugar—to enable storage for very long periods of time because of the securing of the optimal food-storage environment, which is dark, dry, cool, oxygen-free, and protected from pests.

Ensuring these conditions can allow the viable storage of beans, grains, salt, and sugar for at least five years and up to twenty years. Research is still being conducted on these approaches, but evidence shows reliable storage of these foods for at least ten years under ideal conditions. Salt, of course, can be stored indefinitely, and many salts are already millions of years old at time of purchase (for example, Himalayan or Andean salt).

Storing food for very long periods of time using Mylar and food-grade buckets is simple and involves achieving the optimal conditions by (1) drying the food to be stored, (2) placing food and oxygen absorbers into a Mylar bag, (3) sealing the top with an iron, and (4) placing the bag into a tightly closing five-gallon bucket with a strong lid and storing it in a cool, dry, dark environment. I find that buying foods in bulk from the local co-op is a good way to find fresh, large-quantity dry foods at a good price.

I wait for a warm, very low humidity day on which to do the Mylar bagging, using grains and beans I have spread out in the sunshine during the middle part of the day. It's easy to get behind on the process and end up attempting to bag foods as the sun gets low. This is dangerous because the dew often starts to set well before sunset on such a day—rendering the whole drying approach ineffective and likely destroying any possibility for such food to last years in storage. I have not measured moisture content with precision but find that a couple of hours in direct sunshine on very low humidity days (here, that's 40 to 60 percent, which is relatively high for drier climates) does the trick. The beans, grains, or sugar are spread out thinly across screens or dry canvas so that sun access is high. A light breeze can help but is not necessary.

I transfer the food quickly into Mylar with two to three oxygen absorbers added into the bag as the food goes in. Holding the bag tightly so that as little air is inside as possible, I use a hot iron to make the top seal—mine is an electric model, but one could be fashioned at home and heated via a woodstove if necessary. Labeling each bag, of course, is very important. Though Mylar bags are available in full five- and six-gallon sizes to fill a bucket, I prefer to store at least half my long-term insurance foods in smaller one-gallon bags so I can open smaller quantities at a time, and in the event that a seal was not properly performed or the food was not adequately dried, less food is spoiled.

Within one hour after each bag is packed and sealed, you should see the bag tightly crinkled around the contents such that an outline of each bean, seed, or grain is visible on the outside—it should look like it is vacuum sealed. If it does not, you should consider that bag short- to midterm storage at best and eat it within a handful of months to a year. Some have brought up the concern that Mylar could be released from the bag into the food, as is the case with many flexible materials, such as plastic food wrap. This could definitely be a health issue to be sure. My take on the concern is simply that, while it's a possibility—even a likely one—the need to store food for very long periods is important enough to warrant the risk. As with most things, our exposure to artificial contaminants is high and continuous in the modern world—we must counter that with equal consistency through daily food-medicine and other health-enhancing tools.

The longest lasting storage options are of particular value because they allow us to extend harvests across years, not just months. This multiyear storability is crucial when acute events happen—like the Year Without a Summer. While such events are unlikely to happen often, they are inevitable, so a continuous backdrop of preparation for them is foundational. The longest-lasting storage approaches combine the right foods and methods, which yield a stable calorie and nutrient package that can be consumed more than one year from harvest. These food/storage combinations should be used as the baseline to one's food security. These include, in general order of value, the following:

► Live animals for milk, meat, fiber, hide
► Hay
► Dried fruits, vegetables, mushrooms, certain nuts and seeds in their shell
► Grains and dry beans (unhulled, ideally)
► Canned and frozen foods: long storage but high initial and operational inputs limit their usefulness

KEEPING A HOMESTEAD/FARM JOURNAL

Today the ice on our ponds is 6″ thick, it's 28°F out and snowing sideways. A quick look at our farm journal reveals that on this day last year the ponds had been free of ice for a week and the first spring peepers were heard. By checking the journal I also see that we had been eating arugula for weeks already last year while those same beds are now frozen solid. The earliest perennials were leafing out at this time last year—a far cry from this year. Our memories are poor and having a written record going back now almost five years has made me realize this to an acute degree. When it's nearing time to sow a specific veggie seed, look for a certain pest, or think about harvesting a crop I turn to the farm journal. I find an increasingly long span of records that show me the average time the same action was done in years past and the extremes on both early and late ends of the season. I try to record all migrations, new pests, leaf-out dates, ice out, ripenings and harvests, sowings, birthings of an animal, completion of projects, and dozens of other markers that can serve as both seasonal guides and reference points in the future about significant events. It is always an enlightening experience to leaf back through the years and see that whether something on the farm seems productive, early, unhealthy, or late, it all seems to even out by the end of the year.

Chapter Six

Adaptive Fuel and Shelter

On the coldest morning of the year, the thermometer read –17°F. I was pleasantly surprised—actually, almost elated—because you wouldn't know it inside my small home. It was still totally cozy, not a draft and comfortable enough to be barefoot on a concrete floor while making tea. The building had dropped three degrees all night from 66°F to 63°F, with the only heat source being a small (~30,000 Btu/hour) firebox running for maybe a third of the night. This was the building's fourth winter, so it is still in its initial testing phase. But, so far the results have been pleasantly surprising.

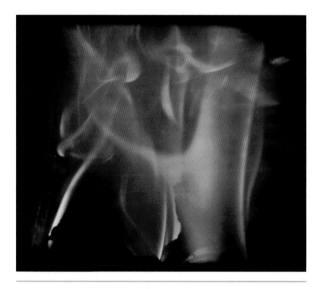

How the inside of a woodstove should look during an efficient slow but complete burn

I've lived nearly my entire life in the very cold climate of the northeastern United States, and for about fifteen years of this time, I've heated with wood. I am familiar with the deep cold of winter, and being concerned with energy use, I've made it a habit for many years to observe the heat requirements of the various buildings I've lived in. So far, spending time in this building has been a novel experience. It's astounding just how little energy a building can get by on in a cold-climate winter relative to what we consider normal.

For instance, an average home in Vermont that heats with wood requires about five to seven cords of dry hardwood per year (if using wood alone) to maintain a fluctuating temperature of 50 to 65°F, with prevalent drafts and uncomfortably cold corners throughout the home. For the past two winters the building I write this from has been heated on less than one cord of red maple and birch (relatively poor firewood) to maintain an average temperature of about 62°F with zero drafts and no cold corners. Granted, this building is smaller than your average home at fifteen hundred square feet—call it 40 percent smaller (twenty-five hundred square feet being the average used above).

So comparing apples to apples it's safe to say that a well-designed and -constructed home can be heated for about a third of that of a typical house. If fuelwood costs about $200 per cord (the current going rate in my area), figure you'll save two cords per year or $400. In ten years that's $4,000. Can one build a well-insulated tight home for $4,000 more than a typical leaky drafty home? No, but that's at current fuelwood (energy) rates—and they're only on the rise. If you are not using wood, you would be saving twice that amount (oil, around here, costs about twice as much for the same amount of heat), or about $8,000 over ten years. Again, these are conservative numbers and don't take into account price spikes, which will continue to happen and to which oil and gas are vulnerable but wood, from one's own woodlot, is not. The value of your comfort is another matter entirely, not easily represented in dollars.

The amount of work required to heat the Whole Systems Design studio and workshop used in the above example has been stunningly low. Inhabiting the building has been equally high in enjoyment and comfort. The systems we used in its design and construction were relatively simple, and the management needs of the building have also been low. But as with any wood-heated building, managing it requires more mindfulness than is generally needed by your average home.

Using Wood for Your Main Heat Source

In the forested cool- and cold-climate regions of the world, wood is the only sustainable and seasonally reliable source of heat that most of us can afford. Superinsulated passive solar homes are great, and if you live in one, you're exceptionally fortunate, but you'll still need *some* wood. And then there are the rest of us, who live in 99.9 percent of the other homes and can't afford a $10,000 to $75,000 complete thermal retrofit.

Step 1 is to secure a firewood source—either your own woodlot or that of a close friend and neighbor with a large one. Remember that not all wood is created equal—see table 6.1 for an overview of the actual heat offered by a given volume of wood across various species. Growing and harvesting the densest, fastest-producing fuels is the most effective way to reach fuelwood self-reliance. Step 2 is to process your firewood—fell, haul, buck, split, stack. This last step is where most of us tend to go wrong. Traveling around Vermont, I see more people slowly rotting their wood than quickly drying it. A stack of wood against the north side of a house with a tarp over it is an ideal way to grow mushrooms, but it won't yield wood fit for your stove, although many people burn such wood year after smoky year. I will now give an overview of why it's hard to dry wood well and what it actually requires, putting you one step closer to local self-reliance.

SELECTING, PROCESSING, AND DRYING WOOD

In selecting the best fuel for heating, it's important to remember the basic goal: getting the most energy from the forest (or field), dried, next to the woodstove with the least amount of energy expenditure and frustration. This is not a simple equation and varies from site to site because of the wood available, its condition (how knot-free it is), its accessibility, and the tools at hand (whether you have a wood splitter, for instance). On most locations the following considerations should be

A young hardwood stand typical of about three acres at the homestead—thus the need to plant more fuelwood

Table 6.1: Species Heat Values

Species	Lbs/Cord	MBtu/Cord	Species	Lbs/Cord	MBtu/Cord
Osage Orange (Hedge)	4,728	32.9	Ash, Green	2,880	19.9
Hickory, Shagbark	4,327	27.7	Cherry, Black	2,880	19.9
Eastern Hornbeam	4,267	27.3	Elm, American	3,052	19.5
Ironwood	4,016	27.1	Sycamore	2,808	19.5
Beech, Blue	3,890	26.8	Ash, Black	2,992	19.1
Birch, Black	3,890	26.8	Maple, Red	2,924	18.7
Locust, Black	3,890	26.8	Fir, Douglas	2,900	18.1
Hickory, Bitternut	3,832	26.7	Boxelder	2,797	17.9
Locust, Honey	3,832	26.7	Alder, Red	2,710	17.2
Apple	4,100	26.5	Pine, Jack	2,669	17.1
Mulberry	3,712	25.8	Pine, Norway	2,669	17.1
Oak, White	4,012	25.7	Pine, Pitch	2,669	17.1
Beech, High	3,757	24.0	Catalpa	2,360	16.4
Maple, Sugar	3,757	24.0	Hemlock	2,482	15.9
Oak, Red	3,757	24.0	Spruce, Black	2,482	15.9
Ash, White	3,689	23.6	Pine, Ponderosa	2,380	15.2
Birch, Yellow	3,689	23.6	Aspen	2,290	14.7
Elm, Red	3,112	21.6	Butternut	2,100	14.5
Coffeetree, Kentucky	3,112	21.6	Spruce	2,100	14.5
Hackberry	3,247	20.8	Willow	2,100	14.5
Tamarack	3,247	20.8	Fir, Balsam	2,236	14.3
Birch, Gray	3,179	20.3	Pine, White (Eastern, Western)	2,236	14.3
Birch, Paper	3,179	20.3	Fir, Concolor (White)	2,104	14.1
Birch, White	3,179	20.3	Basswood	2,108	13.8
Walnut, Black	3,192	20.2	Buckeye, Ohio	1,984	13.8
Cherry	3,120	20.0	Cottonwood	2,108	13.5
			Cedar, White	1,913	12.2

taken into account to determine how you can transform the solar energy stored in standing trees into heat for your home as effectively as possible.

FOREST MANAGEMENT

If you are cutting your own firewood, the question starts with silviculture. What trees should be cut to promote the health of the forest at large and the long-term growth of the most fuelwood, building materials, and other desired yields, such as wildlife habitat? This first part of the firewood chain of processing is the silvicultural end, where your decision as to which tree to fell in the woodlot determines the future development of that forest.

For firewood we generally want to choose trees that are unsuitable for building (sawlogs or poles), as those straight and "clear" trees are less common in the forest.

MANAGING FUELS ON THE HOMESTEAD/FARM

If only the rarity of a material was an accurate representation of the toxicity of a material, the world would be a very different place. Unfortunately today, literally thousands upon thousands of common household products litter the industrial home, landscape, and body. These include primarily cleaners and fuels. Toxic cleaners, being simply unnecessary, have no role in a resilient, regenerative lifestyle and should be left behind in the transition into such a life, along with much of the other sludge filling the daily life of *Homo consumeris*. Although millions of Americans nonchalantly store gas, diesel, propane, and other ubiquitous materials around the home in various ways, these materials are highly toxic substances treated with the respect they deserve by people aware of their power.

In my home I have no toxic cleaners to manage, so the only materials that require particular care, aside from very small quantities of adhesives and paints, are fuels. Such fuels are dealt with often on a small farm, where woodcutting with a chain saw and running a diesel excavator are common occurrences. These fuels, particularly gasoline, require vigilant care in storage and use. The components of a safe fuel system in the home include *procurement, storage, delivery,* and *disposal.*

Procuring the freshest, longest-lasting and highest-quality fuels is the first step to safely and effectively managing fuels and the machines they power. Always go to a reputable fuel station that does brisk business, where stale fuels are less likely to be a problem. When buying gasoline that may last more than a month on-site and/or be used in a small engine, always buy the highest octane rating you can and use ethanol-free fuel if possible. Some people who are very focused on making their engines last or storing fuel for a long time are now buying aviation gas, which is very high octane, above 100, contains no ethanol, and lasts much longer than typical 85-, 87-, or 89-octane gasoline.

With the advent of ethanol in gasoline blends, small engines are experiencing a major surge in problems because of the water-loving and solvent nature of ethanol. I go out of my way to use ethanol-free gas, which, luckily, is available in the valley where I live. Whether I can get ethanol-free gas or not, I always add StarTron fuel additive to help ameliorate the effects of ethanol. I also try to follow the good practice of running the machine dry after each use unless it's going to be used again within a week or two. If I am not going to use the machine for months, I'll run it dry on a heavy blend of additive, then shake out any remaining gas into a cardboard box with newspaper or planer shavings, then immediately light it on fire. There is no safe disposal of gas aside from burning it (and that's not terribly safe for the biosphere either).

Storing fuels must ensure that the materials cannot leak. This sounds simple enough, but almost any gas can will leak if it's tipped over or falls from a shelf. Things fall off shelves—don't put fuel cans where they can fall over. Store near ground level or in very secure higher locations where nothing can knock tanks over. Always use plastic or metal containers in very good condition and protect them from moisture. Moisture is an insidious problem that is the hardest challenge to deal with, because eventually, almost all fuels will take on moisture to some degree—propane being a notable exception and ethanol-containing gasoline being the most problem prone.

Keeping moisture out also means keeping dust, dirt, and organisms out. Organisms, particularly in hot climates, can become an issue in diesel fuel especially. Don't store fuels in areas with high daily temperature swings, which can precipitate moisture in the tank; this is easy to do by accident, so pay close attention to the way sunshine impacts an area before storing fuel there.

Never store fuel in direct sunlight—this is obvious but must be stated. Store in tanks at nearly full or full capacity—just as in a car tank, the more air space available, the more moisture can precipitate out of vapor and into liquid form as water. Water in fuel is one of the primary troubles we are trying to avoid with good storage. Some go out of their way to air-seal fuel storage very carefully in the pursuit of long-term fuel storage so they are prepared for the potential of long disruptions in fuel availability. There is much information available on these strategies online and in books on the subject—an important aspect to think about, given the fragility of fuel availability, the number of steps involved in its processing, and its crucial role in some homestead functions.

Propane is worth discussing separately because of its unusual characteristics. Propane is of immense value if only as an emergency backup source of electrical generation, cooking, and heat because it lasts virtually forever and is almost unaffected by cold weather. Try starting a diesel generator on the –5°F morning when the power goes out, which is always most likely to last a long time in the winter. Diesel presents major cold-weather challenges, and it doesn't last for more than a handful of years, reliably, in storage without accumulating water—unless very special measures are taken. Gas, while good in cold weather, keeps for an even shorter period of time—much less—than diesel, often going bad within months if it contains ethanol and within a year or two if not. If stored impeccably and using additives, one could probably reliably keep gas on hand without water accumulation for a handful of years, but that's risky.

Gas also presents a safety challenge and a risk to infrastructure because of its volatile nature and combustibility. Fumes from a leaking gas can—and all cans can leak, eventually—can travel many feet and ignite. Conversely, you can throw a lit match into a bucket of diesel, and it will extinguish immediately. Do that with gas, and you've got an explosion like that of TNT—hence, no homes have gas furnaces and storage tanks—you'd hear of them blowing up regularly on the nightly news.

Propane is safer given its storage infrastructure, although it is very combustible. The storability aspect of propane alone makes it worth having in the mix of backups we can count on here on the farm. Putting up a few twenty-pound barrels of propane for a propane-fired generator or camp stove gives you the insurance of having a reliable source of heat, hot water, and power or light even in very cold conditions and even ten, twenty, or thirty years from now, so long as you store the fuel containers out of the weather and off moist ground, where they can rust.

I have a portable gas–natural gas–propane trifuel generator available from Central Maine Diesel that can be run in various situations, giving me up to eighty-five hundred watts of power at 120 or 240 volts. Couple that unit with another backup gas generator and a handful of propane canisters, and I am likely to be able to make kilowatts for a long time in various conditions of resources being unavailable or very expensive. We have had some clients that see reason to believe that fuel supplies could be interrupted for long periods of time while the grid is down and aim to be power self-reliant for long periods. Those with the ability to invest toward that end have installed a common five-hundred-gallon propane tank just for backup use—an investment that will last decades upon decades, and if nothing ever goes wrong, they'll end up just saving a lot of money as the cost of propane rises from year to year. The earth isn't making more of it very quickly, after all. Investing in propane is like stocking up on salt—can't go wrong—someone is going to need it at some point, and it's not going bad on you.

Once we've located the cordwood trees—nice curvy, knotty, or otherwise "defective" individuals—we then determine which ones are of a size ready to be processed and which trees, once taken down, will promote the net growth of the forest the most. We always want to manage for overall net productivity—how to get the most Btus per acre (other important variables such as habitat aside for the moment), and in the woodlot that means managing the sunlight entering the system; we can view this as forest-canopy management.

In selecting wood, density is of prime concern, as density gives the best indication as to the amount of heat that can be produced. The denser the wood, the greater the amount of molecules the wood contains; the more energy embodied in the wood, the greater the heat output the wood can create. Generally, the slowest growing tree species produce the densest wood. This is not surprising when you think that an oak, for instance, puts more molecules into the same volume than a poplar. That usually takes more time, with the notable exceptions of black locust and osage orange. (See chapter five for much more on the exceptional utility of black locust.) In the northern forest of the United States, the best fuelwood commonly available in forests includes hickory, hop hornbeam, locust, oak, sugar maple, beech, yellow birch, and ash. See table 6.1 for a comparison of heat values of different species.

FELLING A TREE

Felling is easily one of the most dangerous activities that also happens to be a normal part of homesteading. While felling is not within our scope to cover at any

Silvopasture in action: grazing the fuelwood hedges at the Whole Systems Design testing ground

depth here and is covered very well in books and through workshops, the following are aspects I have found particularly useful in my experiences with logging over the past ten years. I fell trees with a 036 Stihl chain saw wearing Kevlar chaps, mountain boots, and a chain-saw helmet. I use wedges (two to three per large tree, one or none for small trees) and aim to fell the tree with the chain saw off and set aside, using an ax and a wedge to lever the tree over. This is the safest method I've seen and allows for a high degree of control. On occasion, I will throw a rope up in a tree and pull tension via people, my truck, an excavator, or a pulley system/come-along if a tree is heavily leaning in a bad direction.

I like to keep a backup saw on hand in case I run out of fuel amidst a difficult cut, or in the event I pinch a bar, which happens rarely if you are good and careful but eventually happens to everyone. I use mostly waste vegetable oil for my bar-and-chain oil and in five years or so of use have never experienced problems with it such as clogged filters or hot bars. A chain saw is one of the most crucial pieces of equipment in the modern homestead—ranking alongside a screw gun, a drill, a hammer, and an ax, so it pays to be well versed in its use and upkeep, and if you can afford to, own at least two saws and many spare parts.

I am no logging expert and would highly recommend that those new to chain sawing (or who are even experienced but not highly trained) take a Game of Logging or similar hands-on course. **There's no other activity as likely to get you maimed or killed on your land than using a chain saw and felling trees**, so learning the right way is crucial—this is one area where experimenting should be kept to a bare minimum.

CHAIN SAW: THE MOST IMPORTANT GAS-POWERED TOOL?

The most important tools and parts to have on hand so you can keep your rapid wood-cutting capacity going for as long as possible if the global flow of parts stops or ceases for a time

Along with a screw gun, I can't think of a more crucial power tool for general farm and homestead building and operation. Heck, with a hammer, an ax, a chain saw, and a screw gun, you can make most crucial elements and other tools to boot.* In the cold, humid regions of the world, forests dominate the landscape. Cutting and processing trees for opening land, heating, and building is as basic as managing a herd of animals is in the grassland environments of the world. In such a climate we must always be working with trees, whether we'd like to or not. Even if you find yourself with fifty acres of open land and no woods, resilient self-reliant heating needs alone necessitate that one be able to at least buck logs for splitting into firewood, if not fell and haul them as well.

I love processing firewood and being able to provision myself with the basic substance of heat, along with food. Sitting by the woodstove on a long winter night warmed by the radiating heat of the stove and the firelight is as rewarding as being fed by a meal I grew from

seed. I bought my first chain saw, a 025 Stihl, in college thirteen years ago and since have added to the collection as wood-cutting and building needs expanded over the years. I found an old 036 Stihl little used in a neighbor's garage that he was willing to part with for a hundred bucks and later bought a 460 Pro saw, also by Stihl.

I now believe that the collection is well rounded, with a saw for most jobs and good redundancy in three tools to do the same crucial jobs, which is really two, and if things really go bad, at worst, will be one tool to get the job done. I hope it's the 460! Over the years I have also been collecting the needed equipment to maintain the saws efficiently and economically, which largely means the means to keep up saw chain sharpening and replacement. A friend, Kyle Devitt, greatly enhanced those efforts when he began offering me his counsel on this subject a year or so ago, based on his experience as a professional firefighter with the US Forest Service and exposure to top-notch saw use and maintenance. I have built up the following list of supplies under his advice, which I share with you, not to suggest that you repeat the exact same list—though that may be appropriate—but that you see the goal of wood-cutting self-reliance and one approach toward getting there.

---

* I am not speaking in terms of primitive needs here, but of modern homesteads and farms. For a real shit-hit-the-fan scenario or in a wilderness type of setting, the tool list shifts to knife, axe, fire, and the like.

The list of self-reliant chain-saw tool and maintenance needs to keep a saw running for a long time, even through times when parts might be unavailable, is as follows:

The saw: Either a Stihl 460 or larger or 372 Husqvarna or larger. These saws represent the lowest end of the pro classes, which have maintenance features beyond the homeowner class of smaller saws. I didn't buy a 460 because I needed the power of it but because it's the baseline model that allows the kind of long-haul maintenance I am shooting for and because it's a tried and true model that isn't messed with from year to year as the smaller saws are. The 460 is a proven platform; you're not going to get a dud year or batch, in all likelihood.

ACCESSORY TOOLS:

▶ Two bars: 20″ and 24″ or 28″
▶ Spool of chain to fit those: 50′ or 100′. Chain specification: ⅜″ 050 full-skip round ground
▶ Multitool by Stihl: chain breaker and a rivet spinner, a bench-mounted device

▶ Air filter: HD, metal construction w/foam insert, removable cleanable element. These are hard to find. The one coming with a 460 is cleanable but delicate.
▶ Oregon bar (not Stihl), with greasable tip and grease gun with grease

MAINTENANCE PARTS—

THE MOST COMMON PARTS THAT NEED REPLACING:

▶ Fuel filters, a few
▶ Spark plugs, a few; need them with ethanol gas
▶ Air filter replacements for upgraded air filter
▶ Needle bearings
▶ Sprockets
▶ Washers
▶ C-clips
▶ Chain tensioner mechanism and a clutch spring
▶ Chain-break spring
▶ New clutch drum
▶ Package of miscellaneous screws and nuts that hold it together
▶ Bar nuts: three to five

HAULING YOUR WOOD

After the impact on the land system, the next level of decision making involves accessibility and hauling. A cord of wood weighs between four thousand and six thousand pounds wet in the forest. There is no other regular aspect of homesteading or small-scale farming in cold climates that involves the active movement of so much material (aside from construction projects), so you want to be extremely efficient with all aspects of moving this mass from the woodlot to the bucking and splitting area and from there to where it is stacked.

Good firewood management is a refined art of bulk materials hauling and storage. There are numerous methods for moving wood, and each site demands a customized approach. However, some basic principles apply no matter the situation. Hauling full-length logs—twenty to forty feet—from the forest to the processing area is almost always better than bucking into rounds in the woods. This process—skidding logs—can be done

Table 6.2: Wood Consumption—Weight Moved*

Cords burned	lbs moved per day	lbs moved per year
1	22	12,000
2	44	24,000
3	67	36,000
4	89	48,000
5	111	60,000
6	133	72,000
7	156	84,000
8	178	96,000

Given that wood is heavy and that it must be moved multiple times in its journey from standing tree to the woodstove, reducing fuelwood consumption is one of the most strategic ways to reduce labor on the cold-climate homestead.

---

* Pounds per day weight averaged across the entire heating season. Pounds per year weight does not include skidding/hauling/delivery to the processing site, assumed to be moving cordwood 3x per year from stacking to hauling to burning. Data is based on 400 lbs in a cord of wood and 180 days in a heating season.

with horse, skidder, tractor, excavator, or anything that hauls well over usually rough ground.

I skidded my logs for five years with a tractor and no skidding winch (a superhandy but expensive attachment that holds the butt of the logs off the ground so they don't catch). That worked okay if the ground was completely solid, but I always took extreme care on my sloping land to avoid flipping the machine. Now, I haul with my compact excavator, and while it's very slow, it can haul a massive amount of logs at once, which more than makes up for its lack of speed. Also, the digging bucket can be used to lift the butt ends of the logs just enough to avoid their catching on the ground, serving like a skidding winch on a tractor or normal winch on a skidder.

My ideal rig (aside from a draft horse, mule, or ox, which will hopefully be eventual—but requires major skill development), which I have yet to devise, would likely be a cobbed-together forwarder in the form of the compact excavator's hauling a large and beefy trailer on very large and wide flotation tires. The trailer would get loaded with the excavator, then hauled behind it. In Scandinavia, where logging is far more evolved compared to in the States, forwarders are standard practice and work in such a way that they allow logs to remain clean (no rocks and mud from dragging) and, even more importantly, the forest to sustain less damage in the process.

BUCKING AND SPLITTING FIREWOOD

Once the logs have been hauled to the landing/processing area, they must be cut into rounds. Rounds are determined by your stove or burner size, typically fourteen to twenty inches. If you need a mix and are not sure which size because you might change stoves in the future or because varying sizes are needed for multiple stoves on-site, it's always best to cut to the smallest length so it can be used in all stoves.

My cookstoves take fifteen- to sixteen-inch logs or smaller, which is relatively short and more difficult to stack than eighteen or twenty-two-inch-plus cordwood, which the house stove takes, but to be safe and to know that most cordwood on-site will fit the cookstoves, I cut a majority of wood to fifteen inches or slightly smaller. I do not measure with a stick or tape measure but get used to the size needed by using the stove. Measuring is tedious and is unnecessary if you pay attention to the cordwood each stove needs.

When possible, I stack logs parallel well off the ground and cut all the ones I can access easily without cutting the bottom logs. This is one way to avoid having the saw nicking the ground, which is crucial to saving your chain and getting firewood processed quickly. One split second touching a stone, and you need to stop and sharpen your chain, which slows up the rhythm and eats up expensive chain. A peavey is an incredibly handy tool for rolling logs that gets used often when you are managing large logs during bucking. It's common to cut through the log two-thirds of the way, then roll it over and cut the remaining third—thus more easily avoiding the dreaded saw-ground contact.

I love splitting wood by hand. It's second only to scything tall wet grass on a summer morning as far as homestead "work" goes. I can split wood for hours a day on most winter days and feel better off for having done it. If practiced well, it seems to be a meditation that can be good for the body and mind and which, of

Splitting wood: one of the few near-constant jobs in a cold-climate homestead. Fortunately, it's highly enjoyable and good exercise.

A well-made stack (not pile) of wood is crucial to actually having dry wood to burn. And dry wood is crucial to heating one's home optimally.

course, produces a satisfying yield. Having just recently used a very efficient flywheel-based log splitter after splitting by hand for ten years or so, I can now see the advantages of mechanized splitting.

Yet I remain interested only in splitting by hand so long as I am processing wood for a few efficient buildings or less. I can easily buck and split a cord in part of a day, and if I have much of the day and conditions are good (hard ground, clear logs, and nothing breaks), I can get a cord or so, felled off the stump, limbed, hauled, bucked, split and stacked. If the logs are knotty, everything changes, and I am pushing through the splitting process using wedges for many of the rounds. This is not nearly as fun and slows the process down by three to five times, easily. So clear logs are a blessing—and not always something you will have in abundance, so get used to wedges and sledges.

Techniques and preferences for efficient hand-splitting are numerous, but I'll offer the approaches that I have had good results from over the years. When starting with good clear rounds, I split them on large, very short "splitting stumps" using relatively light axes—I like something between a felling ax and a maul in weight and with a fairly fat flair to help open up the round and discourage the head from getting stuck in the wood. Fiskars makes a decent modern cheap version

(though the handle is not replaceable), and Gränsfors Bruks and Mueller make nice more traditional hand-forged versions.

You can swing a light ax all day, whereas a heavy maul tires one easily in short order—though they do have a role in particular woods and grains. I use a 3.5- and a 4-pound ax nearly all the time. I almost always prefer an ax with wedges over a maul and might for tough woods, but again, certain situations allow a maul to be very effective. Please note that "maul" is a misnomer, because it's not intended to crush wood fibers, only to split them. I prefer to call it what it truly is, a very heavy splitting ax or an ax with a widely flared head.

I split what I can from a pile with a light ax and then go at the remaining pieces with a sledge and a wedge, often with an ax that can be pounded (it needs to have a hardened poll for this purpose or you'll mushroom the head). Swinging the ax into the difficult round, then pounding the head through the round (standing at least partly perpendicular, not parallel to the ax handle or you'll bust your knuckles!) works particularly well. Often, this will open the round up, and a wedge can be inserted, which can then be alternately pounded on with the ax.

I also use a hand-powered splitter, purchased from Northern Tool a couple of years ago, which is good for the very difficult rounds. It's slow and has some design flaws, which could get fixed with a welder (the wedge is too short), but it does the trick of busting open superstubborn rounds. Every now and then a log of exceptionally clear easy-splitting wood such as ash will allow you to split before bucking each round. When this happens you can get a ton of work done rapidly.

I discovered this while splitting clear ash in Massachusetts about five years ago. There were four- to six-foot-long logs on the ground, and I had slammed an ax into an end of one, golf-swing style, to rest the ax while taking a break. I noticed a hairline crack shoot down the log as the ax sunk. Instantly, I realized that these logs could get split this way, and I tried. Sure enough, I walked around the pile swinging the ax like I was on a fairway playing golf, each log splitting significantly from the strikes. Later, bucking was magical—each slice of the saw yielded presplit cordwood, which fell away from the saw cut ready to be stacked.

DRYING YOUR FIREWOOD

Wood wants to be wet. In fact it's the only typical raw material that holds more water than good soil (usually 120 to 200 percent of its dry weight). The cells in a tree's wood have such a stubborn grasp on water (it's their life currency) that they only release it fast enough to avoid rotting under specific conditions—and it's in these conditions that you want your fuelwood. To make things even more difficult, these conditions are hard to come by in a humid, cold climate: Throw a hundred pieces of wood from a plane flying over New England, and ninety-nine of them (or probably all one hundred) will begin rotting within a handful of months.

This is why finding a large supply of dead wood to burn in the woods is often impossible—the fungi get to it first. Burning green wood (more than about 20 percent moisture content, depending on species) is a bad idea because it promotes creosote buildup in the chimney, is hard to keep ignited (while at the same time keeping airflow through the stove to a minimum), reduces heat output by 20 to 70 percent (causing one to need one and a half to three times as much wood for the same amount of heat), emits much more air pollution, and is heavier to process. The only tree in our Northern Forest that is burnable in close to its green state is American ash because of its exceptionally low standing moisture content.

It's safe to say that most people where I live burn wood that is far greener than it should be and get a correspondingly low amount of heat value from the wood while also getting more creosote buildup in the chimney, moving more mass than they need to, and tending to a more difficult fire. There are two reasons for such commonplace burning of suboptimal, wet wood: lack of drying time and poor wood storage.

Under average conditions it takes about one year or more to dry sixteen-inch cordwood thoroughly. Under good conditions cordwood will dry within five to seven warm-season months. Under the best of conditions (very sunny; lots of airflow; tall, thin stacks; and stacked with lots of air space between the billets), one could dry wood adequately for efficient burning in three to four warm-season months if the billets (pieces of cordwood) are in very short lengths (fourteen inches

The average wood-heated home in northern New England requires five to six cords of wood per year to heat—that's fifteen to twenty thousand pounds of material to move each year, twice (at least). So ensuring that your wood processing from forest to stove is as efficient as possible is important.

or less) and split on most or all sides. Even small billets that are unsplit take a very long time to dry, as the bark holds moisture in the wood very effectively.

Remember that wood only truly dries in a cold, humid climate between April and November, when temperatures are above 40°F and humidity levels are relatively low. A well-sited and -built wood stack does most of its work from July through September with high heat and low humidity. If, like most people, you find yourself needing to rapidly dry a small amount of wood, piling it near the stove for a week or two before burning it can remove as much moisture (especially in small billets) as months of drying—as well as humidify your house. Having the wood near the stove for even just a few days before burning can polish off the remaining excess moisture of marginally dry wood and is an oft-used strategy.

The soundest approach to properly heating with wood is to put it up well in the autumn or winter a year or more before it will be burned. This requires a surprising amount of space dedicated to wood drying: about 128 square feet for the four cords typical (minimum) of most home needs—that's one stack four feet high by four feet wide by thirty-two feet long per year, two of them at the beginning of winter. In addition to food gardens, the life-after-cheap-oil front yard will be dedicated to wood storage—easily a car parking space's worth of wood—more if your home isn't very well insulated or your stove burns inefficiently.

Skillfully drying firewood (or building lumber) requires managing the moisture factors—precipitation, temperature, and air movement—through the proper location and construction of a wood stack. Optimal wood drying and storage sites are:

- Easily accessible to sled, cart, truck, or tractor
- Off the ground
- In a warm, sunny area
  (against a south-facing wall can be ideal)
- In an area with good airflow
- Near the point of use

Proper construction of your fuelwood stack (you're crafting a stack, not making a pile here) involves the same things as any building: a *stable foundation*, *stable shape* (not too tall for the width), *solid connections* (the way the wood stacks against itself), and a *sound roof*. A solid foundation can usually be made by propping up pallets or 2 × 4s to form a wide, level surface with plenty of support points. Ensure air access underneath the stack. Remember that the bottom layers are most likely going to get wet in snowy weather as it drifts against the pile and by lower airflow volumes and wetter air near the ground. Ideally, you burn the top three-quarters of the pile, then restack the remaining one-quarter on top of another stack for the following year.

PROPERLY STACKED FUELWOOD

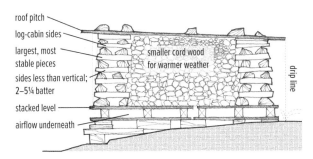

roof pitch
log-cabin sides
largest, most stable pieces
sides less than vertical; 2–5¼ batter
stacked level
airflow underneath
smaller cord wood for warmer weather
drip line

Wood rots if stored carelessly and/or under tarps. A cleanly built and sturdy stack with scrap-plywood or -metal roofing is the best method aside from shed/barn storage.

Stable connections between the layers of cordwood are made by ensuring that the wood is of a uniform length—usually sixteen inches or eighteen inches, that the wood is layered up neatly and flatly, and that any retaining of the walls (see figure 6-08) is rock solid. A sound roof is best made out of anything impervious, large, flat, and rigid, such as scrap plywood or, best of all, scrap metal roofing. Ensure that the roofing is pitched and drains water away from any area that would backsplash onto the wood. Drying wood under a tarp seems like a fine idea until you try it. When you do, you realize how hard it is to keep the wind from removing or misaligning it and snow from forming depressions in it so that water slowly percolates into the pile. If you must use a tarp, heavy canvas or rubber tarps are infinitely more workable than light poly tarps.

BURNING FIREWOOD

Heating with wood is so typical in some areas that many people tend to think it's easy and simple, like jogging. But just like jogging it can be done poorly for a lifetime without your knowing it. Mastering the task of burning firewood involves keen observation and continual experimentation to understand the interaction between your wood and the combustion system you use (stove and chimney). At the Whole Systems Research Farm, all our buildings are heated by woodstoves—all of which are cast iron except the sauna burner, which is steel. While we have wanted to build a masonry oven for years, the recent addition of two wood cookstoves, which also heat water with incredible effectiveness and allow both stovetop cooking and baking along with space heating, has back-burnered the masonry oven plans for a while. So, for the purposes of this chapter the focus is wood heating with a traditional woodstove, covering the tricks and techniques I have found for making the system as optimized as possible. The wood heat system on a homestead can be broken down into the following areas:

1. Fuel properties: density and moisture
2. Firebox arrangement
3. Airflow
4. Burner design and stove selection

COMPOST HOT WATER HEATING SYSTEM (JEAN PAIN MOUND)

The woody water-heating compost mound steaming away on a winter's morning with 155°F temperatures inside on a 10°F day

After studying this method for a few years, we finally got around to making a woody-debris water-heating compost mound last autumn. The concept is simple: As organic matter decays large amounts of heat are released—anyone who's made compost has noticed this. Amazingly, Jean Pain, a Belgian innovator who pioneered this system, found that you could make a pile of completely composed woody debris (carbon) and get up to 18 months of 120°F water from the mound at one gallon per minute or more, with no added nitrogen, such as manure. Since so much woody debris goes to waste in parts of the world, this is an exciting discovery. With a large enough pile, heat release can be captured practically by building the compost pile around long lengths of plastic-water tubing. Pumping cold water into one end of the tubing produces warm to hot water at the other end—the compost pile acts like a furnace. This system's beauty is in its multifunctionality—you get hot water and fertile soil at the same time—and in its lack of pollution—it's a combustion-free furnace. The design details and engineering of this system can get complex and are out of the scope of this work, but the basic approach is simple enough and practical for people with some background in plumbing or composting to pursue a system

like this. We made a large at least 12-yard-wide pile of compostable materials—fresh, fast-rotting woodchips are the basis of this in the Jean Pain method though you could make a "normal" compost pile with much more nitrogen and get soil more quickly, but likely less hot water for a shorter duration. We used 10 to 12 yards of white pine chips, about 2 yards of white pine sawdust, 1 yard of spruce planer shavings, and about 3 yards of horse manure (with bedding). After laying down 100′ of perforated flexible drain pipe, we built the mound inside of hay bales mixing all the material together, 5 layers high to form a cylindrical pile about 16′ in diameter including the bales and 6 to 7′ tall in the middle. Within one month the pile was 120°F and within two months it was 120 to 155°F throughout. We're still in the early testing of it, only four months in, but it seems clear that we can heat about one gallon/minute 24 hours/day to about 115°F. This is enough water to heat a small home with a radiant floor or plenty to heat an in-soil bed greenhouse, which is how we are applying the heat. We used 800′ of three-quarter-inch poly tubing in the mound and 400′ of the same tubing beneath 14″ of soil in a greenhouse-raised bed. We will continue reporting on this impressively productive system via our website, videos, and workshops.

The characteristics of your fuel are fundamental to the performance of the entire system. You can save yourself about 30 percent on time if you process very dense wood such as hickory, hop hornbeam, or locust. Starting with dry, dense wood, the next order of operations is optimal arrangement in your burner. The ideal arrangement depends on the following factors: amount of coals present, draft of stove and chimney and airflow entering the stove, moisture content of the wood, size of the wood, and surface area of the wood. For instance, when starting a fire with no coals present, you want an open, high-airflow configuration, which promotes the fire as much as possible. This setup burns down quickly, of course. So once the fire is going and you have a bed of coals, you can start adding wood in such a way as to reduce surface area and slow down burning—by placing firewood parallel with one another in the firebox.

This is all very intuitive, of course, but is worth pointing out for those new to the art. An all-night burn is most often achieved by loading in a small number of very large billets and placing them parallel with one another so the whole fuel charge is most like a single block of wood. Dialing down the stove in combination with that arrangement will yield a slow burn in all but the leakiest of stoves. The airflow in your system is most heavily determined by the draft created. The draft is the movement of air from the interior of the building through the stove and up the chimney. The fastest drafts are created by straight, hot (interior or insulated) chimneys. The longer they are, the more draft is created, within reason. Poor draft situations are common and often are the result of 90- or 45-degree bends in the system and exterior or cold chimneys.

These situations can be dangerous in the long run because of the way they promote creosote buildup and the resulting chimney fires. A good draft with very dry wood that is burned very hot from time to time in a firebox that is not smoldered often can reliably create a situation where chimney cleaning is never necessary. You should always check your chimney each year, but if you burn effectively in a well-designed setup, you will find your stovepipe surprisingly clean. I check once per year and have never needed to clean

The homestead's most important power plant is our wood cookstove. It is pictured here in typical midwinter action performing multiple functions simultaneously: boiling tea water; cooking a multiday meal of venison, lamb, squash, potato, seaweed, shiitake, sunflower seed, kale, and garlic; boiling gone-by squash for the ducks; baking cookies; simmering chaga-reishi chai for desert; heating all the hot water needed by two people for bathing and dishes; and heating fifteen-hundred square feet of space to 72°F on a 20°F day.

my stovepipes, but I am very careful to almost never smolder the fire. I also do a "cleaning burn" every few weeks, in which I let a lot of air into the stove and rage the fire with small dry pieces of kindling for a few minutes while the stovepipe gets very hot (400° to 600°F). I then damp the stove down before things get dangerously hot. This tends to burn off any creosote that may have started to form on the interior of the chimney before it gets to dangerous levels.

When you are selecting a wood burner, the following points should be kept in mind:

▶ **Airtightness:** Is it nearly 100 percent airtight? It should be. You can check by hooking the stove up to a chimney and, with a fire going, dialing the stove completely down, then holding a candle or lighter next to all possible air inlets. If any indication of air being drawn into the stove is present, it's not airtight. The problem with having a leaky stove (which many of them are, especially old ones) is twofold. You end up with a stove that you can't keep a long burn in—maybe three to five hours at best, and you end up burning inefficiently, always feeding more air through the stove than necessary.

It's important to note that the most efficient fire is one in which the minimal amount of air is allowed through the stove to keep a very hot flame (bluish) alive. Any more airflow beyond this is cold air you've unnecessarily brought through your heated building. Unfortunately, it's hard to check how airtight a stove is in the showroom, so ideally, you would get to a stove in action beforehand in a friend's or neighbor's home.

► **Construction:** Cast iron lasts the longest and should provide a serviceable stove for lifetimes if it's taken care of. Steel tends to weaken and rust over time. A glass door in the firebox is more than just an aesthetic consideration; it allows you to see how the fire is behaving and adjust airflow and wood arrangement accordingly. Without a window into the firebox, it's nearly impossible to achieve optimal burning conditions because you're guessing to some extent as to the condition of the fire. You can't simply check by opening the door of the stove, because as soon as you do, the airflow situation is immediately changed.

Aside from airtightness, insulation and reburning abilities are important. Modern stove makers have begun to understand that a hot firebox is crucial for a complete burn—at least half of the heat in wood is contained in gases that only ignite at 1,000°F or higher. To achieve this crucial hot firebox temperature, insulation is very helpful. I would not consider a new stove that is not well insulated. I have been extremely impressed with my Morsø small stove of about 35,000 Btus, which can hold coals for literally sixteen hours if it's dialed down well.

However, there are some pretty efficient old stoves out there that, because of their design, can achieve very hot firebox temps—the Jøtul 118 is a good example of a very simple stove that does so.

> **The most efficient fire is one in which the minimal amount of air is allowed through the stove to keep a very hot flame alive.**

My wood cookstove, a Waterford Stanley, was not airtight from the factory but with some additions of ceramic insulation can hold coals for a short night if loaded fully before bedtime.

Most modern stoves feature ways of reburning the volatile gases mentioned above. This approach has replaced the once-common catalytic converter. Having owned a Hearthstone Harvest, which cost over $300 in catalytic maintenance and never ran very well to begin with, I would caution you to avoid catalytic stoves at all costs. Catalysts are notorious for breaking and going bad from ash contamination. Luckily, the new stoves (or some old ones) that can reburn volatile gases (via recirculation) are commonly available.

► **Size:** It's very common to oversize a woodstove, and there's no better way to waste wood—aside from drying it poorly—than to use a stove that kicks out more heat than you need. It's like a vehicle—an eight-cylinder truck is great for hauling something, but if you want to coast along with high miles per gallon, you want a very small engine. Think of your home as something that should coast. I have friends who have an enormous stove that can easily get their home to 80 degrees. That's nice when you need to heat the home up quickly after being away, but the rest of the time it's very inefficient. You want to go with the smallest stove you can to heat your space.

Calculating the heat loss of the home can help you do this, as can using the advice of someone well versed in homes and stoves. My office/shop heats with a 30,000-Btu stove (max), and it's fifteen hundred square feet of well-insulated passive solar space but is not superinsulated or superpassively designed. The 1970s house on this property heats, barely, with a 55,000-Btu stove, and it is eighteen hundred square feet.

MAKING BIOCHAR IN YOUR WOODSTOVE

Biochar is an emerging but also very old soil amendment that is likely to be an extremely important addition to the quiver of soil-building techniques for the next century and beyond (see chapter four for more on this strategy). Making biochar can take many forms, and I have been experimenting with some success in making biochar while running my woodstove during the heating season. I have found two ways that seem reliable and easy to do: (1) remove coals before they burn down to ash, and (2) place a baking pan or other heat-tolerant, relatively airtight container filled with sawdust into the firebox during use of the stove. Both techniques work well, but the former is a lot less complicated and involves no extra materials to achieve.

With the former method you want to quench the coals in water immediately—this shatters their structure, creating more surface area—then crush them into powder once they have cooled. The latter strategy yields a fine char powder. With both strategies you then need to inoculate the char to make it *bio*char. As the need for biochar in soils is increasingly being understood, the possible synergy of making soil while heating our homes all winter is enormous and provides what is perhaps the only regenerative yield of the woodstove. This is also an important way for a home to connect with and benefit the nutrient cycle of the whole human habitat.

Biochar coals

Woodstove coals

Biochar crushed

Adaptive Shelter

It's fitting that much of this section (and much of the book) is written in my favorite setting for office work—the balcony of the Whole Systems Design studio. It's a southwest-facing nook about 4′–10′ in size. A third of it is set into the interior of the studio, with two-thirds of it cantilevered out into the sunny, south-facing

The Whole Systems Design,LLC studio and workshop Photograph courtesy of Whole Systems Design, LLC

side of the building. This configuration provides solar access and wind protection simultaneously, providing an enhanced microclimate that extends my enjoyment of the outdoors across the year.

Today it's a brisk 14°F, with light winds and mostly sunny skies. I am warm enough to type comfortably, so long as the sun misses the clouds, while enjoying views of the Mad River Valley and soaking up the rare mid-January sunshine. It's a perfect place in which to ponder what's behind a highly functional dwelling that is built and managed to be as resilient as possible for a future of continual and often rapid change.

The first point to note in the process of understanding, designing, making, and managing shelter is the amorphous nature of it. Shelter is composed of varying degrees of protection from outdoor elements including wind, rain, snow, light, and darkness—it is not, necessarily, a building. The most successful buildings are designed and built around existing and constructed (or planted) spatial elements that offer sheltering values. In other words, a new building should not have to produce 100 percent of its sheltering value from blank space, providing all of that value from its walls and roof alone.

A well-designed building is set into existing sheltering elements in the landscape—a hillside, a south-facing wall of trees, a bedrock outcrop—and extends those sheltering influences farther into the interior of the structure. "Passive solar landscaping," in addition to passive solar house design, is one term that helps us understand this concept. A passive solar house embedded within a passive solar landscape will outperform the same house constructed in a typical suboptimized setting. The point here is that context should be the

> The most successful buildings are designed and built around existing and constructed (or planted) spatial elements that offer sheltering values.

The sauna and an outdoor bedroom at Whole Systems Design's farm
Photograph courtesy of Whole Systems Design, LLC

starting point for all building design, just as it is for the development of biological systems such as a vegetable garden, an orchard, or a fish pond. It's often easier to forget this when locating and making a building than when laying out a vegetable garden.

This section provides an overview of how we design and develop highly functional adaptable shelter. It is broken down into the following components: design and construction; siting and orientation; foundation, frame, walls, roof; mechanical systems; and nutrient cycling.

DESIGN AND CONSTRUCTION

The WSD studio/shop was designed to be a high-functioning space but also to balance the necessary technical components of a high-performance building made for a cold climate. By "balance" I mean that we sought to achieve a result that enhanced the occupants' experience across all aspects of building design. For instance, the building should lend a relaxed feeling of light and warmth and harmony with its surroundings while also being thermally efficient. It should improve the outdoor space around it while also itself being "green" in materials. It

Horse logging was used to harvest all the timbers for the Whole Systems Design studio from the farm property.

should be easy to heat and cool, and not feel like a heating or cooling machine. In other words, the building's materials and form need to enhance all variables related to its use, not be oriented too heavily toward one alone.

That's the fatal error in many modern approaches to "green" buildings: A building needs to be more than an efficient machine—after all, no one wants to live in a mechanical contraption. I have been studying and walking through highly engineered eco-groovy homes for more than a decade, and almost none of them have felt like high-quality space to me. Certainly, the most insulated or most solar or most lighting-efficient among them did not. The few gems that I have experienced were shelters that balanced their design goals, turning them from potential competing aspects into synergistic ones. Those buildings did this harmoniously and create spaces that are unified within themselves and within the site. Remember, while a building requires engineering, it is not simply an engineering feat alone.

There was another goal beyond the balancing of all goals that also organized our design: to make a highly durable, very long-lasting structure that needs as little maintenance and renovation work as possible for as long as possible. "Build it once" was our mantra. See the sidebar "Entropy, Resiliency, and Regeneration" in this chapter for the reasoning behind the need to make our construction projects based on durability above most other goals.

BUILDING IS AN ECOLOGICAL ACT

The Whole Systems Design studio was also conceived as an outgrowth of the site. Its design seeks to connect spaces within the site, to enhance those spaces, and to serve as a room within the landscape itself. Look around, and you'll notice that all good buildings do this. Many famous modern buildings—even if they represent a novel work in and of themselves—often do not do this, however. And those structures don't last over time as a result—cultures don't maintain buildings that do not improve the space in which they are embedded (see Stewart Brand's *How Buildings Learn,* among other works, for more on this idea).

In studying and making buildings over the last half of my life, I've come to believe that there are two basic

The timber frames on-site were hand-raised when practical—always an incredibly rewarding experience.

approaches to architectural design: (1) work from the inside out, using an idea or image as the organizing framework (most "architecture"), or (2) work from the place—the site, inward toward the space composing the building itself. A brief story illustrates the former approach well and why I quit architecture school.

I had started graduate school in architecture and was sitting in my freshman year studio listening to our second assignment. The professor described the design challenge: "Conceive of a home and yard for a writer. The house needs to be a specific size, between twelve hundred and sixteen hundred square feet, it can only be two floors, it must have a main entrance on the road side and a fence of six feet tall around the yard. Everything else is up to your design discretion."

He then opened the room to questions. After a few others spoke up, I raised my hand. "You didn't say where this building was located. Is it here in the cold Northeast or in the tropics? Also, which way is south, and is it in a dry or wet climate?" The professor, visibly annoyed, replied, "That stuff doesn't matter! It's the space that matters, the relationship between figure and ground; just worry about that, about how you develop the positive and negative space and the *architecture* of the solution. Don't worry about where it is; that's irrelevant." I quit the program within a few weeks.

And "where it is" has become the underlying factor of our design work from which all solutions stem, married with the factors of the client: place and people and how they interact in synergy. This is no different a design approach from the way in which we approach design problems in the landscape; only the medium is different—it's dead instead of living—and that makes certain aspects of the system work far more easily; the building doesn't grow in size over time, for instance. And certain aspects are less inspiring: The day a building is finished is the day it begins to decay.

Siting, Orientation, and Layout

After reasonable access, good water, and solar access is established, the task of locating the dwelling within a homestead is the linchpin. Get this wrong, and the rest of the site is forever constrained. Get it right, and the site can work in synergy with highly functional

interaction between spaces, all pivoting off the central zone 1 anchor that is the dwelling. The primary challenge addressed on many of my site consultations is house locating. Several patterns emerge repeatedly here, no matter the property; I have noted these below. We can think of these as a designer's checklist for locating a dwelling optimally on a piece of land. As you read through them, please remember that on many locations, especially in hilly or mountainous regions, there are often only one or two good dwelling sites, at best. More often you are choosing between several not-so-great spots for building. However, on a good property you will be able to sift through most (usually not all) of the following criteria during the process of elimination that represents sound decision making for siting a building.

Whole Systems Design's office nearing completion with the "human solar oven" balcony facing south-southwest over the pond

ENTROPY, RESILIENCY, AND REGENERATION

This entire book is about successfully adapting to what appears to be the only law of the universe: change. Resiliency is the domain of living systems because it is only biological systems that are capable of responding to shifting conditions and can alter form and composition to bounce back from a disturbance. We must remember Bill Mollison's remark about "life being the only organizing force in this part of the universe."

Dead systems can't respond, adapt, undergo value-adding change quickly. They are entropic; the day a building is finished it begins to decay—just the opposite of a tree, which only accrues value as the years roll on. How, then, can a dead, abiotic system—like a building—be adaptive and resilient? At best, the human managers of a building need to adjust it throughout its life cycle to help it respond to shifting exterior conditions of climate and resources, and to the changing interior conditions of occupant needs. This is an important way to understand the possibilities for regenerative action as well.

Regeneration requires the growing of value over time—biomass and biodiversity being two key indicators.

In this view buildings can indeed only be "less bad," while actually doing "good"—performing regenerative action is the sole domain of living systems. Abiotic systems cannot be regenerative; they are in a constant state of erosion and disorganization (entropy). Hence, the resilient homesteader/designer's goal is to replace, wherever possible, nonliving components with biological elements; for example, a hedgerow instead of a fence, a horse instead of a skidder, a dog instead of an electronic alarm system. We must move beyond the concept of green and sustainable, these meaning merely "doing less bad." This requires the prolific involvement of living systems wherever possible.

> The day a building is finished it begins to decay—just the opposite of a tree, which only accrues value as the years roll on.

REASONABLE ACCESS

Can you get to it? Sounds simple enough, but it's not. There are innumerable great spots on which to build if you could only get to them reasonably. If getting to one "reasonably" means making a road up with a steep—15 percent or so—grade, you should seriously consider another location. If it requires vehicle access across a 20 percent slope, forget it. Yes, the landscape is filled with such driveways, and yes, they will be unaffordable to maintain in a world where oil (read gravel and machine hauling) is expensive or less available. Such steep driveways become riddled with impassable ruts in a few years and deep gullies after that. Many an old roadbed is now an intermittent stream.

Can you maintain the driveway by hand with a wheelbarrow? It's worth asking that question—even though it's a tough one. In general you need to be well below a 10 percent grade to achieve a perpetually maintainable accessway that can be relatively erosion-free—not to mention the snow and ice shenanigans

that a slope over 10 percent causes. Note: In a very dry climate, steeper grades become more reasonable—the wetter it is and the more intense the rain events that occur, the lower the grade that should be developed.

After slope comes the consideration of length. The shorter the driveway you need to build and maintain, the better—all things being equal, of course. But I'd rather have a four-hundred-yard-long 2 percent access than a fifty-yard 12 percent grade in a heartbeat—the former would be easier to negotiate in winter and a fraction of the maintenance time and costs. My driveway ranges from 1 to 4 percent grade and has never needed a gravel addition to it in the ten years I've lived on-site. Driving carefully to avoid rut creation is also important.

DRIVEWAY TO THE NORTH

Save the best sun-drenched zones for gardens and people, not cars. A garage on the south side of a house is a type I error, which often makes it worth ditching an otherwise good house buy. You cannot fix this level of

EXPONENTIAL EROSION

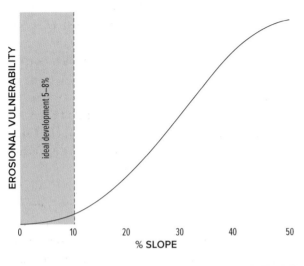

It pays to stay below 10 to 15 percent grade when possible with all driveway and road developments.

OUTDOOR LIVING SPACE MICROCLIMATES

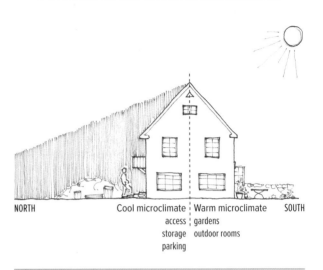

The general pattern for optimizing outdoor uses on all sides of a building in cold climates. Prioritize your south-facing outdoor spaces very carefully.

suboptimization without major renovation, and not renovating is an ongoing problem, because walking far from the house to reach the sunshine is a heavy cost to pay. The ideal is to have your sunny, most intensive zone 1 gardens beginning immediately next to the home on the south side, adjacent to a door from the kitchen. If that's not possible, you want it to be as close to this layout as you can develop.

MICROCLIMATE DEVELOPMENT

How will the location, form, and materials to be used affect the climate of the site around the proposed structure? All buildings make certain areas cold and certain areas warmer. They protect certain areas from wind and sometimes increase airflow in other zones. Buildings always influence the lighting and moisture variations on a site as well. How you intentionally harness these influences is crucial to the successful integration of a building within a landscape.

All successful buildings in cold climates create positive microclimates by (a) accessing sunlight, (b) storing that heat in massive materials such as stone and water, and (c) protecting spaces from wind. Primary questions within microclimate enhancement include these: Where is the easiest place to create a strong sun trap?

Are there knolls or large banks of mature trees that can shelter the building? Windbreaks are easy to plant but take time to develop and are crucial to have solidly established to the north of most buildings; certainly, this includes dwelling and animal barns.

Where do I want to increase the warmth of a space, and what space do I want to cool? A pattern always worth applying can be termed "southern hot spot, northern cold spot." It's safe to say that in almost all situations you want your zone 1 gardens to begin immediately south of the dwelling, utilizing that warmest of locations intensively. South-facing corners of built elements that are protected from wind are the most powerful warm microclimates created on a site: Place the most important and intensive food-production and outdoor human-use spaces there.

Additionally, designing a dwelling such that it allows "living across the solar day" is crucial to human productivity, enjoyment, and health. I follow the sun during my days working the homestead: breakfast or tea on the east-facing entrance steps that have warmed early; lunch in the south-facing midday spaces (or in shade in the high summer); dinner on the west-facing house deck. Remember that in cold climates you only need shade for a small proportion of the year, and cool shade is usually much easier to come by than protected sun-drenched, warm outdoor spaces. It's the latter spaces that need to be thought about from the very beginning of site layout if you want enjoyable outdoor use across the seasons.

LOW ENOUGH TO GRAVITY-FEED WATER

Often buildings are located above key water sources, thus eliminating the possibility for perpetual, free access to water in the home, barn, shed. Don't make this mistake if you can avoid it. And yes, this often means locating your water source before siting your building.

Cornelius Murphy enjoying the view from the Whole Systems Design balcony, an intentionally warm microclimate that is also rain protected. Note how each level of doors is protected by the roof above them—these double as summer shade providers when the sun is high.

UTILIZE AND ALTER SLOPE

If the site is a mix of flat and slope, use the slope for a dwelling (or barn). Having ground-level access on two floors is hugely valuable. Saving the often-rare flat ground for gardens is crucial. A building doesn't need flat ground—get out of the looking-for-a-tent-site mentality that most of us are conditioned to thinking within. I've been on countless site consultations with clients looking for good house locations where I was walked from one flat spot to the next, each time being asked, "What do you think of this?"

Find a slope to build your place into if you have the opportunity, and you'll also have a major solar-thermal and earth-thermal advantage, as well as ease of dual floor-grade access. I've also seen the same error played out over and over again by bulldozer-loving contractors flattening building sites before construction. A little terracing and a sloping site can be just as accessible as a flat one, while being higher performing, more beautiful, and optimally fitting within a site. Additionally, most house construction entails the use of heavy machinery and grade changes. Position the building such that its immediate outdoor spaces benefit from these slope alterations. Most often this comes in the form of allowing the building excavation to form a terrace on which south-facing garden space and an outdoor room is developed and on which a north-facing accessway is developed.

SIGHTLINE LEVERAGED AND BUFFERED

Will the location of the building offer good surveillance of the rest of the site? This can be a hard criterion to judge before construction begins but is an important consideration when multiple good home sites are present on a property (not often the case!). A small rise or knolltop is an obvious choice and can be a good one, if all other aspects are taken into account. Views *from* the building to the surrounding area, as well as views *to* the structure from the surrounding area, can be considered under this heading as well.

> The more you can see at once on your homestead the better.

The Whole Systems Design barn-intergrated greenhouse nearing completion Photograph courtesy of Whole Systems Design, LLC

Consider sightlines within the property to be views having a value beyond aesthetics: Who's eating the chickens? What birds are stealing the fruit off the peaches? Where's the dog or the child? Seeing provides instant awareness—the more you can see at once on your homestead, the better. In some locations, especially near roads and other buildings, sightline buffering is crucial. We all have seen the classic bad example of a house at a T-intersection that gets lit up with each passing car's headlights. This can easily happen even in a rural area if a house is located just in the wrong spot in relation to a road bend or rise.

Be sure to experience the site and building locations within the site at various times of day and, ideally, across the season before finalizing decisions. I once conducted a property evaluation in Vermont for a couple from New York City who was about to put a binding offer on a large piece of land. The property was at least three-quarters of a mile from an interstate highway that you could not see from the property during the day. I visited the site without them and was impressed by its beauty and features. It was perfect for their goals—a better fit, actually, than almost any other property I've evaluated for people.

On my way back from the site, I looked in my rearview mirror while getting on the highway and noticed a rise about half a mile back, in a spot that could, if things lined up just right, pitch headlights upward and straight at the property I had just visited. I called up the client when I got home and told them the property was as ideal as they come—except the one concern about the highway headlight potential. "Have you been there at night?" I asked them. They had not.

The next day I got an e-mail from the wife, who said she had a strong feeling that she needed to see the place at night before placing the offer, so she drove up from New York City after work, arriving in the middle of the night. As she walked into the field above the house, the pasture above her was lit up intermittently with each passing northbound vehicle on Interstate 89, more than three-quarters of a mile away. They never placed the offer and found land a few months later in a different part of the state.

FACE THE MIDDAY SUN—NOT ALWAYS SOUTH

Another mistake made over and over again: architects orienting the building dead solar south. Sounds counterintuitive, I know. But think about the effects of slope and topography. In many hilly locations the *effective* solar

> Solar south is not always the optimal solar orientation—that must be found for each site on a case-by-case basis through observation.

day is not exactly between sunrise and sunset times. Put another way, the sun comes up late on a west-facing slope and goes down late as well. Such a slope has a late solar day. The opposite is true for an easterly-oriented site. This can be affected by both the general slope and by local topographic features such as knolls, which hide the sun early or late in the day. You must design around these if you want to optimize the sun's precious value.

Solar south is not always the optimal solar orientation—that must be found for each site on a case-by-case basis through observation. At our research farm site, we orient buildings south-southwest for optimal solar gain and immediate outdoor space optimization, given our west-facing slope in general. Also worth noting is that western sun is warmer than eastern sun, so if it's a toss-up, orient farther toward the west for solar gain.

Soundscape

This is not usually a major criterion to be sure, but it should be considered on every site. "What influences the soundscape of this exact spot versus this one here? Are they positive sounds? *When* do they occur: often, according to season, or at each rush hour?" These and other questions used to identify what audible influences on the building location can be crucial and sometimes vary greatly across a site of less than five or ten acres, most certainly on larger parcels.

Road noise is easy to underestimate; I did not understand this when I bought my place ten years ago and figured it would be relatively quiet in terms of car noise. Not even close. A cliff on the far side of a quiet state highway, which is nearly half a mile away, acts as a sounding board, sending vehicle noise back at my site. Small sheltered depressions within the site, behind bedrock outcroppings to the west, offer some reprieve from this noise. I can tell if it's a weekday or weekend every morning by this noise, even though most people arriving on this site would think they are in the middle of nowhere.

Interstate highways generate noise that travels incredible distances when the topography allows for it and conversely can be buffered in very short distances when a knoll or ridge deflects the sound. Only earth, rock, and wooden fences or similar materials significantly affect sound transmission; tree planting does

little unless it is very dense and wide—although it can help a lot when covered in snow.

Positive soundscape should also be considered for its ability to mask negative sounds—examples include waterfalls, brooks, frogs on a pond, and wind in coniferous trees. It's worth recognizing that when inside a very well built and insulated home the outside soundscape is often muted heavily and of little consequence. It's more often important to consider these effects in the outdoor living environment, where "white noise" can make a huge difference.

Spatial Design for Security

Spatial design can greatly create or reduce one's security—especially from an awareness perspective, meaning one does not necessarily need to erect blockades; that is, knowing that someone is coming onto your property can be more effective than a huge fence. Most of the design aspects below are hard to achieve all the time, especially in a site or house as is, not built from scratch.

Site: Leverage a Strong Position

The areas we use the most on-site should be higher in elevation than areas where people who would do harm to us or the site are most likely to come from. Land above your zones 1 and 2 should be "hardened" to the extent practical: people deterred from circulation through it, via built fence, guard animals, brushy areas, live fences, rocks, steep ledge, and so on.

You should be able to hear activity at the site boundaries from high-use areas. This is especially true where people are most likely to enter, such as your driveway. Here's an example of a bad situation: Your kitchen or workshop has a wall between you and the driveway entrance; because of noise in those spaces, someone could enter in a diesel truck and you wouldn't know until he knocked on or bashed in your front door.

Foundation, Roof, Frame, Walls

There is much available on this topic in print and via the web, so this will be brief and limited to the areas where our experimenting has added new information to the common literature or where our experience conflicts with commonly understood theory about these aspects of construction. Our approach is always one of making sure built components are absolutely as durable and modular as possible; in no area of construction is this overarching approach more important or consequential than in the primary building elements of foundation, roof, frame, and walls.

Foundation

In making the foundation for a building, I would add two specific pieces of advice to the immense amount of information available on the subject: (1) Extend the foundation walls (stem walls) higher above grade than is often done, and (2) go deeper than is typical in your area for frost stability.* These two approaches will help ensure that your foundation supports as durable a building as possible.

Let's start this discussion with a simple fact: The most common and serious problems in cold-humid-climate buildings have occurred at the interface of the foundation and frame. I have worked as a carpenter doing demolition and reconstruction work and have evaluated dozens of old buildings in my consulting work for clients. I cannot emphasize enough the need to focus on this area first. It doesn't make sense to evaluate other aspects of the building, though important they may be, before this; go into the basement right off the bat, and look at the corners of the building and the tops of foundation walls. Rot in this location is all too common even on modern buildings. The result of this problem is expensive and labor intensive in repair since it's at the base of the building.

Constructing a foundation wall that extends high (called a *stem wall* when above grade) above the ground level upon which snow piles up is the easiest way to ensure a long-lasting structure. This costs more—that's why it's not done in our disposable world of "get it up as fast and cheaply as possible." But compromising on this aspect of construction will cost you far more in the long run. When involved with building design, I like to specify stem walls of at least fourteen inches, preferably eighteen inches or more. The stem wall of

* Insulating foundation walls by frost protecting with foam board is often as effective and less expensive than digging deeper foundations.

> **The most common and serious problems in cold-humid-climate buildings have occurred at the interface of the foundation and frame.**

our studio-workshop is about fourteen inches, which is the minimum I would recommend—there are areas where snow does pile against the wall above this point, but the water detailing in the wall itself should preclude this problem. Insulating against the cold of this stem wall is a challenge, but a surmountable one, and involves stepping the wall back toward the interior of the building to achieve full insulation value in this area.

Going deeper than the typical frost penetration depth (four feet in my area) is another form of cheap insurance against aberrant winters in which cold is as severe as it has ever been but snow cover (insulation) is nonexistent. Nature plays by few rules, and there's no guarantee that frost won't penetrate five feet deep in one winter with typical cold and lack of snow cover. That's all it would take to heave a building—and a building heaved by frost is the worst thing (barring freak events) to happen to a building aside from fire or roof collapse. Foundation movement is to be avoided at all costs.

There are many options to explore for protecting a foundation from frost as well, beyond just going deeply. We have only experimented with two of these approaches, and I would leave this aspect to the many good resources and experts available for shallow frost-protected and other foundation strategies. The point remains: Whichever method you choose, be very conservative about reducing the possibility for frost penetration, and plan for weather patterns that may not be normal for your area. You can't easily redo the foundation later in the building's life, so building it to be adaptive to all possible climate and other scenarios is good prudence.

YOUR ROOF:
STEEP, SIMPLE, STRONG, STEEL (OR SLATE)

The same goes for the roof: Be conservative, and plan for abnormality in weather patterns. Next to the foundation, the roof is the most common source of problems and the scene of many needless issues. There are many resources in the literature on high-performance, long-term roof systems, but the following points are often missing from the discussion and are based on our own direct experience:

A good roof boils down to this approach: Keep it steep to shed snow in cold climates (unless you want to bet the farm that the past hundred years of snowfall patterns will stay the same for the next hundred?). Getting the snow off the roof is crucial to avoid dangerous and damaging snow loading. The idea that snow is good insulation is asinine, because it implies that enough heat is escaping the roof for the snow to function as insulation. A well-insulated roof is cold and does not let enough heat out to allow for snow to serve as insulation.

To shed snow reliably in all conditions (including the dreaded icing-then-snowing event, which is very sticky), you need to make a roof that is at least 12:12 pitch (45 degrees) or steeper and composed of steel or slate. An asphalt roof will hold snow reliably on nearly all angles but vertical (and then it's a wall, not a roof!). A simple roof means avoiding valleys if at all possible. Valleys are always the weakest point in a roof—hence, the vernacular tradition in all cold snow climates is to avoid them.

Take heed of this wisdom. Some building designs will necessitate a valley, but minimizing them is important, and if you do choose to make one, detail it impeccably with wide flashing and careful lapping of all roofing. "Strong" is pretty self-explanatory, but if in doubt, beef it up. Steepness always adds strength, all things being equal, as rafters act more like posts (in compression) with steepening of angles. A simple roofline is also, of course, easier to build and repair. Every change in pitch, ridge, or especially valley is an opportunity for leaks and snow holding that are not present in a single steep plane.

Single steep planes are your friend in snowy country. Using steel or slate—it's a no brainer: Slate will last easily a hundred years or two hundred if it's good quality and maintained well. Steel can last fifty easily, if you spot paint the rusty spots every few years. Both shed snow very well. Both are immensely serviceable after they are done functioning on the structure they were originally installed onto—they can be reused as shed roofing, tile, or to cover cordwood stacks. Asphalt as roofing should simply be against code in all climates. Avoid asphalt—there's simply no need for it! If your budget or time frame or labor do not allow for slate (the best roof choice) or standing seam metal (next best), then choose steel roofing. Basic metal roofing will last at least as long as asphalt, be more leakproof, and shed snow, and instead of filling a dumpster when it's riddled with rust or holes, it can serve crucial functions as wood stack roofing, shed roofing, or animal pen fence mending, among others.

Wood shingles can make some sense if you have a resource of splittable rot-resistant wood such as cedar or white oak (large, clear white oak) and can be highly sustainable as a locally available resource. However, it holds snow very well and is prone to leaking. In the long term locally sourced and made shingles will again return as a primary roofing, but while slate and steel are available, you might as well take advantage of their superiority.

FRAME AND WALLS

Wall and structure framing options are many and, again, covered well in the available literature. I would add the following to the deep and broad information available on the subject for this climate.

- ► Sensibility of a timber frame
- ► Larsen truss framing
- ► Insulating choices and tips

The debate on the function, sustainability, and overall practicality of a post-and-beam frame in this climate continues to rage on, for good reason. A stud-wall-framed building is usually cheaper and as fast to build as a comparable one made with a timber frame. After all, in a cold climate one must balloon frame (surround) the timber frame with an insulated wall, rather than fill in between the timbers with insulation. The latter is a viable strategy only in a much warmer climate. So a timber frame here means building two frames, one for structure and one to hold the insulation. It's easy to see the merits of avoiding the timber frame as outlined above, so I want to point out reasons to actually choose a timber frame.

The first is aesthetics; seeing the structure at work is almost universally desirable by building occupants and for good reason. This value cannot be quantified, but its influence on the enjoyment of the building is undeniable and potent. The second is also somewhat indirect in terms of the actual building value it offers and has to do with process. Felling trees, milling (or hewing or working in the round), then cutting joinery and raising a frame is an ages-old process that can utilize material closer to its raw state than constructing a house from sticks milled one after another, each the same as the last. The rewards of this process, the skills it requires, and the result achieved cannot be attained by the stick-built approach.

My take on the timber-frame versus stick-built debate is that if you are determined to do the most economical thing but still love the values of a timber frame, then stick-frame the exterior walls but use posts and beams for the interior structure. This approach is eminently practical and yields a highly functional, beautiful result.

A light, gusseted frame (Larsen truss) within a timber frame structure has worked pretty well for us. A Larsen truss is a small-member framing system in which an inner and outer post are joined with gussets. In our case we used two 2″ × 3″ posts of softwood separated by a two-inch air gap (thermal break), held together every handful of feet by half-inch plywood gussets. It has performed very well thermally but is a bit light structurally for holding the windows and doors—they rattle a bit more than desired when the wall is bumped into or doors are slammed. It was also very difficult to fill the crucial thermal break (air space) within the wall itself, leading us to feel that a staggered-stud wall of, say, 2″–4″ posts used to form an eight-inch or greater wall would be more ideal and easier to spray cellulose into.

The big advantage to a Larsen truss is the ability to use very small dimensioned material to make a wall assembly with a solid thermal break—allowing a landscape with nonmature trees to offer a yield of dimensioned lumber that's usable for walls. You can

The Whole Systems Design studio before move-in, with roughly 95 percent of the materials by volume used in construction sourced within seventy-five miles of the site and more than 75 percent of the wood sourced on-site.

mill 2″ × 3″ studs from tiny trees—slightly tedious maybe, but doable. Again, staggered studs would probably be our choice next time, and we would likely shoot for a ten-inch wall instead of an eight-inch assembly.

Cellulose was our insulation of choice for the shop/studio, and we've had no problems so far. I like cellulose for its ease of use, relatively low toxicity, sustainability, resistance to mice, and, perhaps most importantly, its ease of modification. It's guaranteed that you'll need to modify an insulated wall at some point in the building's lifetime. You'll need to send a cable or waterline through, add a vent or window, expand the structure, fix a leaking pipe, or retrieve a wire. When you do so with cellulose, it's easy: Pull out the material, put it in a bag, and stuff it back in when done.

Spray foam? Forget about it. You'll have to wreck most of the materials used in the wall when modifying a wall that is foamed. Additionally, foam is not a viable material in my mind purely from a toxicity standpoint and lack of sustainability in production. Foam board is one of the few instances when a toxic material serves a seriously important function, such as when insulating spaces below grade, as in foundation work or root cellars.

I would not choose to use straw in this climate, but I know some good builders who seem to have success with it. The best straw-bale buildings I have seen in this region are smartly done, with the bale walls starting very high on top of stem walls that are at least eighteen inches to two feet high. Foam or cellulose is used lower down in the wall. One issue I have with bales is the need for very thick walls to accommodate both the relatively low R-value of straw (in bale form) and the dimension of a bale. The thinner the wall, the better from a building livability standpoint, as they allow much more light into the interior. A thick stone or bale wall of, say, sixteen inches must have bigger windows and more glazing to be equally bright in comparison with a cellulose- or wool-insulated building with walls of, say, ten or twelve inches. We chose to go for a relatively thin wall (eight inches) and still achieve very high performance of about one cord per year to heat fifteen hundred square feet of space in a very cold climate with negligible passive solar gain.

An important additional point needs to be made about windows. All things being equal, it is more efficient and economical to use a smaller number of large windows than to use many small windows. This is an often-neglected fact, I think, because people fail to realize the insulation value lost in areas around each window frame. The window, of course, has a very low R-value, but so does the framing around each window; you give up R-value with a much wider area than the window itself. Proportionally, you will get more light for less heat loss with bigger windows in smaller quantity. It's also much less expensive if you are paying for labor, as windows are time intensive to install. Larger windows can often provide deeper light penetration into a building as well (if they are vertically elongated); that's why schoolhouses have very tall windows and can often offer a better connection to the outdoor environment.

Mechanical Systems

From a resiliency standpoint, the fewer mechanical systems the better. It's that simple. Every system with moving parts will break and will need to be repaired, or in a time when specialized parts may not be available easily, the system may need to be abandoned for a simpler one. In a well-constructed home, the weak links, resiliency-wise, are the mechanical systems. A maximally resilient home reduces the number and complexity of mechanical systems and uses the most fixable systems when mechanisms are necessary. Let's start with why we need mechanical systems. Briefly, they are commonly related to the following:

▶ Lighting, communications, computing
▶ Heating and cooking
▶ Air movement (including ventilation)
▶ Water distribution

In a well-constructed home, the weak links, resiliency-wise, are the mechanical systems.

Table 6.3: Comparing Mechanical Systems

Service	Typical System	Resilient System
Light, daytime	Electric lights	Daylit via windows
Light, nighttime	CFL/incandescent	LED, task-oriented
Communications, written	E-mail	E-mail?
Communications, verbal	Cell phone or landline	Landline, spoken
Heat, space	Automated furnace	Woodstove
Heat, water	Furnace	Woodstove
Cooking, stovetop	Gas or electric range	Wood cookstove
Cooking, baking	Gas or electric oven	Wood cookstove
Air circulation	Electric fan	Heat source location optimization, open floor plan
Ventilation	Electric fan	Passive vents (high and low)

Achieving a high level of resiliency involves assessing each of these independently and identifying ways of providing the services we need from each of these systems through the simplest, most fixable, passive, and durable means possible.

LIGHTING AND COMMUNICATIONS

These are corollary building functions and not limited to the built environment on a site, but given their technical nature and inherent need for protection from the elements, they fit within the building resiliency focus of this chapter. It is easy to downplay the need for lighting and communications in hard-core resiliency thinking and planning: Under rapidly changing and sometimes emergency conditions, one can easily assume that food, clothing, heat, shelter, water, security are the most basic needs; I sure have until recently.

However, the last time we went without power for several days, I realized how important artificial lighting and communications are, and how much better positioned we are to help our neighbors, if we have power when most people have none. Nighttime lighting allows one to work, fix things, house and support people in ways not possible in the dark. If you're in the dark when the sun sets, all cooking, fixing, socializing, and other crucial functions cease. Doing these activities under a headlamp or with a flashlight is possible but limiting.

In our studio/shop we have developed a baseline of lighting systems in LED bulbs that have a low enough load such that when the grid is down a small backup power system of two deep-cycle batteries can illuminate the building's interior for at least a week between battery chargings if they are used frugally and selectively. Wiring the building with more rather than fewer lighting circuits allows controlling what areas are lit selectively, correlating to a reduced electric use by allowing the user to choose to light only those areas in which she will be working.

It's also important to note that LEDs are extremely focused in their light output, so task-oriented lighting strategies are the directive in their use. You do not get generalized space lighting with most LEDs but can get high-quality focused light with the screw-in Sylvania bulbs we use. For more generalized light we find that a string or two of six- to nine-watt LED Christmas lights are just the ticket to provide general light in a space so we can walk around and see where we are going.

To lower the electric load further, we can supplement lighting needs by running Coleman white-gas lanterns, which throw off a range of light according to need, from low task-based amounts to an intense quantity that allows a large area of work to be conducted. White gas stores for very long periods of time, so it is a good choice for such a use. Supplementing these sources are a few boxes of long-burning candles, which, of course, last indefinitely and offer low-intensity light for general space illumination.

The same two-battery backup system we use for lighting has two other uses: power for communications and cordless tool (drill, impact driver, cell phone) charging. The last time we lost power was from a snowstorm. Such climate-related events often knock out power but leave phone and cable lines in working order. That means for those of us on cable or DSL, the Internet is still accessible *if* you have the power to run the modem and computer. Fortunately, these are very low loads and can easily be met for days to weeks with small backup power systems.

Why is maintaining Internet access such an important thing, given that you can't eat it and it doesn't keep you fed or watered? The web is an easy and fast source of weather and other news events. Knowing the latest developments concerning a storm track, a flu outbreak, a nuclear release or other toxic spill, or many other likely events that affect our lives is crucial. The only ways to find these out are through direct word of mouth, the radio, the phone, or the Internet. Since redundancy is key to resiliency, having multiple lines of such communication and information access is key. During the latest power outage, I was able to plug my laptop, modem, and router into my backup power system and access the Internet for days with no battery charging from the generator necessary before the grid came back up. During this time I could see what other weather events we could expect and plan around them accordingly. In various other scenarios this information could be more valuable than we might imagine.

The other two aspects of communications related to the building system we use are a multiband solar- and hand-powered radio and phones. Both of these can operate independently of a functioning electric grid, and the radio can operate independently of any external electric power source as well. We have replaced our two cordless phones with corded versions that need no plug-in (just a working phone line). These have the added advantage of much lower electromagnetic radiation (EMR) emission that reduces their negative impact on human health. Between this communications triad of Internet, phone (landline and cell), and radio, we are likely to be able to at least receive (if not send) information from the larger world about current events that are necessary to plan around in various disturbances to fragile systems such as the electric grid.

We have a fourth communications source that's usable only for local transmission of information—handheld two-way radios by Motorola, which are handy for us around the farm. In addition to the Motorolas, I also have one CB radio, which can pick up regional signals of information. I plan to supplement the CB with a second handheld device so that if necessary I could communicate with another person in this general area (within a mile or two) in a cell-tower-down situation.

All of these hold charges for long periods of time and can be recharged via generator, deep-cycle battery, or solar-panel fence-charging backup source.

THE WOODSTOVE

The woodstove has become the logical power center of my own resilient homestead. It's nearly impossible to break a wood-powered heating system, and if it does break down, it's easily repairable by low-tech, often on-site means. Aside from stored food and potable water access, it is a dry stack of fuelwood under cover and a woodstove in a well-insulated building that covers your most basic needs for survival and thrival across a wide range of bumpy conditions. I cannot see a scenario in which a cold-climate home should be without the simple power and utility of a wood-burning stove.

> **Our wood cookstove not only heats fifteen-hundred square feet of space, but heats all of our hot water, bakes and boils our meals, dries clothes, and provides firelight—all on less than two cords per year.**

Over the last handful of years, we have taken the basic woodstove to some of the logical extents of this technology's capability, including heating all of our domestic hot water. Our firebox is rated to about 35,000 Btus of which about 25 to 35 percent goes to heating the hot water in the stainless steel jacket at the back of the firebox. This heated water thermosiphons into the hot water tank requiring no pump—if the power fails we can still run the stove and have hot water. Our wood cookstove not only heats fifteen-hundred square feet of space, but heats all of our hot water, bakes and boils our meals, dries clothes, and provides firelight—all on less than two cords per year. The colder it gets outside the more hot water we have. During warmer times of year in the fall and spring we put one charge of wood in the stove just to make enough hot water for bathing; this also helps keep the space from being too humid and gives us a clothes-drying option while taking the

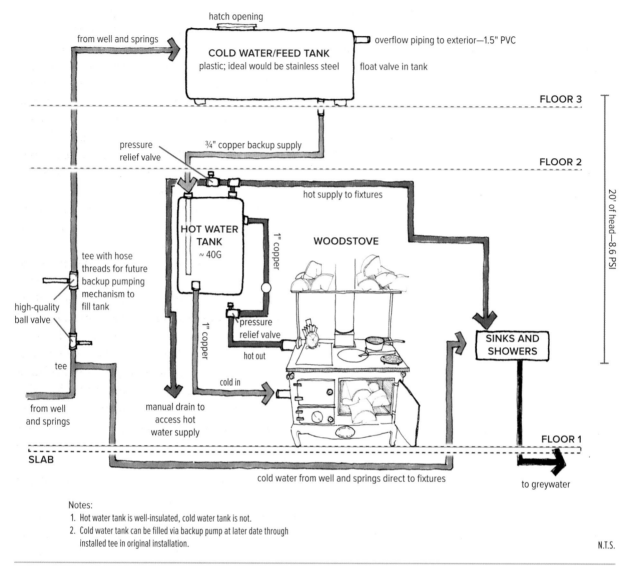

WOOD-HEATED HOT WATER SYSTEM

hatch opening

from well and springs

COLD WATER/FEED TANK
plastic; ideal would be stainless steel

overflow piping to exterior—1.5" PVC

float valve in tank

FLOOR 3

pressure
relief valve

¾" copper backup supply

FLOOR 2

hot supply to fixtures

**HOT WATER
TANK**
~ 40G

1" copper

WOODSTOVE

tee with hose
threads for future
backup pumping
mechanism to
fill tank

high-quality
ball valve

tee

1" copper

pressure
relief valve

hot out

cold in

**SINKS AND
SHOWERS**

from well
and springs

manual drain to
access hot
water supply

FLOOR 1

SLAB

cold water from well and springs direct to fixtures

to greywater

20' of head—8.6 PSI

Notes:
1. Hot water tank is well-insulated, cold water tank is not.
2. Cold water tank can be filled via backup pump at later date through
   installed tee in original installation.

N.T.S.

The generation 1.0 as-built plan for our woodstove-powered hot water systems—working exceedingly well, though improvements are likely; for instance, by adding some cold-water-outlet at varied heights in the tank

chill off. We use simple solar hot water showers and the pond and river for bathing in the warmest four to five months of the year.

We also have solar hot water on site, which works well when we have cloudless days (rare in the winter). But solar hot water with it's electric requirements, pumps, and many parts (in addition to its large cost) does not come close to the resiliency of a wood-powered water heating system. In a much sunnier climate,

solar hot water starts to make a lot of sense if you have the money and a consistent electrical supply—though you can make very inexpensive solar hot water systems if you are in a very hot climate.

Ventilation and Moisture Management
Tightly constructed, well-insulated buildings need significant venting of humid indoor air to avoid moisture (and future rot) problems associated with a buildup of

moisture into liquid water within wall and roof cavities. Indoor air humidity must stay below 40 percent and preferably closer to 25 to 35 percent to avoid water buildup in the insulated cavities, as humid air moves outward toward drier, colder air and condenses.

This condensation is the hidden killer of many tight buildings. You can get your foundation, walls, and roof all correctly detailed (and that's difficult enough in the cold-humid regions of the world) but still lose the battle against moisture management in a long-term durable building if moisture is allowed to condense inside your walls and roof. The typical approach to achieving this ventilation is to install heat recovery ventilation units (HRVs) during construction. These units send humid, warm indoor air to the exterior while bringing in cold, dry outdoor air. In the process they exchange some heat in the indoor air with the outdoor, air providing an efficient way to vent the building—without losing all heat present in the indoor air.

We have chosen to forgo this active fan-based ventilation in the WSD studio/shop in favor of a non-electric, passive, and manual approach to ventilating the building. Now in our third winter, we have performed enough moisture testing and heat-consumption measuring to determine that this approach is working very well. In our second winter we measured interior cavity moisture by drilling holes at different elevations in likely problem areas (where wall plates meet the roof, tops of walls, peaks of the roof, and so on) and inserted a moisture-sensing probe at three depths across the wall profile. We found all three readings in every test hole (over twenty of them) to result in 5.9 to 9.8 percent moisture content, well within acceptability.

Our ventilation approach is simply to open windows slightly at the top of the building and provide some measure of air infiltration at the lowest levels of the building.* This creates a passive air movement ("stack effect") that quickly moves air through the building from low to high

locations. We apply this technique whenever we are in the building and generating humidity (breathing, cooking, washing, and so forth). While we give up a measure of efficiency by losing all the heat present in the air leaving the building, we gain the advantage of having an unbreakable, completely no-electric-load means of ventilating the structure. Given that we use about one and a half to two cords of wood per year to heat the fifteen-hundred-square-foot interior *and* all of the domestic hot water, we do not feel as if the heat consumption disadvantage is worth spending the money, electricity, and time (in replacing the fan or other breakable aspects of the unit) on. If we did use an HRV, we could probably knock that number down maybe a third of a cord or so. Would that be worth the trade?

Figure that it's about $1,500 to $2,000 in parts and installation to get the HRV in. Figure another $50 a year in electric usage to operate it and another $100 to $500 every ten years to fix broken parts associated with it. That one-third cord of wood is worth $50 to $70 in current value. Then there's the time factor of dealing with another set of moving parts in the building. There's also the input and impact of manufacturing the HRV and its components along with the nuclear, coal, or other power production to run it. The answer to whether an HRV is worth it or not is ultimately a judgment call, and one that in my opinion seems less worthwhile than I thought it might be, after living in the building for three years.

I was skeptical about the decision to forgo an HRV during construction, so we designed in the ducting necessary to install a unit later without having to rip open a wall and incur fairly major time or money expense. Fortunately, the need for an HRV seems less each year I use the structure. As a rule, the simplest solution is always the best, and this is a great example of this principle in action.

One other moisture-management challenge is worth focusing on: foundation-frame interfaces. As I inspect landscapes and buildings around New England, I always see buildings in various stages of decay. Nearly always, the area of structures in the worst condition (unless the roof has failed) is where the foundation and walls or first-floor plane interact.

---

* We open windows at the gable end of the east and west side of the buildings. While these are near the top of the building, they are not at the exact highest point. In retrospect there should also be a cupola in the building that can vent air from the absolutely highest point in the structure, which, I am realizing over time, is very important and the basis of that design element in vernacular architecture the world over.

There are two challenges that are exceptionally tricky to get right for the long haul in this regard. First is wood framing and concrete or stone connections. These almost always rot faster than the rest of the building, especially in the case of concrete as opposed to stone, when used as a foundation, because concrete constantly wicks any moisture it is in contact with, whereas stone does not convey it nearly as well. So a concrete footing sitting below the seasonally and massively fluctuating water table will bring that moisture up vertically many feet if it extends that far. That moisture then comes into contact with the building framed atop this footing. That's where you look first when inspecting the structural integrity of a building in a cold climate—and work there is especially expensive.

The second, just as common, problem area—and one much harder to avoid—is a wood-framed wall next to a stone or concrete retaining wall, or a wood-framed floor above a basement or cellar. In both cases the problem arises between cool, humid areas and drier, warmer conditioned space. Such a juxtaposition of spaces is impossible to avoid in cold climates, and it raises a very difficult situation where, at the very least, mold is likely to form and where, more commonly, full-on rot happens. It's only in the most well-designed and detailed buildings that no problems emerge here over time, and even those require vigilant inspection to ensure that this stays the case from year to year.

Inspection is best done during winter and in early spring, when the contrast between temperatures and humidities is highest and when the water table rises the most (surging humidity in the foundation-contact areas). Old buildings often avoided disastrous consequences in this area of buildings, not because they were necessarily designed or built well but mostly because ventilation levels were so high because of poor insulation. Older buildings also stacked the odds in their favor by use of higher-quality materials than we have today, at least in terms of wood—such as chestnut and old-growth, rot-resistant woods. New, second, third, and fourth-growth pine, for instance, is far inferior to old-growth pine, which is actually quite strong and rot resistant. Today you'd never consider pine to have *any* rot resistance, which is true for today's pine, but not of the pine our grandfathers used.

Much of the mold, rot, and attendant air-quality problems of modern buildings are because of their tightness, coupled with poor detailing. Tight buildings necessitate impeccable design and detailing when it comes to dealing with moisture, and in no place is moisture more commonly a problem than at the earth-building interface (roofs are probably the second most common areas—I always start at the bottom and top of buildings when inspecting, then look at the middle). Getting around problems at the ground-structure interface in a tight, well-insulated building requires that we keep as much moisture out as possible via site and foundation drainage, use of concrete sealer, plastic sheeting, and foam board, then promote adequate ventilation to remove any excess humidity that still remains or is generated.

This one-two approach is crucial, because you can never control water with 100 percent assurance; you can only put the odds in your favor via multiple backups. To drive this point home, I've known at least two instances in Vermont where springs formed just uphill and underneath a building during the first ten years of its lifespan, both times creating major moisture problems for the structure. This climate is incredibly wet during certain periods of time; couple that with enormous contrasts in both temperature and humidity from indoors to outdoors, and you've got one of the most challenging climates in the world in which to make a long-lasting building.

It's not in the scope of this book to go into design detail as to how exactly to get through this moisture challenge—there is some good information, amidst the significant misinformation out there on this topic. I simply want to alert the reader to this fundamental challenge in creating resilient, long-term buildings that won't need massive work to repair in the future and offer the general strategies for dealing with this challenge.

Water: Passive Supply for the Resilient Home

Before electricity and the means to drill a deep well, no one went to the trouble of developing a whole farmstead before locating a free-flowing and quality water source uphill of the development location (main house

HOME WATER SYSTEMS

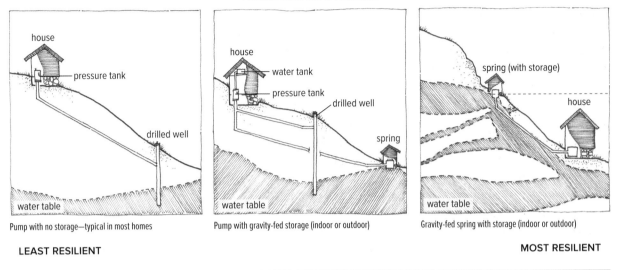

Pump with no storage—typical in most homes

Pump with gravity-fed storage (indoor or outdoor)

Gravity-fed spring with storage (indoor or outdoor)

LEAST RESILIENT

MOST RESILIENT

The most resilient home water systems are often only possible if the property's features allow for it—springs uphill of a home site being a top priority in land selection.

and barn/animals/gardens). When you find an old cellar hole in New England, you know there is—or was—a good spring uphill of that location. There are still pieces of land available that have such a basic resource in place, but they are increasingly hard to find.

In more modern times drilled wells and electric pumps have allowed us to develop almost anywhere in the landscape, even where the water supply is three to five hundred feet, sometimes even six hundred or more feet, below the ground. Our lives are then beholden to the complicated and fragile chain of events, parts, and energy required to constantly bring that water to the surface and pressurize it in our homes.

The WSD studio/shop building was constructed with access to the typical conventional water source in rural locations—a drilled well with a submersible pump located at the bottom of it. As mentioned, this is a brittle system vulnerable at numerous failure points, ranging from pump breakage to grid failure to waterline rupture. You can't make parts for the pump locally, and the entire chain of distribution for replacing a pump or its parts is convoluted, global, and fragile. It will be disrupted, you can count on that. I hope this won't be when your well pump fails, but who wants to make such a gamble when it comes to such a basic need as water?

As in all other systems, we seek a more durable and resilient approach. So in keeping with the principle of replacing a complicated active system with a simpler, passive one wherever possible, we have provided a backup gravity water source to the WSD studio/shop over the past year, consisting of a water tank that is maintained at full and is only used in a grid-down situation. We will also be adding a full spring-fed source to this system in the coming summer. When this spring-fed source comes online, the backup tank (in the attic of the building) still functions as a double-backup gravity-fed source in the event that the spring dries up for a period of time or freezes.

Water supply to the home can be divided into three approaches:

1. Pump-based: from a drilled or hand-dug well with no storage
2. Pump-based: from a well but with storage providing gravity feed for building
3. Gravity-fed: from a spring or shallow well uphill of the building with enough flow for constant supply to building

Of course, approach 3 is the most resilient approach and is the ideal scenario you want to develop if you are

This copper-gutter "drop" connects to a copper downspout that feeds a pond and can be connected to drinking-water storage. Slate and copper combine to offer completely potable water from our roof and also last a century or two.

looking to create a home from a new piece of land. For most of us on an existing piece of land that does not happen to be a good old farm, however, we have a retrofit job on our hands with the goal of moving from approach 1 to approach 2 or, ideally, approach 3 when possible. Getting to a full gravity-fed system is not always possible, as it requires a significant water source uphill of the building. However, the second approach is always possible to some extent and is most often the kind of system we consult clients on in retrofitting into their current homes and farms.

NUTRIENT CYCLING

The typical approach to handling nutrients in modern homes is as senseless as it is unresilient. And as with all nonadaptive systems, it represents a one-way flow of materials, nutrients, and energy. In the case of the rural modern home, it looks like this: food bought at the grocery store—food consumed—food excreted into drinking water—nutrients flushed under the front yard to grow grass.

In the more typical suburban or urban setting, you can replace the front yard with an even more wasteful situation: the sewage treatment plant. Here, we apply enormous amounts of energy and chemicals to the nutrients before flushing them once again, this time into the nearest lake or ocean. In ten thousand years

of civilization, one would think we would have come further than this take-make-waste conveyor belt. The nutrient-flow system in the modern home is a legally enforced and highly efficient system for transporting topsoil and drinking water into the world's rivers, lakes, and oceans using fossil fuel and chemicals every step of the way. Along with the use of the automobile and consumption of nonfood products, this system of nutrient flow through our homes represents the most rapid method ever invented of replacing valuable resources with waste.

Of consequence to the modern resiliency-focused individual is not only the destructiveness of this system but the wasted values it represents. The nutrients we consume in the form of food need to become food again as rapidly as possible if we are to perpetuate a cycle of fertility and health on both our landscape and the planet. How to cycle these nutrients rapidly back into soil, plants, animals, and ourselves is the design challenge. The following strategies distill methods for capturing and harnessing this fertility as potently as possible:

▶ Composting solid nutrients: humanure
▶ Fertigation through urination
▶ Composting food scraps
▶ Potentizing the results

We have found that the easiest and highest-leverage actions for aligning our buildings with an optimized cycle of fertility on the homestead is to use composting toilets and whenever possible divert urine (watered down) into plant roots in the garden, orchard, paddies, and pasture. The best intersecting technologies between a home and the nutrient flows in the larger landscape of the home are a composting toilet and a five-gallon bucket or similar, in which to urinate when the weather is cold. When the weather is nicer, it is best to urinate outdoors either directly at the base of plants (assuming significant soil moisture is present) that need a fertility boost or into a container, then dilute the mix before watering plants. Watering your compost piles with urine-rich liquid is also a great way to accelerate the decomposition action in most piles that run short of nitrogen at times.

The woodstove enters the nutrient flow web because of its carbon-cycling functions and ability to produce ash and char. Both of these products are carbon rich and thus pair perfectly as sponges to absorb and carry the nitrogen-rich effluent generated by the building's occupants. By combining the N and C nutrient streams from people and wood burner(s), we can create a durable, nutrient-rich soil that is optimized to promote plant health in the landscape.

I will not go into composting toilet details here because of the immense amount of literature available on the subject. I would only add that we are developing ways to eliminate the common need for a constant electrical load in the form of a fan for ventilation. We use an old Clivus Multrum that requires no moving parts. We are in the process of developing a stack-effect (convection loop) ventilation system where the pile can be slightly aerated and the air vented to the top of the building without backflowing up through the bowl. A fan that turns on only when the lid is lifted will probably play into the final design, because it is very difficult to achieve a correctly ventilating system (in through the bowl, not out through the bowl) while the lid of the toilet is open.

There are many makes and models of composting toilets on the market with varying degrees of breakability and complexity, from electrical fans to hand-cranking aeration systems. As with all technical systems, we are interested in the simplest method that will get the job done, even if it takes more time to do so—just as the pile takes longer to compost. Who wants to go into the humanure pile to fix the hand crank when it breaks, which is guaranteed to happen? The same goes for the fan or other moving parts. They will all break, so plan on that.

We must remember that resilient homes and farms require highly complex biological systems but are burdened by and made brittle by complex technical systems. The abiotic systems must be as simple as possible.

The Whole Systems Design studio shop with a 150-year-life-span slate roof

Chapter Seven

Resilience and Regeneration for the Long Haul

Over the years I have learned that there is truly only one limiting factor to ensuring a regenerative and resilient endeavor that is consistent from one site to another: human beings. It's not lack of space or the lack of tools or money that keep a regenerative and resilient process from happening. It is people (along with the skills and time they have available) who seem to be the limiting factor to land-based value creation in all examples I have encountered—this seems to hold true across multiple continents and nations that I have visited in my resiliency studies. And as resiliency-seekers we must always look for examples of systems in which people and land have coexisted and evolved with relative stability and sustainability for multiple human generations.

The human system is a subset of the land system and vice versa; they "intermake" one another. These are our reference points in imagining, designing, implementing, and maintaining our systems over time.

> No culture that has figured out the age-old challenge of sustaining itself in an enjoyable manner without depleting the biological capital of their place has ever done so without closely tying themselves to the results of their actions upon the land.

In examining such examples across the globe, one pattern has become clear to me: In those relatively rare instances when people carried out a sustainable livelihood over multiple generations, long-term land tenancy and investment in the landscape at highly localized levels with "direct responsibility" has almost always been the case. I say "direct responsibility" rather than "ownership" because land ownership is a somewhat new idea in relation to many peoples who have lived with land for thousands of years.

The example of the Seneca peoples in what is now the Finger Lakes region of western New York State comes to mind; here was a society that lived in one area for three to four thousand years without destroying the land's ability to sustain their culture. Yet they did not think of land as something that could be owned. Today, our mind-set is very different. How can a rooted culture of a people deeply connected to and aware of a place be sustained over long periods of time? I think this certainly requires an arrangement in which each member of society is closely connected to the results of their actions such that the cycle of cause and effect is tight.

Humans tend to soil in their garden beds most poorly when those beds are out of sight, when it's the commons, when it's removed from their daily interaction—indeed, when it's the infamous "away." No culture that has figured out the age-old challenge of sustaining itself in an enjoyable manner without depleting the biological capital of their place has ever done so without closely tying themselves to the results of their actions upon the land.

Modern-day technology has allowed us to export to great distances, in both space and time, our true impact on the land. This, of course, has quickly become a primary failure point in modern societies. If we are to succeed and live well in a place beyond very brief time frames, we need to take responsibility for our "wastes," cycling them back into the system, connect ourselves with the sources of our food and material needs, know these sinks and sources—where they are, who is involved with them, how their health is being affected by our actions in the system. Achieving this understanding and resulting skillfulness in decision making requires that we find a place, settle in, and put

down roots. Simply because a challenge is immensely difficult does not mean the imperative of meeting the challenge is any less.

Successful human-land and human-human relationships have never been short-term endeavors. It is one thing to imagine such relationships happening in a group of people who "own" land and benefit directly from its health. But how can this be achieved in a world full of landless people, in a world of populations that cannot afford a stake in a piece of land? I have no clear answers except to highlight the connections and mutual dependencies between land and community. Landowning classes have all the advantage in achieving a sustainable situation, it would seem; however there's one rub: Those owning land cannot, usually, afford the

An elderberry harvest being processed; when people are not actively engaged in the system, yields are hypothetical.

Planting out the rice crop during a farm tour with a local college

time or have the skill to work the land in such a way as to cultivate long-term yield. Other people are, and have long been, needed.

For much of human history landowning classes knew this, but they did not give up their stake in land; they simply rented acreage or offered protection or other values to those who worked the land, in part, for the landowners' own survival. Throughout history the lord has needed the peasant, just as the peasant has needed the lord. Though this pattern is perennial, it is not an equitable situation. Can we do better than the medieval fiefdoms of yesterday?

What incentive exists for one to plant an oak tree, a walnut, a nut pine—systems that yield across a half-dozen human lifetimes? How can we cultivate a culture for the long haul—a culture of nutteries? How are people empowered to think about and act for their grandchildren? Acting in this way, as fully developed responsible adults, most often happens when people become rooted in a place. Cultures of displacement, so common today, and the rootlessness they beget are unlikely to support long-term peaceable and enjoyable societies; they never have and probably never will—for a multitude of obvious reasons—mostly having to do with lack of accountability. It seems that the long-term intergenerational success of human cultures that can activate the most value-creating results of land depend upon several basic principles. While not exhaustive or applicable to all people in all places, the following list of conditions seems to be true in my own experience in the intersection between land and people:

- **Land-based tenancy:** Though it almost goes without saying, it is foundational for those on the land to be working in direct contact with the landscape for a long-term functional relationship to emerge. Residents cannot simply be working for money 100 percent of their time (or likely even 80 percent)—a proportion of their investment in time and effort needs to be land-based; keeping a garden, at a minimum, for instance.
- **Long-term tenancy:** Residents are interested in being on the given landscape for more than a period of months; ideally, for at least a few years.
- **Mutually enhancing tenancy:** Users of the landscape receive direct benefit for increases in the skillfulness and labor that they apply to their environment.
- **Self-directed tenancy:** Those using the landscape may be directed or encouraged to do so in particular ways by those with a greater stake in or ownership of the land; however, they have a degree of free agency within this relationship wherein they can choose independently what and how they would like to participate in land management and community activities. There is a very difficult balance that emerges in all communities in this aspect, and it seems only successful when the needs of the group are not forgotten in the face of individual short-term self-interest.
- **Transience-permanence balance:** There are synergies created from short-term users of the landscape—visitors who bring new skills and knowledge to a place, but where the place maintains its roots and cultural memory by a strong group of people who are there for the long haul. Many schools and learning-oriented farms face this particular challenge in acute ways today. Who lives on and in the landscape?
- **Ownership:** In today's world of relating to land both legally and socially, it is hard to imagine achieving the above effects without individuals' actually holding title to the land upon which they dwell. It does

seem, however, that some examples exist where this is not the case—but these seem limited to closely tied communities of people where common values are so high that individuals can function collectively over long periods of time. I know of no examples of this happening in truly functional ways outside of a religious context, such as the kibbutzim of Israel, although I am sure there are some.

Enhancing Vitality in a Time of Biospheric Toxicity

We are guinea pigs in a massive, uncontrolled chemical experiment, the disastrous outcome of which is measured in disease and death.
—**DR. RICK SMITH**, Canada

Life is rare. And as far as we can tell, from Pluto to Venus it finds refuge in less than 0.0000001 percent of this solar system: in the thin, green, watery biosphere beginning at the earth's surface and extending just a few miles upward into this planet's atmosphere. The degree to which humanity and the web of species survive and thrive depends largely on the rate at which we dig up, drill out, and otherwise stir up toxins from the deep recesses of the planet into its atmosphere and onto the surface of the planet, and on the extent to which the development of new chemicals and novel genetic structures outpace the capacity for evolution to adapt to them, and to the extent to which we mount an adaptive response.

Ours is a time of unprecedented toxicity. Since the nineteenth century we've been increasingly digging up, drilling out, and generally unearthing the primary feedstocks needed to operate industrial civilization,

Planting, processing, and consuming nutrient-dense foods revitalizes our bodies to cope with the unprecedented levels of toxicity that exist today.

aside from water and wood. In essence we've engaged a process based on moving toxic matter locked up in the relative safety of the geosphere into the thin and fragile biosphere. To understand the human health challenge we face today, it is important to recall the full context of Earth's inhabitability. From our planet's inception it took about four billion years for life to emerge. Earth spent much of its early life cooling radiologically before complex life could emerge. Toxic substances spread across the surface of the planet required the influence of millions upon millions of rainstorms, hundreds of millions of years of sunshine, and other effects of weathering to become inert enough to be a habitable medium upon which human life could emerge.

Today we find ourselves in the strange and precarious position of spoiling our water, food, and nest at large as we pursue ever-longer lives and more stuff. The problems we have been leaving in our wake, such as persistent atmospheric and soil toxicity (the rain today contains many times more mercury than it has in all of the rest of human existence), started to catch up with us sometime in the latter half of the twentieth century, and most of us have been running from them ever since. The final analysis of the workability of the relationship between humans and our planetary home can be found in the state of our health. Our own bodies are the canary in the coal mine of Planet Earth, and the health of these bodies is a direct indicator of the health of Planet Earth. And working to heal our bodies is working to heal Earth.

On Earth Day 2011, about a month after the hole in the Earth began spewing oil and gas across the Gulf of Mexico, I was swimming at one of my favorite spots in the Mad River—a stretch of mild rapids and green-blue water weaving around boulders through a small gorge. Though it was still May, it was the sixth or seventh day in a string of balmy days in the 80s, a spring that was becoming almost unbelievable with tons of sunshine, warmth, and light. Though the world news seemed to darken with each week that spring, this grotto of bedrock, hardwoods, and freshwater seemed strangely distant from catastrophe. Here was an almost giggling piece of the planet, flowing with a startling grace in spite of everything not well in this world.

I dove from a boulder into the current in a spot where I had done so a dozen times before. But this time, just after entering the water, I could see a metal rod pass below my stomach as I slid by, inches above its rusty tip. I arched upward quickly and surfaced with the sensation you get when an eighteen-wheeler almost sideswipes you on your bicycle. I surfaced, my gut clenched from the near skewering. A beautiful spot indeed, yet not without its man-made hazards. In fact, this spot was no different from dozens of other former mill sites on this river. Abandoned iron gearing and rotting cement work littered the banks. Eroding stone walls that held mill buildings were being reclaimed by sumac, maple, and ash. This now-beautiful place was not that long ago an active industrial site. Yet now it seems to be a largely "natural" place, its legacy fading into the landscape with each passing decade. People swim and play around this abandoned power plant and dozens like it across New England, pondering their past and their quaint legacy.

How benign such industry was back then! Our children may swim the Connecticut River at Vernon, Vermont, looking upon the rotted-out infrastructure of the Vermont Yankee nuclear power plant. Or in the Winooski next to the guaranteed-to-leak-eventually landfill in Moretown, Vermont,[*] with its batteries, paints, glues, and electronics leaching into the toxic muck at the bottom of the landfill, rain trickling through the hundreds of feet of trash, washing this ooze into the river and the lake downstream. But there won't be anything quaint about these legacies.

Our parents and grandparents abandoned to us their mills, mines, and factories. We've done the same for our descendants, but far worse, for what we've left to them is orders of magnitude more toxic, abundant, and

> Our own bodies are the canary
> in the coal mine of Planet Earth,
> and the health of these bodies is
> a direct indicator of the health of
> Planet Earth.

---

[*] At the time of publication multiple studies of wells near Moretown, Vermont, landfill confirm groundwater contamination.

persistent. In the United States alone, we are leaving our children with*:

- ► 104 nuclear reactors in 31 states, operated by 30 different companies. Every single one "temporarily" storing high-level waste that will be lethal for 10,000 to 24,000 years
- ► 40,000 to 80,000 (exact number unknown) chemical factories producing or processing materials with multiple "compounds known to be carcinogenic and/or mutagenic"
- ► More than 40 weapons-testing facilities and 70,000 nuclear bombs and missiles
- ► 104,000,000 cubic meters of high-level radioactive waste from weapons-testing activities alone
- ► 925 operating uranium mines
- ► 20 to 30 times the average historical background rates of mercury in rain
- ► 2,200 square miles of excavated valleys and leveled mountains in Appalachia alone
- ► 478,562 active natural gas mines in the United States in 2008, with 1,800 expected to be drilled in the Marcellus Shale of Pennsylvania alone in 2010
- ► 18,433,779,281 cubic feet of trash per year, or 100,000 acres of trash one-foot deep per year, or about 250 square miles, with trash 400 feet deep

Offshore drilling, natural gas mining, oil shale open-pit mining, coal mountaintop-removal mining, metal and mineral ore mining (and the chemical industries to support these operations, from surfactants to metal-separating solvents) are all accelerating rapidly. It took eons for the earth to cool and for the array of toxic substances native to this planet to safely migrate out of the fragile biosphere and into the deeper layers of the planet. Only then could life emerge. The viability of Earth's thin biosphere, and our ability to live within it, will correlate directly with how we manage this interaction between geosphere and biosphere.

But American industry (and China's and India's) is just now beginning to stoke the engines that will drive

*Many statistics from: http://www.brookings.edu/projects/archive/nucweapons/50.aspx

Seaberry oxymel, with its high-potency fats on the surface

the next century of extraction on this continent. Barring a 180-degree change in direction from the status quo, we will scavenge the American landscape (along with the newly accessible frontiers of the deep ocean and melting arctic) for the remaining dirtiest and deepest fossil resources we haven't been able to access—and haven't needed to—until now. It should not be surprising that America would turn its own landscape into a wasteland similar to those we have already facilitated abroad. What empire has not eventually consumed its own internal resources after exhausting those outside its borders? Dealing with this fact seems to present two options: (1) Disaggregating empire systems before they can consume themselves and their people, and/or (2) creating vibrant societies that do not depend upon the destructive means of today's mainstream society for their basic needs.

The Take-Make-Waste Operating System of industrial society has amassed an unprecedented challenge for humanity: Invigorate or Devolve. Avoiding a future of mass cancer, mutation, obesity, ADHD, apathy, depression, and a general perversion of the human condition will rely upon a human-ecosystem response harnessing the most potently regenerative land, water, and human-health-promoting systems and species. This response can be found in allying ourselves not with any political party or version of Newness or Bigness, but with Smallness, Oldness, and the particulars of living places and their eons of evolutionary heritage—soil, plants, animals, fungi, water, and other forces in the web of life we have been given by this still-breathing planet.

So what can we do at the home and local-community levels to deal squarely with the fact that your world and your children's is likely to continue becoming many times more toxic than it is today for the foreseeable future? The accumulating planetary contamination from the past century of pollution is likely to dictate with increasing significance where we live, what foods we eat, how they are produced, how we die, and the quality of our lives. Our ability to resist toxicity, to maintain and develop ever-deeper levels of personal health will be as crucial as any other way of being resilient in the face of existing and future challenges.

Growing Health and Body-Mind Resilience, Not Just Calories

Let food be thy medicine, and let thy medicine be food.

—**HIPPOCRATES**, 460 BC

As the local response to global resource system failures gains momentum, the need to grow food at the household and neighborhood level is quickly being realized. This is important, but recognizing it is only half the work. Each year more of us fall victim to cancer, MS, ALS, Alzheimer's, asthma, and the hundreds of other diseases that have causal factors, and oftentimes direct

Plums, blackberries, seaberries, elderberries, schisandra berries, and currants: part of a nutrient-dense diet that uses food as medicine

The health of soil and water is the foundation upon which the health of the land-human community is built.

origins, in environmental sources. There will likely come a time when we will look back and realize that being fed was one thing, maintaining our health and vigor another—that calories and nourishment do not go hand in hand. As author Michael Pollan has noted, we are the first generation to have obesity and malnourishment simultaneously.

The medicinal value of our foods (and water) is easy to underestimate, but in an increasingly toxic world, we are coming to realize that these provisions are our first and last line of defense against forces that erode our health. Food has long been humanity's primary medicine and will likely be so again out of sheer necessity. How, then, do we obtain the most potent foodmedicine? Probably, we will need to grow the plants, animals, and fungi ourselves.

The productivity and vigor of a plant or animal is largely determined by the health of the soil in which it is grown. The same applies to health-giving properties of that plant (or of an animal product). The health of soil and water is the foundation upon which the health of the land-human community is built, so attending to their vitalization is always the starting point and backdrop of human-health work. The long and short of it is this: We must increase organic matter (carbon), increase biological activity, and remineralize. All land-management practices in a health-giving ecosystem are rooted in promoting these three aspects of soil health. Such activities come naturally, in the form of composting plant residues and returning them to the soil, minimizing tillage, adding and balancing minerals in the soil, promoting mycelial and bacterial communities, and promoting deep-rooting nitrogen-fixing and dynamic accumulating plants and root dieback events.

Each of us is mostly water. Water is washed through our system on a daily basis and is the medium through which we extract and absorb our nutrients in food.

Erica Koch, ND, with a giant Reishi mushroom (*Ganoderma tsugae*) that we found on a nearby walk. This one mushroom was sliced and dried in the rafters above the woodstove, yielding enough immune-enhancing tea to support us through the long winter.

The basis of life on Earth more than any other single compound, water and its quality are no less important than all the food in our diet. The best living water is spring water, followed by well water, all from protected watersheds (at least locally protected; none are protected from, and all are significantly contaminated by, atmospheric pollutants such as mercury and radioactive isotopes). Securing the best possible water source (ideally, more than one) and ensuring its protection is a top priority for every person wishing for personal health and long-term security.

Nutrient-dense food and water, along with herbs, fungi, plants, and animals that are most potently

Garlic flowers: an easy and potent home medicine made from leaving the scapes on some of the garlic, then letting them flower to make a water-based essence

antioxidant, along with the energetics* of the food-medicine, are primary defenses against increasing toxins in our biosphere from a food and medicine standpoint. Levels of erosive chemicals—those that oxidize and mutate our cells, disturb endocrine system functioning, and in other ways erode our health at the cellular and macro level—continue to accumulate in our biosphere as we enter the twenty-first century. Reducing the effects of these toxins is possible by harnessing the tonifying chemicals and influences of plant, animals, fungal and other life-produced foods, and medicines.

From a land-use perspective, biologically active, mineralized soil and living water are the foundation of this health, as their vigor begets the health-promoting properties of what is grown on and in the land, be it vegetables, grains, meats, fungi, or fruit. It is unlikely

that the produce of land would be healthier than the soil and water from which it is grown, just as it is unlikely that the quality of our own health would be greater than the quality of the foods we subsist upon. Maintaining and enhancing our health therefore begins and ends in large part with land-management practices that restore and develop ever-healthier biological communities that compose the land system.

We can think of healthy ecosystems as the front line of toxic resistance. For every coal or nuclear power plant, we need, say, a million acres of vigorous ecosystem from which to cultivate health and resilience infringed upon by said power plant. One might call this "the restorative ratio."

At the homestead level this toxic resistance is rooted in soil- and water-enhancing activities. Built upon this foundation of healthy water and soil are foods that are particularly powerful at helping us maintain and enhance our bodies' immune and toxin-resisting responses. These are foods that are nutrient dense, loaded with living organisms (cultures), antioxidants,

---

* Energetics of plants, animals, and fungi include influences such as warming, cooling, astringing, relaxing, drying, constricting, and many others. These are important foodmedicine qualities that transcend the mere nutrient quantity of the food or medicine.

Erica Koch transforming our flowered garlic scape harvest into a potent tick and parasite deterrent.

essential fatty acids, amino acids, phytochemicals, bioflavonoids, vitamins, minerals, and micronutrients that support the mind and body in optimal health maintenance. These are, of course, whole foods eaten in the freshest forms possible and prepared in ways that preserve the enzymes present within them at harvest time, or in ways that actually increase the biological activity of the foods—as with kimchi and sauerkraut and other live-cultured foods. One can think of these "superfoods" as sources of "good chemicals," countering the influence of oxidizing and mutagenic chemicals, which degrade—rather than bolster—our bodies' functions.

Money: One (Important) Means to Get Work Done

It's easy to get confused about what money is in today's world. The lack of equity in its distribution, accompanied by its many forms of abuse, can lead one to thinking unclearly about value. More than any other group I interact with, I hear a lot of ideas about money

Lauren Marra brix testing for nutrient density during the Whole Systems Permaculture Design Course.

NUTRITARIANISM

A nutrient-dense meal—think "the opposite of Wonder Bread"

Nutritarianism is an important and foundational concept to understand in relation to human health. We eat for various reasons—for calories (to be "fed"); for nutrients (to maintain healthy organism functions beyond simple caloric needs); for social, psychological, and spiritual reasons. Nutritarianism is the approach of eating based on food quality—specifically, the ratio between nutrients and calories. Nutrient density is defined as the amount of nutrients per calorie of food. This is a crucial concept to understanding health and acting in ways that foster it. Witness the vast amount of overfed but undernourished people in the world today—a clear example of the antithesis of nutrient density.

The term "nutritarian" was first used, it's thought, by Dr. Joel Fuhrman in his book, *Eat to Live,* where he offered the following definition: Health = Nutrients/Calories (or H = N/C, for short). Nutrients include vitamins, minerals, and phytochemicals. Calories include fat, carbohydrate, and protein. A high N/C diet is also called "nutrient dense" or "nutrient rich." While most Americans eat a diet dominated by calorie-rich but nutrient-deficient foods, nutritarians focus on those richest in nutrients per calorie; specifically, the foods found in this chapter, such as leafy greens; dark-colored and specific "super" fruits, nuts, and seeds; certain meats and other food grown in high-nutrient, healthy soils.

This last aspect is often forgotten about in the food-selection thrust practiced by many using the term "nutritarians" and following Dr. Fuhrman's prescriptions. Many foods that should be nutrient dense are not if grown in poor soils, and conversely, many foods that are normally not considered nutrient rich can be if grown in rich soils or in ways that cause the plant to mine subsoil minerals. Stress on a plant—especially drought-stress, is often a great way to build potency in the plant; roots must dive deeper to find moisture and simultaneously tap into more deep-soil nutrients, especially minerals. Grape growers know this and manage accordingly to enhance their wine crop.

Plant medicine: hawthorn, elderberry, cranberry, and seaberry syrups, shiitake mushrooms, garlic, and kombucha

from students in our permaculture design courses and other workshops. They are, understandably, upset with the current version of society and how money is typically used in the world today. However, they sometimes get very confused about the nature of money and, thus, how to use it well. Money, like all other forms of power, can be and is abused often. It is applied poorly, wastefully, and with no regard for its consequences on long-term health.

So, too, can money be used to accomplish a fantastic amount of work when applied correctly. Money is not evil, nor is it beneficent—it is simply one form of energy. Energy is the ability to complete work. When we see money for what it is, we are positioned to ask how best to apply its force for the greatest effect on the regeneration and resiliency of ourselves, our home places, our communities, and the world at large.

Today, there's an intense concentration of money in the same way there is an intense concentration of other forms of energy and power. We have tapped into ancient energy stores, brought them up into the world on the surface of the planet, and fight over who has access to this energy and the wealth and power it enables. All our permaculture work should be rooted in this understanding of present-day society—and it's no different from the ecological realities of all places for all time. Energy tends to concentrate, stagnate, and become ill distributed. Our job as a resilient homesteader is to activate the most direct mechanism for releasing these energies and spreading them across a

One of the most essential foods and medicines we grow at the homestead—garlic—truly foodmedicine

landscape, a community, a region, and the world. This applies to money the same way it applies to water, electricity, fertility, biomass, or anything else.

So our task, in part, is to find the money—to find all concentrations of energy. I know this runs counter to many permaculture approaches around the world where people are seeking to create systems without money, to remake cultural exchanges without the medium of paper currency, to distance themselves from the negative baggage that money often represents. While I respect this, I would caution against taking this approach too literally. Of course we need a world in which money and its power is not abused. Indeed, we may even need a world in which currency is fundamentally very different in form from what it

is today—in which it, again, has some inherent value. I am not sure if that will be necessary or not. But clearly, there are immense concentrations of money today. One can either ignore this reality, dislike it, and distance themselves from it, or ask themselves, "How can I harvest these concentrations of money (energy) to help perform the work I would like to complete?" Taking the opposite route—rebuilding a system with very different forms of money—needs to happen, no doubt. But just like trying to avoid using any and all fossil fuels, this approach, while admirable in its consistency, seems unable to utilize the horrendous wealth concentrations that exist, whether we like it or not. We tap into those concentrations the same way we choose to use some fossil fuel to build terraces, ponds, and swales. We could build them by hand but what would have the greatest change for maximum affect? The concentrations are there to be utilized and doing so does not mean we cannot simultaneously build a different type of economy. We work on that end by bartering and trading for direct goods and services wherever we can.

Few people I work for in my design work could be considered anything but "well off" financially. Their accumulated energy storage (wealth) is utilized by us to help them achieve their goals—usually establishing a sound homestead that will be adaptable and enjoyable for the long haul—while we spread that energy out into our landscape of swales, terraces, ponds, biodiversity, community, educational opportunities, research, and other manifestations or "work."

Staying

This book could have been aptly called "The Empowering Lifestyle." Truly, that's the biggest reason I stay in it—not to fix the world (it might be broken beyond repair, who really knows?), not even to build fertile soil and plant a new forest. Those are big motivations to be sure, but perhaps the most consistent day-to-day fulfillment comes from having a central role to play in my own survival and thrival, from keeping myself warm, to feeding myself, to enlivening each day with a swim in the pond, a ski through the rice paddies, a night on the rock under the silent stars. Enlivening the land around me and the person within me has been the most dependable outcome of this lifestyle. And that's been a surprise. The shape my life has taken here does not stem from some grand design but from a series of small actions—trying a pond here, a rice paddy there, a seaberry plant here, a swale there, some mushroom mulch over here. With each passing season the outcome of these tiny experiments becomes visible. I start to see what works, adjust, and try more new things. Each of these things unfolds. Being open to that unfolding is key. It's so easy to expect specific outcomes, but that hides possibilities. About four years ago a student asked me, "If you could offer one piece of advice about how to make a

landscape work, what would it be?" My answer was as simple then as it is now, "Try stuff." It's as simple and complex as that.

I moved to this piece of land with the intent to live here for three years, but I have stayed for ten. My plan was to finish a master's degree program in architecture that I had enrolled in. I was going to renovate the small home and sell it, hopefully at a profit, when I left. A homestead and small farm was *not* what I had intended. Looking back at that time now, I am very surprised. How did this all happen?

I suppose the idea of homesteading and farming—of living close to the things that sustain me and to the beauty of this earth—was always on my mind. Yet, it was always in the future—something I would *eventually* get to. And then it just started. In retrospect, I realize that staying here—not leaving as I had planned—was what allowed this project and lifestyle to take shape. I began to plant some plum trees and walnuts, dig a pond, turn up an area for a vegetable garden. A few years in and these trees had started to get pretty big. Then we started to harvest big salads and potatoes and plums. The pond brimmed with frogs and salamanders. New birds arrived in the spring. This home around me was growing into something new, something that I was a part of, something that I had a role in. I realized that the idea of "home" had always been incomplete in my life. This was new to me, the feeling of being a co-creator in the world around me—of taking a walk on the land and seeing that with each passing season some new form of life was taking shape. I could see where my actions helped the life process along and where I could help more. Without realizing it at the time, I was slowly becoming a member in the community of this small hillside. And slowly (there weren't many eureka moments) the meaning of this filtered into me. The weekly walk around the property began to be one of the most reliably fulfilling experiences in my life. I don't have children yet, but I suppose it's not too different from watching them grow up, learn new things, and become themselves. For ten years now I have had the good fortune of being a part of this place become itself, and I become myself, increasingly an extension of this larger thing. It is almost as if my "self"

The bottom pond on a typical summer evening Photograph by Brian Mohr/EmberPhoto

has been expanding to meet this land, growing out-ward, becoming three, five, ten acres. Calibrating myself to this place takes time and it's been full of sur-prises. Largely, the process has been a reminder to never let my concepts get in the way of possibilities. Having been to college and graduate school before I got here, I was confusedly full of concepts. One of them had to do with how I give to and take from this land through planting trees.

When we consider a tree planting job, my clients almost always ask the same question that I have asked many times as well, "How long will it take for this tree to give me fruit?" Last autumn I harvested the first apple on this land. I had planted the tree that bore this

> **Calibrating myself to this place takes time and it's been full of surprises.**

fruit about seven years earlier and it has grown into a beautiful young sapling with widely spaced horizontal branches, a thick base, and a strong sense about it. Each year I diligently mulched its base, offering it a wheelbarrow full of woodchips, sawdust, and some manured duck bedding. I had weeded the base, pulling back any grass that was competing with it and a few times gave the leaves a foliar feed on early summer mornings. With each passing year the tree kept on growing and I kept on pruning, feeding and weeding. Over time I began to admire the tree greatly—it was such a beauty and the largest apple tree I had ever planted. I was proud of it and it made me happy to walk by it. Apparently, I had come to appreciate having the tree so much, that it was strangely surprising when I walked by it one day in its seventh year and found a ripe apple ready to be eaten. I remember the moment very clearly. It was a new sensation and not one I ever expected. Instead of feeling like "Finally, an apple,

The author and permaculture design course students taking a log-rolling break in the lower pond. Photograph by Costa Boutsikaris

victory at last! This is what I've been working so hard for!" I was quieted, humbled. The apple I had pulled from the tree and was now feeling in my mouth was clearly not a reward I deserved, but a gift. You see, the tree had already fed me. The tree had already given me so much and this all became crystal clear in that moment. Somehow, in tending to the tree I had come to appreciate it in and of itself, not for some future thing it could offer me, but just because it was there. Quiet, serene, beautiful in form, patient. This tree had already yielded value to me and to this place. So when it finally gave something else—an apple—the offering seemed extra, a bonus, a gift. And as I stood there admiring the apple and the tree from which it came, I realized that I had lost my ambition about the tree—that the mental place in which I had planted the tree had faded. The apples I was dreaming of then had been replaced with an appreciation for the mere presence of the tree and for the opportunity to care for it.

As I looked across the landscape filled with dozens of other food trees, I realized for the first time how much there truly was here. "Abundance" has never meant the same thing since.

When the bad news of the world starts to overwhelm me, I always remember one piece of overwhelmingly good news; that in this work of facilitating abundance, life wants to live. Spread seed and plants will grow. Nourish a patch of ground and it will live. When we try, when we take a stand, root down, and begin to build a relationship with the world at our doorstep, life responds. The clues in this journey are there to discover each and everyday. And finding these clues provides more than enough inspiration to keep on going. We are all the discoverers, each day, on our own little piece of Earth. The real solutions are in front of each of us—they cannot be outsourced to "experts." Disempowerment has no place in a living future. So become your own expert; alas, there's no one else to turn to who can know your life and your land as well as you can. You already have what you need to enliven your own place and your own life. Inaction quickly consumes a lifetime. Be curious, be bold, pay close attention to the world in front of you. And start trying stuff.

Acknowledgments

This book has resulted in no small part from the influences and hard work of many people over the years. Their efforts have contributed to this homestead and my life here—all of which are interwoven throughout this book.

I am grateful in particular to Cornelius Murphy for his hard work, skillful hand, problem-solver's mind, huge heart, and steady partnership for more than five years. You have helped make Whole Systems Design, LLC, what is today, and I look forward to many decades ahead with you.

To my family for their endless support in helping me manifest my life—my parents, Marcia and Stephen, and sister, Elizabeth, who have been the most supportive and loving family anyone could ever ask for. None of this work would likely have emerged without their continuous nourishing of my life from day one. To my grandparents Annette and Martin Kamisher, for providing a foundation on which the generation before me could stand, immigrants in a new land—I can only imagine the difficulties they encountered and their courage. And for my grandfather Martin Kamisher, especially, whose love affair with plants, the water and winds, sunsets, and all forms of the sublime was an early and continuous inspiration. I have never met anyone so attuned to beauty and so thoroughly enraptured by the simple joys of life; his example still serves to guide my life.

To my loving wife, Erica Koch, who has contributed immeasurably to my understanding of health and healing, and the vital connections between the body, mind, and land. And for her loving partnership in the ever-unfolding story of this place and my own life; nourished with tender care each and every day.

I am indebted to those who have provided a foundation for this work to be built upon, and to those who have inspired and influenced my thinking, whether in person or through their example, their writing, or speaking. These include Wendell Berry, Aldo Leopold, John Todd, David Orr, Amory Lovins, Bill McDonough, Chris Shanks, Sepp Holzer, Jack Spirko, Richard Czaplinski, Bill Mollison, Eric Sloane, Buckminster Fuller, Christopher Alexander, Allan Savory, Masanobu Fukuoka, Paul Stamets, and Mark Shepard, among others.

To the entire team at Chelsea Green for their interest in and hard work on this book.

With gratitude for those who have believed in me and have supported my company's design and building work repeatedly, including Melissa Hoffman, Josh Hahn, Jack Kenworthy, Stan and Helen Ward, Anne Burling, Chris Maxey and Christian Henry, Peter Forbes and Helen Whybrow, Dave Brodrick, Ted Blood, Amy Seidl, Shawn Smith, Mari Omland, and Laura Olsen, among others.

I am also grateful to Kristen Getler, who helped cultivate this farm as her own. I hold a particular gratitude for Dave Johnson, craftsman of the highest order and deeply dedicated human being; you've been an inspiration and have left the lasting mark of quality on this place and my own life here. To Buzz Ferver, Kyle Devitt, Micah Whitman, Jackie Pitts, and Chris Eaton—thank you for dedicating your insight, craft, and hard work into creating lasting beauty here. For Erica Koch, Vic Guadagno, Joe Bossen, and Eileen Shine, who are continuing to bring forth new life from the once worn-out slopes of this hillside and for whose presence here I am grateful.

To old and dear friends who have helped make the story of my own life what it is—inseparable from the work described on these pages: Leigh Axelrod, Marty Nolan, Adam Maker, Ludvig Thor, Greg Koskinas, Kevin Natapow, Hadley Clark, Michael Blazewicz, Chris Shanks, Josh Hahn, Brian Wade, Sean Gaffney, Carsten Homestead, Jack Kenworthy, Neil Ryan, and Ralph Tursini.

To those who have been fellow adventurers in the vertical and backcountry world, where the skills and awareness in resiliency were first born in me: Brian Wade, Chris Shanks, Jack Kenworthy, Brian Mohr, and Marty Nolan. And to Dan Sobol.

Assessment is always an important, albeit imperfect, subjective, and incomplete tool. In order to understand one's skill in living a resilient lifestyle, I have developed the following assessment tool. This test is useful in identifying strong points—where one can help others most directly, and weak areas—where the lowest hanging fruit is. Developing skills as rapidly and thoroughly as possible requires that we focus on the weakest links in ourselves, which raises the function of the whole system most easily. Since we only have so much time and energy, being strategic with these precious resources is key.

The results of the test below should not be taken literally but as an *indicator of patterns*. As you go through the test, notice in what areas you are strongest and in what areas you are weakest. Think about how these strengths can help others around you. How can you share them? In what areas do you need someone else to learn from? Please note the value placed not on hard skills per se but on the aptitude to develop them when necessary. Also note that these skills, like the rest of this book, are specific to the author's lifestyle and setting— a rural cold-climate homestead. This test is useful in other contexts, but it must be modified accordingly. In that regard, think of the test as a template from which to make your own assessment tools.

RANKING (OUT OF A POSSIBLE 4,685 POINTS)

4,000+: Likely adaptable to major change, likely an asset to any community, should likely be facilitating other people's learning and sharing skills and resources

3,000–3,999: Probably adaptable to changing conditions, a likely asset to most communities with much to share

2,000–2,999: Adaptive patterns to work from, positioned to become highly resilient

1,000–1,999: Some resilient tendencies to build on

0–999: Average American—a liability until major changes are undertaken

As with all tests, the breadth and depth of what can be measured by this evaluation is very limited. The point of this "test" is to help you identify areas in which you have sound skills and those areas that would be most strategic to work on.

Scoring your evaluation should be done in the following manner:

1. Read the question, and think about how competent you are at the skill described.
2. Mark a number corresponding with that competence. This gets subjective, but do your best. For instance, say the points available are 10 for the question of "Can you weld?" If you could probably cob together a poor weld because you've tried it once, mark 3 to 5. If you can do a satisfactory job with basic welding tools, mark a 10. If you can weld with an array of welding equipment very well, mark a 15.

The scoring should be done in a weighted manner, with a maximum of 50 percent more points possible than shown as a baseline for each skill area.

The Test

Please answer the following questions in each skill area:

SOCIAL-ECONOMIC	Potential	Actual
I am useful to neighbors—I make or do something they need	200	
I can generate a surplus	50	
I am well liked	150	
I am a long-time resident of this area with many social connections and am well liked here	50	
I am financially well off, without debt, can purchase most tools or other resources I need	200	
I can organize people well and/or work well within a group setting	200	
SUBTOTAL	**850**	

PERSONAL and PSYCHOLOGICAL	Potential	Actual
Aptitude		
I can quickly figure out solutions to challenges I have never been trained in and enjoy doing so	500	
Attitude		
I have a positive outlook in difficult situations and experience in adverse conditions which demand calm, calculated, effective action in the face of emergency	150	
I do not give up when encountering difficulty	100	
I am patient when it comes to dealing with challenges	50	
I enjoy challenging situations	50	
I am confident, believe in myself, and act decisively and with poise	200	
I know what I do not know	200	
Mental Health		
I am in sound mental health, stable, and happy; I enjoy my life, like to engage with others	300	
SUBTOTAL	**1,550**	

FOOD, FUEL, and HEAT	Potential	Actual
I can do everything needed to heat a home efficiently with minimal wood	50	
I can haul wood from woodlot to house with on-site resources	20	
I can process firewood with an ax	20	
I can make and tend to a fire	20	
I can cook on a woodstove	20	
I can clean a chimney	10	
I can install a woodstove and chimney	10	
I can stack and cover wood very well	10	
I am able to grow vegetables, grains, fruits, nuts, pulses, and meat	50	
Two-year supply each season	50	
One-year supply each season	25	

Three-quarters of a year's supply each season	20	
Half a year's supply each season	15	
One-quarter of a year's supply each season	10	
I can lactoferment	5	
I can dry/dehydrate	5	
I can smoke, cure	5	
I can slaughter and butcher	10	
I can plant a variety of food trees properly	30	
I can propagate trees and other plants	30	
I can raise seedlings and transplant well	100	
I can keep a vegetable garden in good condition all growing season	100	
I can save seed in a vegetable garden	5	
I can breed animals	5	
I can lamb/kid/birth animals	5	
I can shear, shoe, and perform various animal maintenance in general	5	
SUBTOTAL	**635**	

PHYSICAL HEALTH	Potential	Actual
I can cook using local and seasonal ingredients	50	
I can make a variety of potent medicine	30	
I can identify and treat various ailments with local medicine/approaches	30	
I can treat someone for emergency/acute trauma if given the right tools, and I know how to use them (emergency medicine, including CPR)	30	
I am currently in good physical health	250	
I am likely to have a long life ahead of me	50	
I am not addicted to any foods or drugs	250	
I know how to care holistically for my particular body type and mental habits, and I have learned to maintain overall health to a high degree—I am healthy and know how to heal when sick	150	
SUBTOTAL	**840**	

VARIOUS SKILLS	Potential	Actual
Electrical		
I know basic wiring of switches, outlets, batteries, fencing, lighting, etc.	15	
I know how to use a multimeter and charge vehicle batteries	15	
Plumbing		
I can set up a gravity-fed domestic water system without freeze problems	5	
I can capture and store roof water	10	

I know irrigation systems	5
I can sweat copper and do basic indoor plumbing	5
I can work with pex tubing	5
Crafting	
I can make some clothes and repair them	10
Ropework: I know basic home and farm knots and rigging—bowline, clove hitch, trucker's hitch, fisherman's knot	10
Vehicles	
I can change a wheel	10
I can fix a flat tire	5
I can check and change oil	5
I can check and change other fluids	5
Construction	
I can frame a wall efficiently	15
I can frame a house	20
I can use a full woodshop's array of tools	25
I can design, cut, and raise a timber frame	15
I can do concrete work: forming and pouring	10
I can build a proper dry-stack stonewall of 4 feet in height	15
I can weld	5
I can repair and maintain buildings in general from rot, leakage, and mechanical problems	50
Ecological Awareness and Literacy	
I can identify 10 common medicinal plants in my immediate area and make medicine from them	25
I can identify 10 edible plants in my immediate area and make food from them	25
I can eat from and live in local wild areas for one week	50
I can eat from and live in local wild areas for a year	100
I can eat from and live in local wild areas for a warm season	25
I can stalk, hunt, trap effectively	30
Safety	
I am highly aware of my surroundings	200
I can defend myself well from a physical threat	50
I don't get injured often	45
I am trained to help others in an emergency	50
SUBTOTAL	860
TOTAL	4,735

To offer transition literacy, awareness building, and general human development opportunities to children through the development of a landscape and home, the following strategies have been used in some of our (Whole Systems Design, LLC) projects. This curriculum-connection outline is aimed at helping ensure that children are offered enough real-world challenges to develop high levels of skill, knowledge, and "usefulness/fitness" that will benefit them in the coming century.

Areas of growth in particular to focus on:

► Critical thinking: *design* as interdisciplinary problem solving for multiple variable challenges
► Kinesthesia, handwork, crafting, tool literacy
► Ecological: working with natural systems
► Social: working with people
► Discipline-specific skills: Math, science, writing, art, physical education

CURRICULAR AREAS ACROSS THE TIMELINE OF DEVELOPING A HOME OR FARM

Project Phase	Sample Course Work
Planning & Design • House layout, components, details • House design	Make three models of house designs using wood blocks; photograph or write about them. Why are things laid out the way they are? What are you trying to achieve with each layout? Draw plan-view and cross-section depictions of your bedroom.
Site Analysis & Land Planning • Vegetation, soils, microclimate, views, wildlife, inventories, etc. • House, solar panel, garden and road location and orientation	Inventory all plant and animal species on-site—who lives there and where? Why? Research if any are endangered; if some are, what do they need to ensure they are preserved/enhanced? Dig soil-test pits around the site, show their locations on a map, test all of them, and document the results. What does each finding mean? What steps need to be taken to help "fix" the soil?
Site development: *Home/outbuilding construction,* *fencing, earthworks, planting*	Where is the site sunniest (sunshine analysis)? Make a map of the site, show two locations for the house, gardens, driveway, and solar panels. Why is each of them where they are in each drawing? Innumerable hands-on tasks; e.g., frame a wall, site and dig a swale or garden bed, hoe and seed a row . . .

Appendix C
Crucial Skill List for Emergencies

These skills are most crucial, both during emergencies lasting from a week to multiple years and during the rebuilding phase, which could last months or decades. This is not oriented to a strictly wilderness setting, though it applies to a large extent.

THE DURABLE TEN: KNOW THESE COLD, AND YOU CAN SURVIVE ALMOST ANYWHERE

NEED	TOOL/MATERIALS
Making and keeping a fire	Bow drill, lighter, matches
Finding and securing water	Shovel, pick, tubing, clay, mortar
Germinating seeds and raising vegetables	Seeds, soil, hand tools, water, sun
Building soil	Nutrients, shovel
Storing food without refrigeration, electricity, or fuel	Root cellar, ice, salt, sawdust, sunshine, drying racks, fire, pots and pans
Planting and raising trees	Seeds/plants, shovel, fencing/ deer repellent/dog/sprayer, mulch
Felling and bucking trees, splitting wood	Saw, axe
Basic carpentry: wood framing, masonry, structures	Saw, shovel, wheelbarrow, pencil, paper, tape measure, level, ax, chisel, drill or driver, hammer
Making stone walls	Shovel, pick, the tools above, trowels, sledge
Hunting, fishing, and wild foraging	Knife; bow and arrow; gun and ammo; rod, reel, hooks

SKILLS FOR THE LONGER TERM AND FOR GREATER USEFULNESS

The list below should form some of the basis of an educational approach for young people (or old alike) seeking to be resilient and useful members of their family and society in the twenty-first century.

I. Food and Water
Wells and Springs
- Locating, tapping, diverting, and maintaining water sources; spring digging; and spring box construction

Gardening
- Garden locating, site preparation, soil building, planting timing, seed selection, seed starting, plant arrangement, pest identification and management, companion planting, weather predicting, harvesting and timing, seed saving, season extension

Food Processing and Storage
- Dehydrating, canning, cooling, freezing, and other strategies
- Wine, mead, beer, kraut, kimchee, and other fermented foods preparation

Soil Building/Fertility
- Composting, compost tea, inoculating, biodynamics, animal management

Hunting and Fishing
- Gun, bow and arrow mastery; tracking, stalking, habitat identification; skinning, hauling, processing, storing of game
- Deer, turkey, moose, grouse, woodcock, rodent
- Fish habitat, identification, fly tying, trapping, casting, lure and bait selection, cleaning, processing, storing

Animal Husbandry
- Birthing, training, and care of; communication with; slaughtering, processing and storage, and fencing of

chicken, duck, rabbit, pig, goat, sheep, cow, cattle, llama, horse, oxen, alpaca, geese, and other animals
* Saddlery, shoeing, harnessing, driving, plowing with livestock

Grazing
* Pasture management: species selection, seeding timing, rotation timing and stock density, animal tractoring, fencing, keylining, water systems, predator control, soil food web health

Orcharding and Perennials
* Species and variety selection, planting arrangement, soil preparation, inoculant selection and application, planting, disease and pest management, pruning, grafting and propagation, harvesting, mulching, foliar feeding, guild and understory development, groundcover selection and management

Wildcrafting
* Ecology, habitat identification, awareness, fungi, tree and herb ecology and identification for gathering and maintaining populations of mushrooms, elderberry, nettle, ginseng, arrowhead, willow, bearberry, blueberry, blackberry, raspberry, thimbleberry, butternut, beech nut, black walnut, hazelnut, echinacea, mountain ash, Solomon's seal, cattail, chicory, chokecherry, chickweed, cow parsnip, crab apple, apple, cranberry, bilberry, dandelion, dock, ferns (many varieties), wild ginger, wild leek (ramps), wild grape, groundnut, hickory nuts, Labrador tea, lamb's-quarter, bloodroot, blue cohosh, mulberry, New Jersey tea, pickerel weed, plantain, hawthorn, sassafras, sheep sorrel, serviceberry, shepherd's purse, milkweed, sorrel, sweet flag, thistle, yarrow, mullein, rhubarb, burdock, and *many* others (there are over one hundred well-distributed edible and medicinal plants in most cold-climate woodlands)

Forestry
* Tree selection, felling, hauling, bucking, milling, splitting
* Understory crop development, slash and char soil building, mushroom production, managing to grow "wild"life
* Road construction: siting, culverts and drainage, bridges, shaping, maintenance

Ponds
* Siting, construction, succession management, seeding, fish rearing

Mushroom Cultivation
* Siting, choosing and gathering substrate, inoculating, maintenance

II. Shelter

Construction
* Homebuilding and repair, reinsulating, wiring, plumbing, roofing, weatherizing, window and door replacement, siding, deconstruction and materials salvaging, barns, coops, pens, fencing, root cellar retrofitting
* Working with local materials and assemblies: timber framing, masonry, cob, wattle and daub, clay slip, straw, straw-clay, clay plaster, milling, hewing, and so forth

Machinery, Milling, Manufacturing, Tool Making and Maintenance
* Blacksmithing, mechanics, woodworking, small engines, pulleys, gears, hydro power, bearing repacking, lubrication, engine repair and rebuilding, salvaging parts and retrofitting

Clothes-Making
* Sewing, weaving, knitting, darning, felting, cobbling, tanning

III. Wellness

Nutrition
* Whole, live, nutrient-dense and antioxidant-rich food preparation and combining

Medicine
* Herbs, homeopathy

Body Work, Mindfulness, Spirit
* Massage, reiki, acupuncture
* Meditation, "wilderness" immersion, ritual, ceremony, community
* Music, graphic art, other arts

IV. Energy

Biomass
* Wood heating: processing, drying, storage, burning, chimney maintenance and repair

Solar PV, Microhydro, Wind, Solar Thermal
- Siting, installation, wiring, plumbing, maintenance

V. Mobility

Bicycle
- Repair, rebuilding, maintenance, trailer (in my home state of Vermont you can arguably gather more food per calorie expended (at the right time of year) via a mountain bike with a trailer than by any other means)

Foot and Ski
- Long-distance travel, difficult-terrain traveling and navigation

Canoe, Kayak, Sailboat
- Hazard identification, route identification, navigation, operation, maintenance

Vehicle
- Mechanics, communication/coordination, organization, compromise, driving, parking

Bus/Rail
- Flexibility, slowing down, compromise, organizing/coordination, communication

VI. Community: Local Currencies, Cooperatives, Shared Infrastructure, Ritual

Home and neighborhood enterprise, organization, speaking, writing, art, networking, trust, teamwork, determination, vision, value adding, ceremony, music, celebration

Stored Resources

Certain nonperishable strategic investments will likely be hard to get or very expensive during system-failure periods and for potentially long periods after disturbances to food, energy, economic, or other systems are encountered.

Currency

Certain investments actually retain durability during system-failure periods when "normal" investments lose value acutely in credit bubble/finance/currency collapses.

- Precious metals: silver and potentially gold in small units
- Cash, in small bills
- Dry beans, grains, sugar, salt, coffee, livestock, fuel, cigarettes, guns, ammo, alcohol, small tools, generators, cordwood

The following list is not exhaustive but represents some of the more important tools needed for rural self-reliant and community living. It's likely that some of these tools are simply transitional; you can count on needing and being able to use a shovel in a hundred years, less so a backhoe. Heavy machinery, for instance, is incredibly useful in developing resilient post-peak oil systems today, but will be less affordable and potentially less accessible in the future.

HOME AND HOUSING CLUSTER/NEIGHBORHOOD SCALE:

- Knife, chisel, adze, ax
- Rake, hoe, cultivator, flat fork, broad fork, and so on
- Rope/cordage
- Hammer
- Shovel
- Handsaw, pull saw, chain saw
- Scythe, machete
- Barn
- Sockets, wrench, pliers
- Fencing tools
- Wood, engine, metal shop with all basic tools
- Plumbing torch, oxyacetylene torch
- Pipe cutter, pipe wrench
- Wire stripper, cutter, wire nuts, and so on
- Screw gun, drill, backup batteries, and chargers
- Bench grinder
- Hacksaw
- Hoists or jacks, block and tackle
- Tractor and PTO attachments, loader
- Truck
- Horse- or ox-drawn sled, wagon, plow
- Grain drill, reaper, flail, harvester
- Wood-fired oven
- Wood gasifier, pyrolysis/biochar/charcoal-making
- Pumps, sterling/steam engine, wood/masonry stove
- Welder, forge
- Root cellar
- Oil-seed press
- Commercial nut cracker
- Gun, ammo, bow, arrows
- Fishing rod and reel, hooks, line, lures, flies
- Pick or mattock
- Rock bar
- File
- U-bar
- Shears, pruners, clippers, loppers
- Wheelbarrow or cart
- Dibble, planting bar
- Cold frames, small hoop houses
- Loom, sewing machine, needle, thread, yarn
- Musical instruments, art and communication tools
- Computer, Internet, printer?

VILLAGE- AND COMMUNITY-SCALE TOOL SYSTEMS

- Excavator, dump truck, barn, sawmill, tractors, loaders, hydropower-milling
- Nursery, seedbank, library, greenhouses, plant and spore propagation facility
- Yogurt/cheesemaking, distillery, cold storage
- Animal slaughtering and processing, microtextiles
- Methane digestion/biogas, biochar facility
- Wood/metal/engine/machine shop, pottery shop, forge

- Child care, health clinic, theater, gallery, shops, markets
- Biomass-based cogeneration food canning and dehydrating
- Log splitter, chipper
- Generator, solar PV, wind/hydro turbine, inverter, batteries
- Micro–power grid
- Schools, research facilities

SMALL CITY/REGIONAL–SCALE TOOL SYSTEMS

- Wind farm
- Manufacturing and processing of all kinds
- Rail, highway, path, mobility systems
- Education, research

BOUGHT MATERIALS AND TOOLS: SOME RECOMMENDATIONS AND THINGS TO AVOID

Recommended Tools and Materials

- Fencing by Premier Fencing: These fences stand up well, unfurl easily (for electro-net) and seem to be generally made very well. I buy a lot of items like this through Wellscroft Fence Systems in New Hampshire.
- The King of Spades planting shovels: These are fantastically high quality single-piece forged aluminum—worth the $90 or so price tag; should last a lifetime or more. Clean them after use, and treat them well—they deserve it.
- Pruners by Felco and some products by A. M. Leonard for tree work.
- Tool handles by Tennessee Hickory tend to have correct grain patterns and hold up well.
- For off-grid appliances and the like, Backwoods Solar is hard to beat.
- Dripworks is a massive resource that will help you design systems for anything drip-irrigation related.
- The Japan Woodworker, Silky, and Lee Valley for hand-tool and related materials.
- Lehman's nonelectric: Some of their stuff is cheap in recent years, but they stand by everything without fail.
- Scythe Works, formerly Scythe Connection: the only source for a completely great scythe this side of the

Atlantic Ocean. Scythe Supply makes a decent tool, but the handles are weak. Avoid hardware store variety scythes like the plague.

- Frost/Mora knives: hard to beat compared with knives twice the cost.

Tools and Materials to Avoid

- Kencove electro-net: Premier is far superior in every respect and not much more expensive. Kencove's fences have extra plastic bumps on each connection, with little hanging tabs of material that repeatedly become caught each time the fence is moved. Moving a Kencove fence is like wrestling with a knotted up net compared to a Premier fence, which flows off itself when laid out and rarely snags on itself—when it does, a quick snap of the fence usually frees it.

  Kencove seems to have made a good three-joule charger, however, which I own and mated with my own 30W panel (panels via Kencove and most dealers are highly overpriced); I use a Morningstar charge controller to join it with the battery. The power controller on/off dial is defective, however, and spins, such that you have to ignore what it says and simply go by feel for how far high or low it is.

- Most hardware store variety axes and tool handles, especially those made by Truper, which are complete junk: The grain runout is too high on these to make durable handles; you must evaluate each handle at the store to find ones that have continuous grain—often impossible, since most are poorly made. I have not found a good source for these and have even been shipped defective ones by Madsen's, a reputable professional chainsaw–related dealer in Washington State.

- Most low- and midpriced power tools: I have had very short life spans on Porter-Cable tools, such as their pancake compressor, some DeWalt tools (though many are good). Avoid the cheapest end of the power tool line, such as Homelite saws and similar electric tools—the 25 to 35 percent you'll save off the bat will cost you dearly in project time, ease of use, and replacing them when they break quickly. In the past five years Delta power machines have gone way downhill as well, and it is increasingly impossible to find parts for many of their basic tools; for instance, drill presses. The pulley ring on my five-year-old Delta benchtop drill press shattered recently, and I cannot find a part anywhere to replace it. Without custom machining a new one, this otherwise good-condition tool is now effectively worthless.

► Most smoke detectors and motion-detecting lights: Buy these carefully. I have not found a reliable brand of either of these.

► Most hardware store variety plumbing parts (such as Gilmour) and many electrical parts are junk. You have to go out of your way to find durable materials for these applications, looking usually to replace plastic with metal. Buy only the best quality ball valves you can afford.

► Most cheap inverters: Find recommendations on these, and don't go with off-brands. Same with batteries. Steven Harris of Solar1234.com is a great resource on this front, as is Backwoods Solar.

FOOD

Design Conditions

GOAL: Establish production, storage, and processing systems for foods that are most sustaining, and do not need electricity.

▶ Without electricity, most homes in cold climates have no heat or water.

▶ Most homes have a three- to six-day food supply on hand and no access to drinking water aside from the tap.

Production (Maximize for the Long Term)

You want food that is:

▶ Highly nutritious and available in high volume
  • Carbohydrates (potatoes, radishes, other roots, grains)
  • Fats and proteins (meats, butters, milks, fats, and nuts)
  • Micronutrients, antioxidants, vitamins (berries, fruits, leafy greens)

▶ Easy on the soil
  • N-fixers: beans, peas and other legumes

▶ Durable to climate and pest extremes (reliable)
  • Cold-loving root crops primarily! What did people in your area grow 150 years ago?
  • Potato, radish, beets, kale, collards, apples, perennial herbs, nuts, animals, and so on

▶ Uncommon
  • Tree crops, uncommon berries, uncommon perennial vegetables (skirret and others). If other people don't recognize it as food . . .

• "Trade crops": things you are good at producing and have the resources to make that people place high value on: maple sugar, grain, fruits, nuts, animal products, salt, herbs, fibers, tools

Storage (Minimize Energy Intensiveness and Maximize Length of)

▶ No fridge
  • Grains, beans, nuts, roots, canned goods, seed

▶ Very long shelf life, between one and ten-plus years
  • Grains, beans, sugar, salt, seed, canned goods
  • Store via buckets and for longer term in Mylar with O_2 absorbers

Processing (Minimize Energy and Time Intensiveness of)

▶ No need for electricity

▶ No need for water (for severe situations)
  • Canned, jars

▶ No need for cooking (for severe situations)
  • Canned, jars, rehydrated foods

▶ Minimal milling/cutting need (fast)

Among the most important foods to store for long-term use are the following:

▶ Hard winter wheat

▶ Quinoa

▶ Amaranth

▶ Lentils, split pea, rice, mung and various beans

▶ Salt, sugar

▶ Seed—unhulled nuts/seeds

- Live animals—tree crops and perennials—vegetable gardens
- Compost/compost materials, planting trays or soil block-makers, amendments
- Ability to water (via gravity is superior to every other approach)

Grid Failure

Needs to provision in descending order of importance:

1. Heat: space and cooking (should have nonelectric source: wood)
2. Water (should have nonelectric source: spring/shallow well, stream with filter or boil)
3. Light (should have nonelectric source: LED headlamps, white gas, candles)
4. Communications (landline, cell phone, Internet, radios)
5. Power tool operability

Generator Considerations

Ideally, a homestead is equipped with the following:

- A small generator
- A fuel store: five to fifty gallons, sealed well, stored well, ethanol-free (if gas), treated with ethanol fuel stabilizer (if gas). Propane is superior; diesel is better.
- A covered space for generator, removed from main buildings
- Quieted
- Ideally, two of them

See the fuel storage sidebar in chapter six for crucial information regarding how to make highly volatile and perishable gasoline last for more than a couple of months.

Shop (keeps all other aspects operating, and for production)

- Plentiful hand tools (cutting, fixing)
  - Most crucial **tools**:
    - Axe—saws—guns—ammo—knives—scythe—sharpening tools—drivers/drills—chain saw and chain/file—*generator*—sledge—all manner of wrenches/pliers/sockets/

wire cutters/hammers—tape measuring tools—mattock—shovels—candles—flashlights, headlamps
- Most crucial **maintenance/building/fixing materials**:
  - Fuel—tape—batteries—wire—rope/cordage—nails, screws, bolts, nuts, washers, and so on—lumber—fire starter—kindling—wood—buckets—hose—tubing

Storage/Animals

- Plentiful hay/feed storage
- Plentiful wood storage (for home/shop)
- Plentiful fasteners and building materials
- Highly durable and weatherproof

Health

GOAL: Get and stay in top health for as long as possible.

- "Fix" yourself now
  - Teeth especially
  - Major issues
  - **Prevention** (diet, exercise, mental acumen)
- Sustaining homestead crops (see above)
- Fitness
  - Physical
  - Mental
  - Spiritual
- Care: for yourself and others
  - Have basic training
  - Know and be near an EMT, doctor, nurse, midwife, herbalist, and so on
- Medicine—make, have access to, and store
  - Those you already need
  - Antibiotics—various kinds
  - Most important medicines to store:
    - Broad spectrum antibioltics, Compazine suppository (I've seen these save multiple lives when one cannot hold down medicine via mouth and would die of diarrhea), bandages/wraps, pain meds, EpiPen
- Self-defense
  - Awareness, skill, speed, strength (see below)
- Pest control (in a city especially)

COMMUNICATION

Need to maintain sources of regional and national awareness and be able to communicate to others.

In ascending order of priority for dependability and likelihood of usability:

▶ Internet/e-mail
▶ Cell phone
▶ Mail
▶ Landline
▶ Radio FM/AM
▶ Radio short wave

PASSIVE HOME

GOAL: To develop highly flexible and resilient home, work, and storage/animal spaces.

▶ Needs no electricity to perform all vital functions (heat, water, dryness, light, food storage)
  • Systems necessary
    - Gravity-fed water (ideal), close to hand-powered water if not
    - Well-insulated walls, roof, floors
    - Root cellar (or in separate building)
    - Modular: open/close interior spaces
    - High-mass interior
    - Plentiful windows
    - Good solar access (antirot, daylight, passive solar)

Appendix F
Vocabulary and Concepts

❧

This list of terms lends insight into ecologically regenerative and resilient design and development. As in any fundamental change, the shift from an extractive relationship with the Earth to one of mutual health requires the development of new language.

ACCESS: Element in a landscape that allows entry and exit to and from the site. The ability to move to and from a location within the site. The element most directly relating to off-site locations. Usually consisting of paths, roads, trails. Often the aspect of site development and maintenance that is most destructive and expensive.

ACTIVE: An element with moving parts or mechanisms. A pump as opposed to a thermosiphon. *See also* Passive.

AERIAL: A view of a site from above and in perspective. Often a photograph or drawing from a bird's eye view.

ALTITUDE: The position of an object such as the sun in relation to the horizon at 180 degrees; for example, the altitude of the sun in Burlington, Vermont, on December 21 at solar noon is approximately 22 degrees. Perpendicular to the azimuth bearing.

ANALOGUE/ECOLOGICAL ANALOGUE/CLIMATE ANALOGUE: A place or site sharing similar fundamental elements and relationships between elements. Can be broken into cultural analogues or ecological analogues. Other sites sharing similar challenges with a given site. Ecological analogue: A biological environment sharing similar species, climate, processes, and other fundamental traits as they relate to the scene of a design. A site in Vermont has many ecological analogues in Scotland and northern Scandinavia. Many clues to effective design strategies are found by assessing ecologically analogous sites.

ANALYSIS: The systematic review of the existing conditions of a site. Often includes a description and rendering of sun and shadow patterns, microclimates, circulation and movement on site, zones of use, soils, hydrology and drainage, legal conditions, and so forth. Analysis is done through a harder, more systematic process than that of an assessment and is thus more limited in scope.

ANGLE OF INCIDENCE: The angular measure between an incoming light ray striking a surface and the normal (a line perpendicular to that surface). The lower the angle of incidence, the more energy is transmitted from light to surface.

ANGLE OF REPOSE: The angle in degrees relative to horizontal at which a given material will come to rest without unusual disturbance strictly through gravity's influence; for example, the angle of repose of large boulders is higher than that of sand (one can establish a stable slope more steeply with boulders than with sand).

ARBOR: A structure upon which plants can be grown. Often used to support vining plants.

ASPECT: The direction in which a land predominantly faces; for example, southerly aspects are warmer and dryer in this region than northern aspects.

AZIMUTH: The numeric position (bearing) of an object in a 360-degree horizon; for instance, azimuth 22 degrees north-northeast.

BASE MAP: A map used for reference in a plan set that helps the designer understand the positions of existing elements on a site.

BERM: An accumulation of material, usually soil, against an object, such as a building. It varies from a mound, which is a freestanding pile of material.

BIOMASS: Biological material. Often used for thermal storage in a house.

BIOREMEDIATE: To improve the health of an ecosystem by employing biota, usually in the treatment of pollution.

BLACKWATER: Any water carrying human effluents ("wastes") or water from kitchen sinks; distinct from greywater. *See also* Greywater.

CIRCULATION: The flow or movement of people in a landscape.

CLIENT: Usually the person or persons seeking design services. Always a person or group of people with a problem or challenge to be addressed. May be any major stakeholder that has sought and is responsible for funding design services. There are many "hidden clients/stakeholders," such as the students in a schoolyard design, the users of a museum, the visitors to a landscape. The line between "user" and client may blur at times.

CLIMATE: Weather over a period (usually decades or more) of time. The general characteristics of weather patterns in an area.

CLIMATE CHANGE: Also referred to as "global climate change," climate change is a more descriptive and accurate term than climate warming, with respect to Earth's current climate trends. Climate change has always been the case on this planet and is becoming increasingly severe since the emergence of fossil fuel dependence and the results of transferring massive amounts of carbon from within the earth's crust into its atmosphere. Actively designing for changes in the global and local climate will be increasingly critical to ensure the survival and success of human settlements.

CLOSED LOOP: A system in which inputs feed back directly into outputs without leaving the system; for example, vegetable garden to dinner to composting toilet to sheet mulch to soil to edible plant to dinner. There are almost always materials and energy that "leak" into the surrounding larger systems, such as carbon dioxide, oxygen, and other gases in this example. The earth is a closed-loop materials system with incoming solar energy and outgoing radiation.

COGENERATION: The production of one material or service simultaneous with others; that is, in a system with multiple yields, a product or service is cogeneratively produced. Heating a greenhouse with the excess heat from a wood combustion heater would be cogenerative. Growing plants in a courtyard through which "waste" (excess) heat from buildings is piped is cogenerative. "Nature" is everywhere working via cogeneration; there is no growth or production without concurrent feeding of another process in the system.

COLD AIR DRAINAGE: Any low-lying path along a landscape through which denser cold air travels. Usually an element that pulses in and out of existence over a day, a month, a year. Highly related to climate and aspect of the landform.

COLD HOLLOW/POCKET/FROST POCKET: A depression into which air flows and settles. Often the coldest and lowest place in a landscape.

CONCEPT/CONCEPTING/CONCEPTUAL DESIGN PHASE: A general design idea that is broad but has a direction or pattern. Concepting is a part of the design phase, early on, where overall directions, symbols, themes, forms, and strategies of a particular element or a group of elements are revealed. Concepting a garden would reveal overall layout and arrangement of plantings but not the specific species or materials used. Concepting a house would be revealing how it fits to the site; revealing the major forms of the building; possibly some openings, entrances/exits, but not the materials used or exact dimensions. Concepting always connects the element to its larger context.

CONCEPTUAL DRAWING: Any graphic representation of the above process. *See also* Construction drawings/documents.

CONSTRUCTION DRAWINGS/DOCUMENTS: Graphic depictions of building and installation specifics necessary for the implementation of elements in a design. A set of construction documents for a landscape would specify all materials to be used in the hardscape, all species, and dimensions. The layout and arrangement of these pieces is usually described in plan view or cross-sectional drawings.

COPPICE: To harvest wood while maintaining the roots. Harvesting part of a tree, such as a limb, but leaving the main shoot and roots. A system in which this technique is practiced.

COVER (VEGETATIVE): Any surface of the landscape upon which vegetation is growing. Permeable. Solar absorbing. Living.

CRITERIA: A specific standard of judgment. A discrete statement defining a measure of success on which the design or part of a design is based. A direct, brief design statement used to guide design strategies; for example, landscape would require watering only during the driest months of the year; lighting would be less than forty foot-candles at twenty feet; time to walk between the compost and the kitchen is less than one minute; snow on roof would be redirected away from entrance; and so forth. For many designs there could be dozens if not hundreds or more of criteria. Criteria guide designers and help communicate between the designer, the client, and the builder.

CROSS SECTION: A graphic depicting the vertical aspect of a design, usually aligned with a cut line that transects a landscape or hardscape. A "cross section through the pond" would show the arrangement in the vertical plane of everything lying on that cross-section line.

CYBERNETICS: From the Greek Κυβερνήτης (*kubernites*, meaning steersman, governor, pilot, or rudder; the same root as government). Increasingly, we are seeing the design of living systems as necessarily being one of steering, piloting, orchestration.

DEAD SPACE: An area that sees little to no use, usually due to poor design of the particular space or space around it. The best designs have the least amount of dead space, obviously.

DESIGN: A highly intentional problem-solving approach to a specific or broad challenge. Differs from art, in that the purpose of design is to provide solutions to a problem, whereas art may be a process whose result is simply aesthetically pleasing, entertaining, or otherwise useful but does not necessarily overcome a problem. Much of what is called "design" in the modern world is actually only art.

DESIGN DATE: Time of year in which a particular design is optimized for. Often best if it is the limiting factor design time in a year and/or a time when conditions are typical but challenging.

DESIGN FOR CLIMATE CHANGE: The maxim that when we plan and develop places we should do so in anticipation of change; in particular, change of the largest caliber—the long-term weather patterns of the site. This approach is fundamentally different from conventional reactionist approaches to place making. It is forward thinking and progressive, attempting to harness a force for a positive result, rather than react to a challenge after it emerges.

DESIGNER: An arranger of energy, land, and biota whose process is intentional, systematic, and problem solving.

DESTINATION: Any feature of a landscape that one is drawn to for whatever reason. Destinations help draw one into little-used areas of a site. Destinations are sometimes visible from inside a building and draw one out of doors and into the landscape. Generally, a timeless landscape teems with destinations while retaining a certain subtlety.

DISTINCT FEATURES: Anything in a landscape that is of unusual value or significance in guiding the design; often a special tree, rock, water feature, or other sensitive element, something the client has a particular connection with. It is not always a physical feature but sometimes is an aspect, such as a special sightline or sound.

ECOLOGY: The interrelationships of living things to one another and to their environment, and the study of these interrelationships.

ECOSYSTEM SERVICE: An output in a biological system that has direct or indirect human value; for example, oxygen production, erosion control, carbon sequestration, water and air quality enhancement. Ecological designs maximize and utilize ecosystem services.

ECOTONE: The transition zone between distinct natural communities. Often the most biologically active area in a landscape and often the source of much design potential. Often an area of particular human interest and enjoyment; for example, a pond edge or where a field meets a forest. *See also* Edge effect.

Edge effect: A tendency for the interface between species and communities of species to be especially fertile, productive, and interesting. Usually, a zone of increased surface area where energy flow and relationship intensity are heightened.

Embodied energy: The amount of energy transformed in the production of a given material or group of materials. Often used quantitatively but can be just as useful in qualitative terms, such as "type of embodied energy"; for example, the embodied energy of lumber is recent solar energy (transformed by the tree); the embodied energy of a nylon is derived from fossil fuel (transformed by a factory). Choosing materials and systems with the lowest and most biologically based embodied energy is critical in ecological design.

Emergent property: An initially nonexistent condition that is "produced" by the interaction between parts of a system over time. A characteristic of a system that derives from the interaction of its parts and is not observable or inherent in the parts considered separately.

Energy: The potential to do work; it can come in the form of a human body, money, water storage.

Energy flow: Movement of potential or power from one part of a system to another.

Entropy: The second law of thermodynamics: the tendency for energy or matter to erode or become less organized over time in a system. Notably, biological systems seem to break this law and often develop order and organization (org-anism) from less order. *See also* Order/organize.

Erosion: The loss of biological material, especially soil, from weathering. Most often enabled by human mismanagement of a landscape and made especially rapid with use of heavy machinery.

Existing conditions: The state of a site previous to the action of the designer. This includes all elements in the landscape and their arrangement. An existing conditions map, often called a base or index map, identifies and communicates these elements and often notes any particular challenges or opportunities. For design process purposes, the status quo of a place, the baseline situation and raw material that the designer will adjust.

Farm (regenerative): A system in which the flow of energy and materials is managed negentropically, where biological production, ecological structure, and complexity are increasing over time, where yields increase while inputs decrease. Antonym would be mining.

Farming: Choosing which animals one lives with in a landscape (Leopold).

Feedback: Outputs of information, materials, or energy that flow back into the inputs of a common system. Positive (system reinforcing) and negative (system discouraging) feedback loops result. Consideration and ongoing management of feedback in a system is critical to desired results. The incorporation of feedback in a system provides an endless stream of opportunities for optimization.

Fertigate: Using water as fertilizer, always best done as a gravity-fed system.

Fractal: A word coined by Benoit Mandelbrot in 1975 to describe shapes that are "self-similar," shapes that look the same at different magnifications. To create a fractal, you start with a simple shape and duplicate it successively according to a set of fixed rules. A simple formula for creating shapes can produce very complex structures, some of which have a striking resemblance to objects that appear in the real world.

Frost pocket: *See* Cold hollow/pocket/frost pocket.

Gardening: To cultivate biological systems, usually plants, through the management of nutrients and energy in a system. The most productive gardening does this in such a way that labor and energy inputs are minimized, while outputs and yields are maximized. This is approached largely by harnessing the positive relationships between elements in the system.

Grade: Verb: to manipulate earth or other material to achieve a specific design goal, such as access. Noun: the angle of a slope.

Gradient: The gradual transitioning from one quality to another. Gradients are usually productive, enjoyable, and full of life. *See also* Edge effect.

Greywater: Water from bathroom sinks, showers, and any other nonkitchen and toilet fixtures. A particularly valuable source of water on a site. *See also* Blackwater.

GUILD: A mutually beneficial relationship of cultivated plants. A positive plan association. An arrangement of species in such a way that synergy and yields are maximized.

HEALTH (OF LAND): "The capacity of land for self renewal" (Aldo Leopold). The ability of land to recover from stress. Can be measured in the overall amount of biodiversity and biomass in an area.

HOLISM: Holism (from holon or holos) is the idea that the properties of a system cannot be determined or explained by the sum of its components alone. The word, along with the adjective "holistic," was coined in the early 1920s by Jan Smuts. According to the Oxford English Dictionary, Smuts defined holism as "the tendency in nature to form wholes that are greater than the sum of the parts through creative evolution."

HOLON: A holon (from the Greek *holos* = whole and *on* = entity) is something that is simultaneously a whole and a part. The endless "nesting" of elements within one another is holonic. The term was coined by Arthur Koestler in his book *The Ghost in the Machine* (1967).

INSOLATION: The amount of sunlight falling upon a surface, usually measured in Btus per square foot per hour or Btus per square foot per day.

INVASIVE (SPECIES): An organism that thrives in abused and abandoned locations and where humans do not usually want it to proliferate. Often an organism that allows for further successional development of the ecosystem; for example, nitrogen-fixing plants.

LAND: The earth's surface. "By land is meant all of the things on, over, or in the earth." (Aldo Leopold, "The Land Ethic," in *A Sand County Almanac*, 1949).

LANDSCAPE: The traits, patterns, and structure of a specific geographic area, including its biological composition, its physical environment, and its human social patterns.

LIMITING FACTOR: A condition that, at a given time, discourages a desired function or output more than any other identifiable factor; for example, a flat tire on an otherwise functioning automobile would be the limiting factor to speed. Soil erosion in an otherwise healthy landscape would be a limiting factor. There is usually a host of limiting factors in any given situation. Comprehensive, health-promoting land design and management can be seen, in part, as the addressing of existing and emerging limiting factors and reducing, eliminating, or alleviating them.

LIVING MACHINE: A system that completes specific tasks (for instance, a machine) and is designed by humans but is composed of biological components; for instance, a constructed wetland or a biological water treatment facility. Living machines are multiyield and often low input. Living machines have been articulated and developed most notably by Dr. John Todd.

MASTER PLAN: A report or set of recommendations, usually in graphic and textual form, that details the existing conditions of a site and outlines directions of site development. A master plan is the primary and broadest planning action to complete on a site. Master plans should be geared to accept more specific designs as an ongoing process of land use over time.

MATTER: Most fundamentally, it seems to be energy in physical form. *See also* Energy.

MICROCLIMATE: A small-scale climate nested within a larger surrounding climate, usually resulting from a variation in solar access, wind exposure, or human-derived heat sources. Microclimates occur at all scales.

MULTIYIELD: An element that produces more than one intended output. In an integrated design, as in natural systems, most elements are usually multiyield.

NATIVE: Usually used to describe a species that has been in a North American region since pre-European settlement. The idea that an organism is "from" a specific place is often theoretical and inaccurate, as species are in constant movement from both human and nonhuman forces.

NICHE ANALYSIS: The identification, by the designer, of the inputs and outputs in an element in a system. A crucial step in the design process.

OPTIMAL/OPTIMIZE: A condition of being most integrated, beneficial, or desirable. Arranging a system for the most desirable outcomes. Optimize is often not the same as maximize.

ORDER/ORGANIZE: To arrange in a predictable or patterned way. The seemingly nonrandom pattern of systems, especially living systems. Permaculture founders Bill Mollison and David Holmgren noted that "life is the central organizing force in this part of the universe." *See also* Entropy.

Orographic: Weather patterns endemic to mountainous regions. Precipitation often resulting from rising and cooling air masses that are moving over mountains.

Outdoor living space: Open-air spaces that are geared toward spending time sitting, sleeping, gathering, and so on. An outdoor room. Extension of the house into the landscape and the landscape into the house. Outdoor living spaces are prime human habitats taking advantage of the edge effect where the built and biological environments meet.

Passive: Any element or system of elements that function without moving parts or regular user input, as opposed to an active system. For example, a south-facing wall of glass is a passive solar device, whereas a solar photovoltaic panel (PV) is an active solar device.

Pattern: A repeating or in some way predictable set of forms, elements, or events that have coherence relative to one another or to other systems. *See also* Order/organize.

Pattern language: A term coined by Christopher Alexander and colleagues to describe a vocabulary of interacting design strategies that can be used to develop human-scale, enjoyable, and durable spaces, buildings, landscapes, and towns. The original book, *A Pattern Language*, is particularly geared toward building and architecture.

Percent grade: A description of the pitch of a slope. Noted as 5 percent, 20 percent, 135 percent, and so on. A 100 percent grade equals 45 degrees from level.

Permaculture: A conceptual framework and decision-making system, formalized to a large extent initially by Bill Mollison and David Holmgren, that is aimed at the development of human systems fitting into more-than-human ("natural") systems in synergistic ways such that the health of both is increased. Permaculture, in contrast to most gardening or farming views, yields as a logical side benefit of ecosystem partnering, not as a singular goal. In this way permaculture doesn't truly aim to grow "crops" but to promote vigor in whole systems.

Plan/planning (site plan/planning): The long-term arrangement of a site in relation to intended uses and the existing conditions of the site. More broad, less specific and long-ranging than design.

Plantscape: The system of plants on a site.

Plan view (graphic): A two-dimensional graphic that is communicated from the perspective of being directly above the subject. Different from bird's-eye view or aerial view, which are typically rendered as a three-dimensional perspective.

Positive drainage: Drainage away from buildings, gathering areas, roads, and other spaces of high use. Usually achieved by grading or installing subgrade drains.

Positive outdoor space (POS): Spaces near built environments (usually residences) that take advantage of the hardscape for the enhancement of outdoor uses. POS leverages the built environment for privacy, shade, solar gain, windbreak, and other services that help create an enjoyable and productive space near buildings.

Process: A pattern repeating itself over time with physical ramifications.

Program (design/client program): A boiling down of client goals and challenges guiding the designer in her endeavor. Usually, a program is a textual statement of a sentence to a short paragraph.

Rain shadow: Reduced precipitation on the leeward side of a hill or mountain.

Regenerative system: A relationship in which the whole function of any one element in the system is realized and the value of its outputs increases over time. A system in which outputs are more valuable than inputs. A system that is fundamentally economical. Land use in which entropy is reduced and biological stability, integrity, and long-term health are increased for both the local and the global environment. A system whose interest increases while capital inputs decrease.

Renewable: Usually referring to sources of energy or materials that are produced in relatively short time frames. A more specific term would be "rapidly renewable," indicating that all resources are renewed over some period of time, however long it may be.

SCHEMATIC/SCHEMES: A drawing or often a series of drawings describing various approaches to solving a problem. Schematics compare and contrast various approaches to site layout, elements to be included, scale of elements, and other planning questions to empower the design team to sift through options clearly, taking the best and discarding the rest. Schematic design is the "guts" of the design phase, in which the variables in the design are confronted and narrowed down and decisions are made. This is the most empowering part of the planning and design process, and truly, no process is complete or thorough without it—though much "design" is done without ever drawing on schemes.

SEASON: A label applied to climatic or biological patterns occurring in a specific time period. A comprehensive designer is aware of a multitude of seasons that the site may experience. Some of these patterns emerge from features of the site, such as plant foliage changes; others emerge from patterns passing through the site, such as migrating animals. A thorough design highlights particular landscape elements as they constantly shift and come into and out of season. Thorough landscape design takes advantage of the seemingly infinite variety of seasons across the year on a given site; for example, sumac season, chanterelle season, sugar season, wild leek season, trout lily season, aspen season, maple foliage season, mud season, monarch butterfly season. Some seasons last a month, others a day, others perhaps only hours or minutes. Some insect species, for instance, have their season for hours in the entire year. Some seasons occur only a handful of times a decade or less, such as the masting of an oak tree. The more climatic flux a site is exposed to, the more noticeable these seasons usually are. For example, continental locations have intense seasonality.

SELF-ORGANIZE: The tendency for systems, especially biological systems, to develop order and complexity over time. A force offering design and management opportunities.

SHEET DRAINAGE: Water moving across the surface of a landscape.

SIGHTLINE: The discrete area of land under the designer's visible attention.

SLOPE: An area of land angled relative to the horizontal level.

SOIL: The living matrix of materials and organisms in which plants grow and the foundation for much of life on Earth.

SOLAR GAIN: Positive contribution of heat from the sun into a landscape or building

SOLAR NOON: The highest point the sun reaches on a given day. Time during which maximum energy is available to the landscape; 10:00 a.m. to 2:00 p.m. is commonly referred to as the "solar window."

SOLAR SOUTH: The direction on the horizontal plane (azimuth) where the sun lies at solar noon.

SOUNDSCAPE: The audible environment experienced by a user. Should be defined during a given period, as soundscapes shift quickly over an hour, a day, a season.

SPRING (WATER): A location of groundwater emergence because of head pressure (gravity force) developed from the aquifer's connections with higher elevations on the landscape.

SYSTEM: An assemblage of interrelated elements (or holons) composing a unified whole. *See also* holon.

THEME: A unifying framework that holds the overall design together. A set of symbols or meanings, associations, or goals that serve to guide a design; for example, the theme for this children's playground is connection to nature (all elements aim to support that wherever possible).

THERMAL BUFFERING: Reduction in amplitude of temperature swings. Often the result of thermal mass. Usually allows a more productive and enjoyable living environment.

THERMAL LAG: The tendency for temperature change to have inertia. For example, a stone exposed to sun will stay warm after exposure for a time relative to its mass and specific heat (heat-holding capacity). Both spaces in the landscape and the earth experience thermal lag, as evidenced by later-day microclimates in the landscape and hot Julys, after the sun has reached its highest path in the sky.

THERMAL MASS: Heavy materials such as stone, water, earth that store incoming solar heat and reradiate that heat during times of little to no solar gain;

typically, an element in the southern area of a building or on the southern sides of a structure. Materials with the highest specific heat (ability to hold heat) such as water and stone are used for thermal mass. "Massing" of living environments is particularly important where temperature fluctuations are high.

THERMOSIPHON: The passive flow of heat through a system. Often called a convection loop. A particularly powerful way to move heat through a building or landscape because of its passivity and reliability.

TIME: The nonphysical context that contains all processes and elements. The vessel within which all physical developments happen. For the designer it is the most powerful point of leverage in biological systems. One of the most fundamental and oft-forgotten resources in landscape planning and development. *See also* Order/organize.

TRIM TAB (PRINCIPLE): Coined by Buckminster Fuller to indicate the leverage possible when one applies energy to the most strategic part of a system for the maximum effect; from the tiny trim tab on a massive ship causing large changes in direction and stability.

VENTURI: The wind tunnel effect occurring when the flow of a fluid (such as air or water) is increased in speed when a given volume passes through a constriction. A venturi is often unintentionally created by building or vegetation positions. A venturi can be utilized for passive cooling in many locations. Windbreaks and venturis can be used to affect the direction and speed of air currents in a landscape.

VIEWSHED: An area that a user visually experiences from a given location. *See also* Sightline.

WASTE: A resource misplaced. Usually, the unintentional outputs of a system. A biological or technical "nutrient" that is without an optimal match in a system. A sign of design failure—as good design matches inputs with outputs: "food" with "waste."

WATERSHED: The area of land draining into a common basin. Like many other landscape processes, this occurs fractally, at all scales; for example, from a puddle to a pond to an ocean.

WINDBREAK: A natural, biological, or built feature that deflects or slows the flow of air.

WOODLAND: A tree-based environment consisting of fairly large openings in the canopy. More open than a forest, more closed than a savanna.

YIELD: An intentional output of a system. Typically, the unintentional outputs are "waste."

APPENDIX G:
Resources

EARTH ENGAGEMENTS AND DAILY PRACTICES

The following ecological systems and daily/weekly/yearly (or less often) activities have been far more fundamental than books to informing the design approaches I have used in developing my homestead. These include:

1. **Being sick and healing:** Acquiring disease (for me it was a bone infection) and working through the healing process has been one of the primary teachers in my life, and the experiences gained there allow me to do my work each day and year. Immense clarification of the way living systems behave and respond to actions and intentions that occurred during this process of transformation.

2. **Keeping animals and tending to people:** Caring for and learning to be competently responsible for another living being is expansive and awareness building in ways that allow the intimate connection with land systems that is required to do regenerative and resiliency work effectively.

3. **Making things by hand and living close to things you make:** This includes buildings, plantings, gardens, water systems, and anything else that was brought into existence with your own hands that you can then be in a relationship with over time. An empowering and crucial teacher.

4. **Sleeping under the open sky and being outside as much as possible:** Crucial awareness increases of the universe as a whole, of one's place in the expanse of time and space, of the largest cycles affecting our lives and place including sun, moon, and stars and potent health benefits, especially related to sleep.

5. **Any vigorous and endurance-based outdoor exercise, especially in a "wild" setting:** This is a process of centering in the body-mind and has myriad benefits of connecting with the world around us. Risk is especially helpful as well. Leaving the safety net of "civilization" and immersing in the self-reliance necessities of life in the mountains, woods, and the like has been crucial to learning some of the basic aspects needed to do the work here.

6. **Diet, meditation, observation, revelation, music:** Treating myself to the highest quality foods, periods of time, rest, reflection, soundscape, and vigorous outdoor activity, including in vibrant natural systems, has been crucial as well. A weekly walk through the farm during the growing season may be the most reliable source of inspiration and information in my life and never fails to bring new lessons to light. Maintaining curiosity and fascination with the world around me has been of central importance. By "meditation" I do not mean sitting still, for the most part, but engaging with the world beyond one's self.

7. **Surrounding oneself with inspiring people and culture:** To be able to continually learn, inspiration and reciprocity with others is key. Traveling to new locations for a shift in perspective can be key.

8. **Climbing steep rock and negotiating steep terrain:** Engaging in this activity involves intense kinesthetic awareness enhancements, sensory enhancements, beta-state practice, confidence and trust in one's body-mind, boldness, risk negotiation, decisiveness, and body-mind expansion, among other potent values.

9. **Swimming in swift-moving rivers:** I have gained basic lessons in water movement, fluids, and breathing, and kinesthetic awareness increases.
10. **Moving around in the dark:** This promotes fantastic sensory awareness gains, land connections and understanding of land shape, sounds, and other sensory gains.

Books, Essays, Videos

Acres USA magazine. Austin, TX.

Alexander, Christopher. *A Pattern Language and a Timeless Way of Building.* New York: Oxford University Press, 1979.

Bailey, Liberty Hyde: Various works.

Benyus, Janine. *Biomimicry.* New York: William Morrow Paperbacks, 2002.

Berry, Wendell. "Solving for Pattern," "The Economics of Substance," "The Gift of Good Land," among other essays, in *The Gift of Good Land.* Berkeley, CA: Counterpoint Press, 2009.

Brown, Asby. *Just Enough.* New York: Kodansha, 2010.

Deppe, Carol. *The Resilient Gardener.* White River Junction, VT: Chelsea Green Press, 2010.

Environmental Building News. Professional journal—the best, latest information on building science, air quality, building performance, and more.

Fukuoka, Masanobu. *A One Straw Revolution.* Emmaus, PA: Rodale Press, 1978.

Holzer, Sepp. Various videos—his is likely the most sophisticated and integrated hill farm on the planet.

Jacke, Dave, and Toensmier, Eric. *Edible Forest Gardens*, Vol. 2. White River Junction, VT: Chelsea Green Press, 2005.

Jackson, Wes. *Becoming Native to This Place.* Berkeley, CA: Counterpoint, 1996.

Leopold, Aldo. *A Sand County Almanac* and other land management essays. New York: Oxford University Press, 1949.

Lstiburek, Joe: Various presentations—a practical and insightful building scientist for cold climates.

Mann, Charles. *1491.* New York: Vintage Books, 2006.

McDonough, William. *Cradle to Cradle and the Monticello Dialogues.* Santa Rosa, CA: New Dimensions Foundation, 2004.

Mollison, Bill. *Permaculture Designer's Manual.* Tasmania: Tagari Publications, 1988.

Nearing, Helen and Scott. *The Good Life.* New York: Schocken Books, 1973.

Orr, David. *Earth in Mind* and others. Washington, DC: Island Press, 2004.

Salatin, Joel. Various videos—an articulate ambassador and example for sane and smart farming on a large scale.

Sloane, Eric. *A Reverence for Wood* and others. New York: Funk and Wagnalls, 1965.

Smith, J. Russell. *Tree Crops: A Permanent Agriculture.* New York: Harcourt Brace and Co., 1929.

Spirko, Jack. The Survival Podcast—an exhaustive resource for all things preparedness.

Stamets, Paul. Various videos—inspiration from the fungal world.

Steingraber, Sandra. Various essays and books—clear communicator of the toxicity challenge we face.

Theodoropoulos, David. *Invasion Biology: Critique of a Pseudoscience.* Blythe, CA: Avvar Books, 2003.

Todd, John. *From Eco-Cities to Living Machines.* Berkeley, CA: North Atlantic Books, 1994.

Weisman, Alan. *Gaviotas.* White River Junction, VT: Chelsea Green Press, 2008 reprint.

Wessels, Tom. *Reading the Forested Landscape.* Woodstock, VT: Countryman Press, 2005.

Zelov, Chris. *Ecological Design: Inventing the Future.* Documentary; 64 minutes. Easton, PA: Knossus Publishing, 1995.

For More on WSRF

This book is not the entire story of the lessons learned on this farm—it simply cannot be, given the limited form of a book. For additional information on Whole Systems Research Farm, please visit our website, www.whole systemsdesign.com for short workshops, our acclaimed permaculture design course, tours, and other opportunities that are continually offered here. The most direct learning from this system occurs through direct contact on site.

Index

JEB WALLACE-BRODEUR

Ben Falk, M.A.L.D, developed Whole Systems Design, LLC, as a land-based response to biological and cultural extinction and the increasing separation between people and elemental things. Life as a designer, builder, ecologist, tree tender, and backcountry traveler continually informs Ben's integrative approach to developing landscapes and buildings. His home landscape and the Whole Systems Design (WSD) studio site in Vermont's Mad River Valley serve as a proving ground for the regenerative land developments featured in the projects of Whole Systems Design. Ben has studied architecture and landscape architecture at the graduate level and holds a master's degree in land-use planning and design. He has conducted nearly two hundred site-development consultations across New England and facilitated dozens of courses on permaculture design, property selection, microclimate design, and design for climate change.

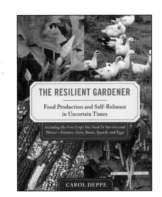
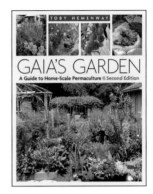

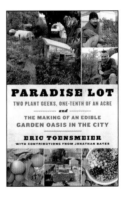

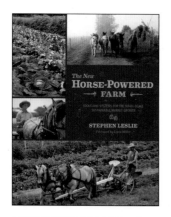
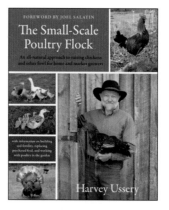